Praise for *Deep Learning for Biology*

This is the book I wish I had when I started my PhD! It demystifies many important aspects of machine learning and how they apply to core issues in biological sciences. Natasha and Charles dedicate outstanding care to these topics, both from the theoretical and the implementational aspect, allowing the reader to immediately dive deep into the action.

—*Petar Veličković, research scientist, Google DeepMind*

A compelling introduction to the transformative power of deep learning in biology—a gateway to the future of technology-driven medicine, agriculture, and biotech. Exploring the depths of machine learning and biology, this is essential reading for anyone eager to be at the forefront of AI-driven biological discovery.

—*Lipi Deepaakshi Patnaik, senior software developer, Zeta Suite*

An excellent guide for students and scientists looking to tackle biological research questions with deep learning. It's hands-on, approachable, and contains lots of valuable tips.

—*Žiga Avsec, research scientist, Google DeepMind*

This is the book I wish I had read before transitioning from computational biology to machine learning engineering. It's pragmatic, opinionated, and filled with great examples that kept me motivated to work my way through.

—*Felix Kruger, NLP scientist specialist, Novo Nordisk*

Deep Learning for Biology
Harness AI to Solve Real-World Biology Problems

Charles Ravarani and Natasha Latysheva

O'REILLY®

Deep Learning for Biology

by Charles Ravarani and Natasha Latysheva

Published by O'Reilly Media, Inc., 141 Stony Circle, Suite 195, Santa Rosa, CA 95401.

O'Reilly books may be purchased for educational, business, or sales promotional use. Online editions are also available for most titles (*https://oreilly.com*). For more information, contact our corporate/institutional sales department: 800-998-9938 or *corporate@oreilly.com*.

Acquisitions Editor: Michelle Smith

Development Editor: Corbin Collins

Production Editor: Ashley Stussy

Copyeditor: Audrey Doyle

Proofreader: Kim Wimpsett

Indexer: WordCo Indexing Services, Inc.

Cover Designer: Karen Montgomery

Cover Illustrator: José Marzan Jr.

Interior Designer: David Futato

Interior Illustrator: Kate Dullea

July 2025:   First Edition

Revision History for the First Edition

2025-07-16:   First Release

See *https://oreilly.com/catalog/errata.csp?isbn=9781098168032* for release details.

978-1-098-16803-2

[LSI]

Table of Contents

Preface

This book introduces you to the fascinating intersection of biology and deep learning. It's written for both biologists eager to acquire computational skills and computational practitioners curious about applying their expertise to biological problems. This fusion of disciplines is already transforming biotechnology and medicine—and is poised to become foundational across the life sciences.

The material in this book is pitched to be introductory, guiding you from the basics to more intermediate concepts. We aim to balance practical code examples with clear explanations, making new terms and ideas accessible. Real-world Python code appears early and often, helping you develop hands-on intuition. While deep learning is a powerful tool, it's not a one-size-fits-all solution—we emphasize the importance of understanding your data and framing your problem thoughtfully before diving into modeling. We encourage you to start simple, build modular and debuggable code, and add complexity only when it serves a clear purpose.

Although this book is designed for beginners, each chapter builds on the last to develop a practical, end-to-end workflow for applying machine learning to biological data. Our goal is to equip you with tools that are robust enough to solve real-world problems and flexible enough to adapt to your own research questions. In the final chapter, for example, we reproduce key results from a recent *Nature Methods* paper that uses deep learning to uncover spatial protein patterns in microscopy images.

This field evolves incredibly rapidly—libraries change, datasets get updated, and model architectures come in and out of fashion. The core ideas you'll learn—how to structure projects, prepare data, build and evaluate models, and connect predictions to biological questions—will remain useful regardless of the framework or trend.

We use the JAX and Flax deep learning ecosystem throughout, helping you build efficient, flexible models using tools that are increasingly popular in machine learning research.

Who This Book Is For

This book is for anyone interested in using deep learning to solve biological problems—whether you're a biologist, a data scientist, or a software engineer. We have aimed to provide an accessible bridge between disciplines and a practical foundation for working in this emerging field. We've designed the book to be user-friendly and interdisciplinary, with the goal of supporting readers from a wide variety of backgrounds.

Why We Wrote This Book

The short answer is that there just aren't many educational resources for people who want to work at the intersection of biology and machine learning. Most materials either focus heavily on one field or assume deep prior knowledge in both. We set out to write the book we wished we had when we were getting started—something approachable, hands-on, and grounded in real biological problems.

While this book is far from perfect (please do send us your suggestions and extensions), we hope it offers a helpful and motivating starting point—and maybe even inspires some readers to pursue a career in this exciting, fast-moving field.

Prerequisites

We assume a basic familiarity with Python and the core scientific Python libraries—namely, NumPy, pandas, and Matplotlib. You don't need to be an expert programmer, but you should feel comfortable reading and writing basic Python code. If you're new to Python, there are many excellent tutorials and courses available online to help you get started. We particularly recommend the Codecademy Python online course (*https://oreil.ly/oLar3*), and the free book *Automate the Boring Stuff with Python* (*https://oreil.ly/uEGbU*) by Al Sweigart.

Some background in machine learning is also helpful. Concepts like optimization, loss functions, training steps, overfitting, and feedforward neural networks will come up frequently. If these are unfamiliar, don't worry—brief explanations are included throughout the book, and you can always pause to brush up using external resources as needed.

Why We Focus on Molecular Biology

This book focuses primarily on the "small" side of biology: molecules, cells, and tissues. There are two main reasons for this. First, our backgrounds are in molecular and cellular biology, which naturally influenced the topics we chose. Second, most deep learning applications in biology so far have concentrated on this space, largely due to the availability of large-scale datasets and the quantitative nature of the field.

That said, the techniques and workflows in this book are broadly applicable, and we hope that readers interested in other biological areas—such as ecology, neuroscience, or zoology—will also find inspiration and useful tools here. We're excited to see deep learning methods continue to spread across the life sciences.

Why All the Hands-on Programming Projects

When it comes to machine learning, we believe programming is one of the most effective ways to truly understand what's going on. Theoretical explanations alone can often feel abstract or confusing—especially when complex notation and dense jargon get involved.

Hands-on projects cut through that by making ideas concrete. They offer several distinct advantages:

Active learning
Writing code helps reinforce concepts through direct experience.

Deeper understanding
Implementing a model forces you to understand each component—there's no room for hand-waving.

Immediate feedback
It can be unforgiving, but code that doesn't work quickly highlights gaps in understanding or setup and invites experimentation.

Practical payoff
By the end of a project, you'll have a functioning model you can analyze, adapt, or apply to new data.

Theoretical understanding is still essential—but for most people, there's no substitute for the clarity that comes from building and debugging something yourself.

Notebook Availability

Executable notebooks for each chapter are available on GitHub and can be launched directly in Google Colab. These notebooks are designed to help you follow along with the material in a hands-on way.

We'll remind you where to find them as you go, but we strongly encourage you to dive in—there's no better way to learn than by experimenting and getting your hands dirty. If you spot bugs or have suggestions as you explore the code, feel free to submit a pull request or open an issue on GitHub.

O'Reilly Online Learning Platform

This book is also available through the O'Reilly online learning platform, where you'll find all figures in full color. The print version is in black and white, and while we've done our best to ensure clarity, some visualizations do rely on color to be fully interpretable.

We recommend referring to the digital version when viewing figures that use heatmaps, overlays, or color-coded labels.

Quick Tour of the Book

This is a highly interdisciplinary book built around end-to-end examples of deep learning applications in biology. Each chapter walks you through a self-contained project—from biological background to dataset construction, model training, evaluation, and possible extensions. While some chapters build on earlier ones, most can be read independently based on your interests.

To support learning, each chapter includes a brief biology and machine learning primer relevant to the task at hand. We cover a wide range of biological modalities (e.g., sequences, images, and graphs) and deep learning architectures (e.g., transformers, convolutional neural networks, graph neural networks, autoencoders), gradually increasing in complexity.

Here's a summary of the five main technical chapters to give you a quick sense of the tasks, data types, and techniques explored:

Chapter	Title	Task	Modality	Task Type	Architecture	Difficulty
2	"Learning the Language of Proteins"	Predict protein function from protein sequence.	Sequence	Multilabel classification	Linear classifier on transformer embeddings	Beginner
3	"Learning the Logic of DNA"	Predict DNA–protein binding events from sequence.	Sequence	Binary and multilabel classification	Convolutional neural network, transformer	Intermediate
4	"Understanding Drug–Drug Interactions Using Graphs"	Predict whether a pair of drugs will interact.	Graph	Graph-based learning, binary classification	Graph neural network	Intermediate
5	"Detecting Skin Cancer in Medical Images"	Classify skin lesions from images.	Image	Multiclass classification	Convolutional neural network	Intermediate
6	"Learning Spatial Organization Patterns Within Cells"	Predict protein subcellular localization.	Image	Representation learning, multilabel classification	Autoencoder, convolutional neural network	Advanced

The chapters generally build in scope and depth, starting with a simple linear classifier on extracted embeddings in Chapter 2 and progressing to more complex models built from scratch on increasingly rich and challenging data.

Chapter Summaries

Chapter 1, "Introduction"
Learn about the promise and challenges of deep learning in biology. We walk through practical questions to consider before launching a new project—like what your model could replace, whether deep learning is even necessary, and how to structure your workflow.

This chapter also includes a short technical introduction covering JAX/Flax, Python patterns common in machine learning, working environments, and practical setup tips.

Chapter 2, "Learning the Language of Proteins"
Use protein language model embeddings to predict protein function. You'll extract embeddings from a pretrained protein language model and train a simple linear classifier to perform multilabel classification. Along the way, you'll learn how to prepare protein function datasets, evaluate predictions, and visualize embeddings to gain biological insight.

Chapter 3, "Learning the Logic of DNA"
Train lightweight convolutional and transformer models to predict where certain regulatory proteins bind to DNA. You'll learn about how to uncover insights from DNA sequences, interpret models, and evaluate predictions across multiple binding factors.

Chapter 4, "Understanding Drug–Drug Interactions Using Graphs"
Use graph neural networks to model drug relationships, predicting potential drug interactions—harmful or beneficial—based on network features.

Chapter 5, "Detecting Skin Cancer in Medical Images"
Train convolutional neural networks to classify types of skin lesions from images. This chapter focuses on image preprocessing, transfer learning, and managing overfitting in small medical datasets.

Chapter 6, "Learning Spatial Organization Patterns Within Cells"
Build a variational autoencoder to analyze protein localization from microscopy images. This advanced chapter explores unsupervised learning, latent space analysis, and extracting structure from high-resolution imaging data.

Chapter 7, "Tips and Tricks for Deep Learning in Biology"

Gain practical advice for designing, debugging, and improving models in real biological workflows. This chapter also provides guidance on avoiding common pitfalls, extending your learning, and exploring new project ideas.

Conventions Used in This Book

The following typographical conventions are used in this book:

Italic

Indicates new terms, URLs, email addresses, filenames, and file extensions.

`Constant width`

Used for program listings, as well as within paragraphs to refer to program elements such as variable or function names, databases, data types, environment variables, statements, and keywords.

This element signifies a tip or suggestion.

This element signifies a general note.

This element indicates a warning or caution.

Using Code Examples

Supplemental material (code examples, exercises, etc.) is available for download at *https://github.com/deep-learning-for-biology*.

If you have a technical question or a problem using the code examples, please open an issue, submit a pull request on GitHub, or email *support@oreilly.com*.

This book is here to help you get your job done. In general, if example code is offered with this book, you may use it in your programs and documentation. You do not need to contact us for permission unless you're reproducing a significant portion of the code. For example, writing a program that uses several chunks of code from this book does not require permission. Selling or distributing examples from O'Reilly books does require permission. Answering a question by citing this book and quoting example code does not require permission. Incorporating a significant amount of example code from this book into your product's documentation does require permission.

We appreciate, but generally do not require, attribution. An attribution usually includes the title, author, publisher, and ISBN. For example: "*Deep Learning for Biology* by Charles Ravarani and Natasha Latysheva (O'Reilly). Copyright 2025 Charles Ravarani and Natasha Latysheva, 978-1-098-16803-2."

If you feel your use of code examples falls outside fair use or the permission given above, feel free to contact us at *permissions@oreilly.com*.

O'Reilly Online Learning

O'REILLY®   For more than 40 years, *O'Reilly Media* has provided technology and business training, knowledge, and insight to help companies succeed.

Our unique network of experts and innovators share their knowledge and expertise through books, articles, and our online learning platform. O'Reilly's online learning platform gives you on-demand access to live training courses, in-depth learning paths, interactive coding environments, and a vast collection of text and video from O'Reilly and 200+ other publishers. For more information, visit *https://oreilly.com*.

How to Contact Us

Please address comments and questions concerning this book to the publisher:

O'Reilly Media, Inc.
141 Stony Circle, Suite 195
Santa Rosa, CA 95401
800-889-8969 (in the United States or Canada)
707-827-7019 (international or local)
707-829-0104 (fax)
support@oreilly.com
https://oreilly.com/about/contact.html

We have a web page for this book, where we list errata, examples, and any additional information. You can access this page at *https://oreil.ly/deep-learning-for-biology-1e*.

For news and information about our books and courses, visit *https://oreilly.com*.

Find us on LinkedIn: *https://linkedin.com/company/oreilly-media*.

Watch us on YouTube: *https://youtube.com/oreillymedia*.

Acknowledgments

This book was a monumental project—one we could not have completed in isolation—and we are deeply grateful to everyone who supported us along the way. We especially thank Melis Kayikcı and Tibet Fonteyne and our families for their unwavering support. This book would not have been possible without them.

We are also grateful to our technical reviewers—Christian Winkler and Lipi Deepaakshi Patnaik—whose thoughtful feedback helped sharpen both the clarity and correctness of the material. We would also like to thank Danny Elfanbaum for developing the `jupyter-book-to-htmlbook` tool, which greatly improved the content development. Another big thank-you to our O'Reilly editor, Corbin Collins, for guiding the project from start to finish; production editor Ashley Stussy, editors Audrey Doyle and Kim Wimpsett, and WordCo Indexing Services, Inc for helping to polish the manuscript; and acquisitions editor Nicole Butterfield, for encouraging us to take on this project in the first place.

Many friends and colleagues offered insightful reviews, advice, and much-needed encouragement throughout the writing process. Our sincere thanks go to Sam Beckbessinger, Petar Veličković, Justin Dollman, Kristofer Linton-Reid, Felix Kruger, Joshua Pan, Richard Tanburn, Sebastian Bodenstein, Arnaud Aillaud, Vaibhav Bhardwaj, Fredrik Eckardt, Hannes Baukmann, Justin Cope, Levent Mengütürk, Murat Mengütürk, James Sullivan, Loïc Royer, Toby Pohlen, Žiga Avsec, and John Jumper.

Finally, Chapter 2 was inspired by the 2023 Deep Learning Indaba practical "Machine Learning for Biology: Learning the Language of Life," co-developed by Tom Makkink, Kevin Michael Eloff, Natasha Latysheva, and Kyle Taylor.

## CHAPTER 1

Introduction

Biology is increasingly becoming a data-driven science, and deep learning—a powerful subfield of machine learning—is opening new ways to uncover patterns in complex, high-dimensional datasets. As these two fields converge, new opportunities are emerging to extract meaningful insights using modern computational tools. This book is a practical introduction to working at that intersection, focused on developing the skills and mindset needed to apply deep learning effectively in biological contexts.

Getting Started

This opening chapter helps you get oriented. Before jumping into code, we walk through how to frame a project, evaluate your data, and avoid common pitfalls. A bit of structure and planning up front will make your work more reproducible, more flexible, and ultimately more useful and impactful.

Deciding What Your Model Will Replace

The success of a deep learning project in biology often hinges on what happens before you write a single line of code. It's easy to get lost in technical details or spend weeks exploring data and architecture variants that don't lead to meaningful outcomes. Especially in a field as interesting as this one, the temptation to tinker is strong. To stay focused, it helps to ask a few grounding questions up front.

One of the most useful is: *What existing process will my model replace or improve?* The most impactful projects in this field often (though not always) have a clear answer. Here are some examples across different domains:

In healthcare and drug discovery:

Skin cancer classification models
> Aim to replicate a dermatologist's ability to visually diagnose melanoma or other lesions from clinical images, offering faster, more scalable screening for at-risk populations

Pathogen detection systems
> Trained on sequencing data or imaging to detect bacterial or viral infections directly from raw clinical samples (e.g., blood, saliva, or tissue), potentially replacing slower, culture-based diagnostics

Brain tumor segmentation models
> Automate or accelerate the process of outlining tumors on MRI scans, a task that radiologists typically perform manually and with great time investment

Drug-target interaction prediction tools
> Aim to prioritize the most promising compound-target pairs, reducing the need for costly wet-lab screening of massive chemical libraries

Antibiotic resistance prediction models
> Forecast whether a given bacterial strain will resist certain treatments, helping clinicians select effective antibiotics more quickly

In molecular biology:

AlphaFold
> This protein structure prediction model, in many cases, replaces the need for experimentally determining 3D protein shapes via expensive lab-based techniques like X-ray crystallography, cryo-EM, or NMR

Gene expression prediction models
> Forecast gene activity from raw genomic sequences, offering a computational alternative to RNA sequencing (RNA-seq) experiments

Variant effect prediction models
> Help automate the interpretation of genetic mutations, supporting clinical decision making by prioritizing likely pathogenic variants for follow-up analysis or experimental validation

And in ecology and environmental science:

Acoustic species classification systems
> Use forest sound recordings to identify animal species present, offering a scalable and less labor-intensive alternative to in-person biodiversity surveys

Crop disease detection
> Via drone or satellite imagery, enables early identification of plant stress, reducing the need for manual scouting across large fields

Animal facial recognition tools
> Track individual animals, reptiles, birds, and mammals over time without the need for tagging, collars, or other invasive methods

Poaching detection systems
> Trained on infrared or motion sensor data, can automatically flag human activity in protected wildlife zones, assisting conservation efforts

Ideally this gives you a flavor of the kinds of workflows deep learning can improve or even replace. Where possible, try to estimate the potential impact of your model—how much time, cost, or manual labor it could save, or what new insights it might enable. This will help you stay grounded in the real-world utility of your work and communicate its value to collaborators, stakeholders, or the public.

> That said, not every valuable model needs to replace an existing process. Some open up entirely new capabilities—like generating novel biological sequences, uncovering hidden patterns in large datasets, or linking data types that were never connected before. These models might not streamline a lab task, but they can enable new kinds of discovery, expand what questions we can ask, or offer fresh ways to interpret complex systems. If your model creates something new, just be clear about what that is and why it matters—and be thoughtful about how you evaluate success when no established benchmark exists.

Determining Your Criteria for Success

It's important to define, as early and explicitly as possible, what success looks like for your project. Research can be time-consuming and open-ended, so clear goals help you stay focused and avoid endless tweaking—repeatedly changing models, architectures, or training settings without a clear hypothesis or evaluation plan. This kind of trial-and-error loop is common in deep learning due to the large number of design choices and hyperparameters. Without structure, it can waste time and produce results that are hard to interpret or reproduce.

Examples of success criteria include:

Performance metric (e.g., accuracy, AUC, F1)
> You might aim to match the performance of a human expert, achieve a correlation with experimental results comparable to a technical replicate, or keep the false-positive rate below a certain number.

Level of interpretability
> In many applications, it's important not only that a model performs well, but also that its decisions can be understood by domain experts. For instance, you

may prioritize well-calibrated uncertainty estimates or interpretable feature attributions, especially when trust and explainability are critical.

Model size or inference latency
If your model needs to operate in a resource-constrained environment (e.g., smartphones or embedded devices) or meet real-time throughput targets (e.g., process 20 frames per second), your success criterion might focus on efficiency—such as achieving high performance per floating point operation (FLOP), which measures how effectively the model uses computational resources. In such cases, metrics like inference time, memory usage, or energy consumption may matter more than raw accuracy.

Training time and efficiency
When compute is limited—or for educational contexts—you may prioritize fast training or minimal hardware requirements. Since training deep learning models typically involves large matrix operations, they are often accelerated using graphics processing units (GPUs). In low-resource settings, developing a simpler model that trains quickly on a CPU may be a more practical goal than maximizing performance.

Generalizability
In some cases, the goal is to build a model that works well across many datasets or tasks, rather than one that is finely tuned to a single benchmark. For example, *foundational models*—large models trained on broad datasets that can be adapted to many downstream applications—prioritize flexibility and reuse. In such settings, broad applicability may be more valuable than squeezing out the best possible performance on a specific task.

Defining these goals up front helps you answer the key question: *When is the project done?* You'll likely need to balance multiple criteria—but having them laid out early will keep your efforts aligned and your scope realistic.

Invest Heavily in Evaluations

Once you've defined your criteria for success, it's time to prioritize *evaluation*. This means thinking carefully about precisely how you'll measure progress—including what metrics you'll use, how you'll validate results, and which baselines you'll compare against. Without a clear, well-designed evaluation strategy, even a technically impressive model can fail to produce meaningful conclusions.

Strong evaluations don't just help you measure progress. They also help you detect bugs, estimate task difficulty, and build intuition. The key idea is simple: you need a known point of comparison to understand if your model is doing anything meaningful.

While no general rule exists, it wouldn't be surprising if successful machine learning projects spent 50% of their time designing evaluation strategies and running baselines, 25% curating or processing data, and only 25% on model architecture. Without good evaluations, you're flying blind: you won't know whether your model is actually improving, what trade-offs you're making, or even whether it's learning anything meaningful at all.

Spend time here. Evaluation isn't just something you do at the end. It's something you design at the beginning, and it guides the entire project.

Designing Baselines

One of the most practical evaluation tools is a strong *baseline*—a simple method that gives you something to beat. Good baselines help you measure progress, catch bugs early, and understand the difficulty of your task. Sometimes they can even be surprisingly hard to beat.

Designing good baselines requires thinking carefully about the task. Here are a few common baseline strategies for classification tasks:

Random prediction
    Assign labels completely at random, with equal probability for each class. This tells you what performance looks like with no information at all.

Random prediction weighted by class frequencies
    Sample labels randomly, but in proportion to how often they occur in the training data. This is useful for imbalanced datasets.

Majority class
    Always predict the most common class. This can be a surprisingly hard baseline to beat in highly class imbalanced settings.

Nearest neighbor
    Predict the label of the most similar example in the training data (e.g., 1-nearest neighbor using Euclidean distance). This is often effective when inputs are low dimensional or well structured.

And for regression tasks:

Mean or median of the target
    Always predict the average or median target value from the training set. This often matches what a model would do if it's not learning anything meaningful.

Linear regression with a single feature
    Fit a line using just the strongest individual predictor (e.g., one biomarker). This helps gauge how much a more complex model improves over a simple signal.

K-nearest neighbor regression
> Predict the target as the average (or weighted average) of the *k* most similar data points. This is simple to implement and often surprisingly competitive on structured datasets.

And for both:

Simple heuristics
> Use straightforward rules based on domain knowledge. For example, in diagnostics, classify a patient as positive if a single biomarker or measurement exceeds a threshold. For skin cancer images, rank lesions by average pixel intensity. In genomics, if the task is to predict which gene a mutation affects, a simple baseline is to assume it affects the nearest gene in the genome.

> If your model can't beat a basic baseline, something is likely off— and that's useful to know. It's a key signal to revisit your data, features, or modeling approach.

Time-Boxing Your Project

It's important to time-box your project—that is, set a fixed amount of time to work on it, after which you pause or stop regardless of the outcome. Many research ideas "fail" in the sense that they don't achieve the desired metrics. That's normal. All projects, even the unsuccessful ones, generate insights that inform future work. Time-boxing helps ensure that failed experiments still move you forward—without consuming unlimited time and energy.

> Time-boxing doesn't mean giving up easily—it means setting boundaries to maintain focus, avoid burnout, and keep making progress.

Here are some tips for time-boxing effectively:

1. Set a clear deadline
> Choose a realistic time frame (e.g., two weeks, three months) and stick to it.

2. Define checkpoints
> Identify intermediate milestones—like completing dataset preprocessing, training a baseline model, or hitting a certain accuracy—to track progress.

3. Reflect at the end
> Take time to evaluate what worked, what didn't, and what you learned.

Time-boxing is also useful *within* a project. For example: "I'll experiment with this new processing or modeling idea for one week, and if it doesn't help, I'll move on."

The biggest risk with time-boxing is yourself. It's easy to justify extensions, add new ideas, or convince yourself you'll strike gold if you just try 10 more things. Scope creep and perfectionism are common traps. In these cases, it can help to talk to someone else—a collaborator, mentor, or friend—to get perspective and avoid spinning your wheels. A quick conversation can often cut through indecision or obsessiveness and help you refocus on the broader context.

Deciding Whether You Really Need Deep Learning

This might seem like strange advice in a deep learning book, but before diving in, take a moment to ask yourself: *Do I actually need deep learning for this problem?* We'll say it again—seriously consider simpler approaches.

Deep learning models are powerful (and undeniably interesting), but they're also resource intensive, complex to train, and difficult to debug. In many cases, traditional methods—like linear regression, decision trees, or basic statistical techniques—can achieve your goals with far less effort. These approaches are often:

Easier to implement
    Quicker to set up and require less expertise

Less computationally demanding
    Can run on standard hardware with minimal training time

More interpretable
    Easier to explain, troubleshoot, and validate

Carefully weigh the trade-offs. If a simpler method delivers the insights or performance you need, it's often the smarter and more efficient path.

Ensuring That You Have Enough Good Data

Deep learning models don't just need a lot of data—they also generally need high-quality data. Models trained on poor data can often fail catastrophically, regardless of their sophistication.

Make sure you have:

Sufficient quantity
    Deep models typically need thousands of examples or more. What counts as "enough" data depends on your problem and architecture. Check relevant

literature for benchmarks. If you're working with a small dataset, consider *transfer learning*, where you start from a model trained on a related task and fine-tune it on your own data. This approach can dramatically reduce the amount of data needed to achieve good performance.

Sufficient quality
Clean, consistent data is critical. Label errors, noise, or inconsistencies can seriously degrade performance. Even large language models—like those powering modern chat-based assistant systems—can benefit substantially from training on carefully curated, high-quality data. Prioritize quality checks and thoughtful curation.

Assembling a Team

Working alone is totally fine—but teaming up can accelerate progress, improve your ideas, and make the process more enjoyable. Here are some tips for finding great collaborators:

Engage with the community
Join relevant forums, online groups, or webinars to connect with others, share ideas, and discover potential collaborators. Communities like Reddit, Discord servers, X, and specialized Slack groups can be great starting points.

Participate in hackathons and competitions
Platforms like Kaggle, Zindi, or local university events offer structured challenges, feedback, and opportunities to meet people with similar interests.

Form an interdisciplinary team
Combining expertise from different areas often leads to stronger projects. If you're a biologist, team up with someone in machine learning, and vice versa.

Collaborate with experts
Domain experts can help shape your approach and identify blind spots early. Look for collaborators at conferences or workshops, or reach out to authors of relevant papers. Strangers are surprisingly responsive to genuine cold requests from interested people.

And here are some tips for once you find people to work with:

Define clear goals and roles
When working with others, it helps to clarify responsibilities early—who's doing what, what success looks like, and how decisions will be made. This avoids misunderstandings and keeps the project moving.

Use shared tools for collaboration
> Version control (like Git), shared notebooks (e.g., Google Colab), and simple task trackers (like Notion or Trello, or just a shared Google Doc with some lists) can make it much easier to coordinate and stay organized.

Support specialization
> Let people lean into the parts of the project they enjoy most—some may focus on infrastructure and software engineering, others on data curation, modeling, or biological interpretation.

Start small
> If you're unsure about long-term compatibility, try a short project or exploration together first. A small, low-pressure collaboration is a great way to test the waters.

Whether you're working solo or as part of a team, the most important thing is to stay curious, keep learning, and take that first step.

You Don't Need a Supercomputer or a PhD

There are a few common misconceptions about working in deep learning for biology that are worth challenging:

You need huge budgets and compute power.
> In an age of multimillion-dollar training runs for massive language models, it's easy to assume you need vast resources. But that's not always the case:

Prototype with small models
> Start small to iterate quickly. You might find that lightweight models are more than enough for your goals.

Use free or affordable compute
> Platforms like Google Colab and Kaggle offer free GPU access for smaller projects. For more demanding workloads, cloud providers such as Amazon Web Services (AWS), Microsoft Azure, and Google Cloud Platform (GCP) offer scalable paid instances.

Not everything is about scale
> Many valuable projects focus on analyzing existing models rather than training new ones. These often require modest compute but can yield deep insight. There's still a lot we don't understand about how deep models behave.

You need deep expertise in machine learning or biology (or both).
Another myth is that only highly trained experts can contribute meaningfully. In reality:

Better tools
Modern frameworks make it easier than ever to build and experiment with powerful models.

Open source culture
Freely available code and pretrained models let you learn from and build on existing work.

Educational resources
There's now no shortage of tutorials, videos, and walkthroughs online to help you get started.

Plenty of untapped problems
Many important biological questions remain unexplored by machine learning. You don't need a PhD or a Kaggle medal to work on them.

While the bleeding edge of research may require specialized knowledge and high-end infrastructure, there's plenty of room in this field for curiosity, creativity, and new perspectives—no supercomputer required.

As you explore this field, you'll almost certainly come across academic papers—whether you're digging into a specific method, reading related work, or looking for project ideas. Both biology and machine learning papers can feel impenetrable at first: the language is dense, the ideas are highly condensed, and there's often a lot of jargon. But remember:

- You're seeing the result of months or years of work by a team of researchers—and encountering it for the first time.
- Reading papers is a skill, and like any skill, it improves with practice.
- Blog posts, YouTube videos, and open source projects can also be great, more accessible ways to learn the same concepts.

With that background in place, let's dive into the technical foundations of this book.

Technical Introduction

We'll be using Python-based deep learning frameworks, in particular, JAX and Flax. JAX is a system for high-performance numerical computing and machine learning, and Flax is a flexible neural network library built on top of JAX. We'll motivate by explaining why we're using JAX and reviewing a few Python features that tend to come up in a lot of machine learning (ML) code. Then, we'll introduce some foundational machine learning concepts recurring throughout, with the main one being how a training loop is structured.

> As mentioned in "Prerequisites" on page xii, this book assumes basic knowledge of Python. If you're new to Python, check out the recommended resources listed there first.

Finally, to avoid repetition across chapters, we've created a small companion library called dlfb (Deep Learning for Biology), which wraps up common utilities and components. We'll reference it throughout the book.

> Don't worry if some parts of this technical introduction feel unfamiliar or challenging at first. You're welcome to skim or skip ahead. Many of the concepts will become clearer when you see them in action later in the book.

Why JAX and Flax?

This book uses the JAX and Flax ecosystem. But why did we make this choice, when other options like PyTorch or Keras are more common?

First, some honesty: there is no one "best" framework. All of them can be used to build effective biological models, and many of the concepts in this book will carry over easily if you're using PyTorch or Keras.

We've chosen JAX/Flax primarily because:

Familiar NumPy API
    JAX's jax.numpy module (commonly imported as jnp) offers an API that so closely mirrors standard NumPy for array manipulation and mathematical operations that NumPy np calls can often be directly substituted with jnp. This means users already proficient with NumPy can transition to JAX with a significantly reduced learning curve, leveraging their existing knowledge to quickly build and adapt code while gaining JAX's powerful transformations and accelerator support.

Functional programming encourages clarity
    JAX's pure function style can reduce hidden state and make training logic more transparent. This fits well with the educational goals of the book—explicit is better than implicit.

Transformations are first-class

JAX provides powerful, composable transformations like `jit` (just-in-time compilation), `grad` (automatic differentiation), and `vmap` (vectorization) that work cleanly on Python functions. These tools simplify and unify many aspects of model training and evaluation.

JAX aligns with cutting-edge research

JAX has gained traction in recent machine learning research, particularly for biology, physics, and large-scale models. Using it here helps you align with newer toolchains and experiment with modern practices.

Speed

JAX uses a compiler that can yield significant performance gains on specialized hardware such as GPUs (e.g., from NVIDIA or AMD) and TPUs (made by Google), making it well-suited for large-scale deep learning workloads. This compiler is based on XLA, a low-level system for optimizing numerical computations on accelerators.

That said, JAX and Flax come with trade-offs: a smaller ecosystem and APIs that evolve quickly (sometimes breaking things along the way). And while JAX can offer impressive speedups, that speed isn't exclusive to JAX/Flax. For example, Keras (*https://keras.io*) now supports a JAX backend, offering another option for users who prefer a higher-level API. If you're already comfortable with PyTorch, Keras, or TensorFlow, you're welcome to implement the ideas in this book using those tools—and even contribute your own version to this book's repository.

It's not necessary when you're just starting out, but over time it can be helpful to become familiar with more than one deep learning framework. Each has strengths in different ecosystems—for example, we use PyTorch in Chapter 2 to extract pretrained embeddings from a Hugging Face model, since many models on Hugging Face are primarily released and maintained in PyTorch.

> The deep learning field moves fast. While we use the `linen` API in Flax throughout this book, a newer API called `nnx` has recently emerged as the recommended way to build models. `linen` remains fully supported, but be aware that you may come across other tutorials or examples that use `nnx`, which has a slightly different syntax.

We'll introduce key JAX concepts as needed throughout the book, but we won't cover the entire library in detail. For more detailed hands-on learning, check out the official JAX tutorials (*https://oreil.ly/Jcqtu*). And if you run into unexpected behavior, the JAX "sharp bits" notebook (*https://oreil.ly/TcOzQ*) is an excellent reference for common gotchas and how to avoid them.

A note on performance

As this is an educational book, our focus is on clarity rather than peak performance. That means we won't cover things like precision tuning, advanced hardware strategies, or distributed training. But these things can matter a lot in real-world setups.

If you're comfortable with the basics and want to go deeper, here are a few areas worth exploring:

Numerical precision and tuning
> Many machine learning operations, especially matrix multiplications (matmuls), benefit from reduced-precision formats like `bfloat16`, which can significantly improve speed and memory usage with minimal impact on model accuracy. JAX lets you control the precision used for matmuls via `jax.default_matmul_preci sion`, helping you take advantage of specialized hardware like Tensor Cores (on NVIDIA GPUs) or matrix units (on TPUs). Lower-precision training is widely used in large-scale setups because it enables training larger models more efficiently and cost-effectively.

Profiling tools like `jax.profiler` or TensorBoard
> Profiling helps you identify where your code is spending time and memory so you can spot bottlenecks in training and optimize the most expensive operations.

Memory-efficient training techniques
> Methods like gradient checkpointing (remat in JAX) let you trade off computation for memory, allowing you to train deeper models without running out of RAM.

Multihost/multidevice training
> Training across multiple GPUs, TPUs, or even machines allows you to scale up models and datasets that wouldn't fit on a single device.

You won't need any of these to follow this book, but they are good to be aware of and are worth exploring as you grow more comfortable with the JAX ecosystem.

Python Tips

While this book doesn't cover Python background knowledge in depth, this section highlights a few helpful Python concepts you're likely to encounter when working with machine learning code in general—and with JAX and Flax in particular.

Type annotations and docstrings

Python is a dynamically typed language, meaning you don't need to declare variable types (such as strings or integers) explicitly. Instead, types are determined at runtime, which makes the language flexible—but this flexibility can also make bugs harder to catch, especially in larger codebases. Adding *type annotations* helps mitigate this by improving readability, enabling static type checks with tools like mypy, and making debugging easier.

Here's a simple function that computes the mean squared error (MSE) between two NumPy arrays:

```python
import numpy as np


def mean_squared_error(y_true, y_pred):
    squared_errors = (y_true - y_pred) ** 2
    return np.mean(squared_errors)
```

Here is an example of its use:

```python
y_true = np.array([1.1, 0.1, 1.0])
y_pred = np.array([0.9, 0.2, 1.2])
mean_squared_error(y_true, y_pred)
```

Output:

```
np.float64(0.030000000000000002)
```

We can improve this function by adding type hints to specify that the inputs are np.ndarray objects and the return type is a float, along with a docstring to explain what the function does:

```python
def mean_squared_error(y_true: np.ndarray, y_pred: np.ndarray) -> float:
    """
    Calculate the Mean Squared Error (MSE) between two NumPy arrays.

    Args:
        y_true (np.ndarray): Ground-truth values.
        y_pred (np.ndarray): Predicted values.
    """
    squared_errors = (y_true - y_pred) ** 2
    return np.mean(squared_errors)
```

These changes don't affect the function's behavior, but they offer several benefits:

Clarify input and output types

It's immediately clear that y_true and y_pred should be NumPy arrays and the return value is a float. Note that some machine learning code goes further by specifying the data type within the array (e.g., arr: NDArray[np.float64]), but we will not do this in this book.

Enhance documentation

IDEs and documentation tools can provide better inline help and autocompletion. This can really improve productivity.

Improve readability

The function is easier to understand for others (or your future self).

Enable static checking
  Tools like mypy can catch type-related errors.

The MSE example is very simple, so adding type hints and a full docstring is arguably overkill—but the principle is important.

We won't necessarily always use typing and docstrings throughout this book due to space constraints, but they're good habits to adopt in your own projects. When we do include docstrings in the main book text, we'll usually keep them to a single line, which will save a few trees in printing.

Decorators

Decorators are functions that modify the behavior of other functions or methods. In machine learning and data science, decorators are often used to enhance performance, cache results, or log function behavior.

Just-in-time (JIT) compilation with JAX. One of the most common decorators when working in JAX is `jax.jit`, which performs JIT compilation to accelerate code execution. The first run of a JIT-compiled function is slower because it is compiled to XLA (Accelerated Linear Algebra) machine code. However, subsequent calls run significantly faster.

Suppose we have a function that takes a JAX array, raises all values to the 10th power, and sums them:

```python
import jax
import jax.numpy as jnp


def compute_ten_power_sum(arr: jax.Array) -> float:
    """Raise values to the power of 10 and then sum."""
    return jnp.sum(arr**10)


arr = jnp.array([1, 2, 3, 4, 5])
compute_ten_power_sum(arr)
```

Output:

```
Array(10874275, dtype=int32)
```

We can speed up this function in one of two ways. Either directly:

```python
jitted_compute_ten_power_sum = jax.jit(compute_ten_power_sum)
jitted_compute_ten_power_sum(arr)
```

Output:

```
Array(10874275, dtype=int32)
```

Or, via the Python syntactic sugar, we can apply it when defining the function with the `@jax.jit` decorator:

```python
@jax.jit
def compute_ten_power_sum(arr: jax.Array) -> float:
    """Raise values to the power of 10 and then sum."""
    return jnp.sum(arr**10)



compute_ten_power_sum(arr)
```

Output:

```
Array(10874275, dtype=int32)
```

The output remains the same for all, but the jitted functions run much faster. If you are working in a Jupyter notebook, you can use `%timeit` to measure the execution time for a line of code (or `%%timeit` for entire cells). Try timing the function with and without `@jax.jit`. On a GPU, you may see an ~20× speedup.

> How does `@jax.jit` work? Briefly, when you apply `@jax.jit`, JAX doesn't run your function like regular Python. Instead, it first traces the function—it runs through it once using special tracer objects (not real data) to build a computation graph. This graph is a static representation of all the numerical operations performed, with control flow unrolled and variable shapes and types fixed.
>
> Once the graph is built, JAX compiles it using XLA (Accelerated Linear Algebra), a backend that generates highly optimized machine code. This compiled version is cached and reused whenever the function is called again with the same input shapes and types—leading to significant speedups.

JIT compilation is powerful, but it comes with trade-offs, especially for debugging. This is because:

- Python debugging tools like `print()` statements or `pdb` don't behave as expected inside jitted functions.
- Side effects (e.g., `print()`, logging, or modifying a list) are not actually executed during tracing, since JAX skips anything that doesn't affect the computation graph.
- Error messages can refer to internal JAX or XLA code instead of your original function and can be quite cryptic.

While you can temporarily disable jitting by commenting out `@jax.jit`, this becomes impractical if many functions rely on JIT. Fortunately, you can globally disable JIT by setting the environment variable `JAX_DISABLE_JIT=True`, which forces all jitted functions to run normally. This is a convenient way to debug without rewriting your code. See the JAX debugging documentation (*https://oreil.ly/oXI97*) for more details.

Preconfiguring JAX jit with partial. There is a common source of confusion in machine learning code regarding usage of `partial`, especially with JAX code. The use of `functools.partial` is to prefill (or "bind") some arguments of a function, returning a new function with those values fixed. This is a general Python utility and is not specific to JAX or ML.

Here, we adapt the `scale` function and create a new function, `scale_by_10`:

```
from functools import partial


def scale(x, scaling_factor):
    return x * scaling_factor


# Create a new function 'scale_by_10' where 'scaling_factor' is fixed to 10.
scale_by_10 = partial(scale, scaling_factor=10)
scale_by_10(3)
```

Output:

```
30
```

Here, `scale_by_10` is a new function that behaves like `scale(x, 10)`.

In the context of JAX, `partial` is often used to customize a decorator before applying it, like this: `@partial(jax.jit, static_argnums=...)`. This is a way to configure the `jax.jit` decorator itself. As mentioned previously, `jax.jit` compiles your Python function for speed. However, JAX needs to know if certain arguments are static. Static arguments are typically non-JAX array types (like integers, strings, or booleans) that control the structure of the computation (e.g., in if/else statements). If a static argument changes, JAX may need to recompile the function.

Let's say we want to compute a summary statistic over an array, choosing either the mean or the median based on a string argument `average_method`. Since this choice affects the control flow, JAX needs to know the value of `average_method` at compile time:

```
from functools import partial

import jax
import jax.numpy as jnp


@partial(jax.jit, static_argnums=(0,))
def summarize(average_method: str, x: jax.Array) -> float:
```

```
if average_method == "mean":
    return jnp.mean(x)
elif average_method == "median":
    return jnp.median(x)
else:
    raise ValueError(f"Unsupported average type: {average_method}")


data_array = jnp.array([1.0, 2.0, 100.0])

# JAX compiles one version of 'summarize' for average_method="mean".
print(f"Mean: {summarize('mean', data_array)}")

# JAX compiles another version for average_method="median".
print(f"Median: {summarize('median', data_array)}")

# Calling with "mean" again uses the cached compiled version.
print(f"Mean again: {summarize('mean', data_array)}")
```

Output:

```
Mean: 34.333335876464844
Median: 2.0
Mean again: 34.333335876464844
```

If we didn't mark `average` as static with `static_argnums=(0,)`, JAX would throw an error, because it can't trace control flow that depends on strings unless it knows their value ahead of time. Marking arguments as static tells JAX to compile a separate, specialized version of the function for each unique value of that static argument it encounters.

> A bit of a clarification on the meaning of "static" versus "dynamic": JAX treats most numerical inputs (like `jax.Array`, `float`, or `int`) as dynamic, meaning they can vary between calls without requiring recompilation, as long as their shapes and types stay the same.
>
> Other inputs, like strings, Python objects, or functions, are static: they affect control flow or can't be traced as part of the computation graph. If you pass them into a jitted function, you must mark them as static using `static_argnums` or close over them using a closure instead (see the next section).

Closures

A closure is a function that "remembers" the environment in which it was created. This means it can access variables from its enclosing (outer) function's scope, even after that scope has finished executing:

```
def outer_function(x):
    def inner_function(y):
        return x + y  # inner_function "closes over" x.
```

```
    return inner_function


add_five = outer_function(5)  # x is 5.
result = add_five(10)  # y is 10.
print(f"Closure result: {result}")
```

Output:

```
Closure result: 15
```

In this example, add_five is a closure. It "remembers" that x was 5 when outer_func
tion was called.

Closures are used extensively in JAX-based machine learning code. Many components—like loss functions, regularizers, and augmentation pipelines—are parameter-
ized by configuration values. Instead of passing these values as arguments (which
might require static_argnums if used inside control flow), they're often closed over.
A little later we will see this in action when defining the JAX training loop.

Generators

Generators are functions that allow you to iterate over data lazily, yielding one item
at a time. They are especially useful for working with large datasets where loading
everything into memory (RAM) at once is impractical or impossible.

Here's a simple generator function that simulates streaming batches of data:

```
from typing import Iterator

def data_generator() -> Iterator[dict]:
    """Yield data samples with features and labels."""
    for i in range(5):
        yield {"feature": i, "label": i % 2}

# Example usage.
generator = data_generator()
next(generator)
```

Output:

```
{'feature': 0, 'label': 0}
```

We will be working with TensorFlow datasets (TFDSs) in some chapters. Because JAX
doesn't include a native data-loading library, it's common to see hybrid setups using
TFDS. If you have your data in NumPy arrays, you can easily create a TensorFlow
dataset using tf.data.Dataset.from_tensor_slices. This allows you to integrate
NumPy data into TensorFlow pipelines for efficient training and preprocessing. It
provides a clean API for batching, shuffling, and prefetching (preloading data before
it's needed to increase training speed), which is helpful when getting started:

```
import tensorflow as tf

features = np.array([1, 2, 3, 4, 5])
labels = np.array([0, 1, 0, 1, 0])

# Create a TensorFlow dataset from the NumPy arrays.
dataset = tf.data.Dataset.from_tensor_slices((features, labels))

# Batch dataset with batch size of 2 and drop the final batch if incomplete.
batched_dataset = dataset.batch(2, drop_remainder=True)

# Create a dataset (ds) iterator and retrieve the first batch using next().
ds = iter(batched_dataset)
next(ds)
```

Output:

```
(<tf.Tensor: shape=(2,), dtype=int64, numpy=array([1, 2])>,
 <tf.Tensor: shape=(2,), dtype=int64, numpy=array([0, 1])>)
```

In later chapters, we will also be writing custom data pipelines that offer more control.

Anatomy of a Training Loop with JAX/Flax

While the details of machine learning projects may vary, the core structure of training a model remains fairly consistent. This core structure will serve as a foundation throughout the chapters of this book. Let's walk through the basic layout of training a model using JAX and Flax—a pattern you can build upon as you explore more complex examples.

Defining a dataset

Let's create some toy data where the target y values are a linear transformation of the x values with a bit of random noise added. We'll use the relationship y = 2x + 1 with Gaussian noise, as you can see in Figure 1-1.

```
import jax
import jax.numpy as jnp
import matplotlib.pyplot as plt
from flax import linen as nn

# In JAX, randomness is handled explicitly by passing a random key.
# We create a key here to seed the random number generator.
rng = jax.random.PRNGKey(42)

# Generate toy data: x values uniformly sampled between 0 and 1.
rng, rng_data, rng_noise = jax.random.split(rng, 3)
x_data = jax.random.uniform(rng_data, shape=(100, 1))

# Add Gaussian noise.
```

```
noise = 0.1 * jax.random.normal(rng_noise, shape=(100, 1))

# Define target: y = 2x + 1 + noise.
y_data = 2 * x_data + 1 + noise

# Visualize the noisy linear relationship.
plt.scatter(x_data, y_data)
plt.xlabel("x")
plt.ylabel("y")
plt.title("Toy Dataset: y = 2x + 1 + noise")
plt.show()
```

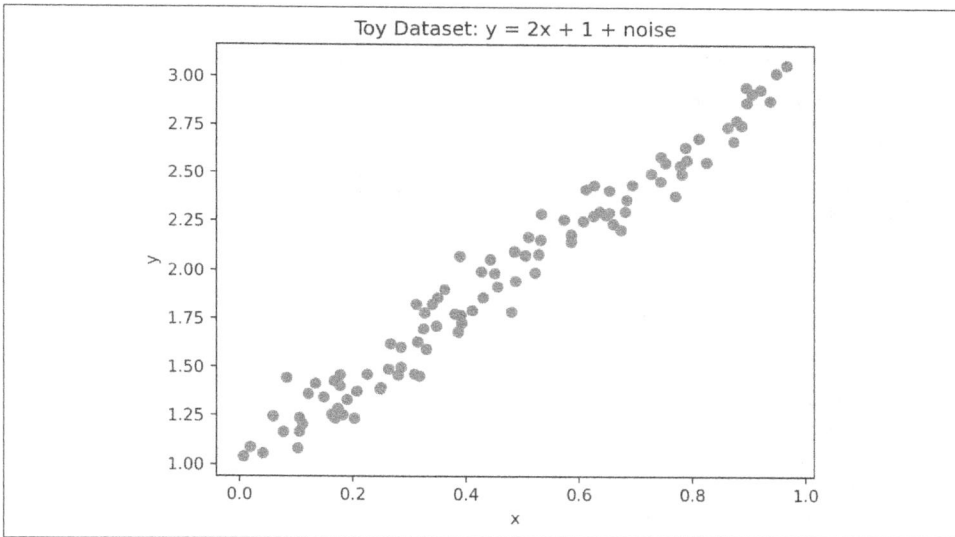

Figure 1-1. This scatterplot visualizes the underlying relationship that we want our model to learn.

Defining a model

In Flax, we define a model by inheriting from nn.Module. The @nn.compact decorator allows us to define layers directly inside the __call__ method, rather than in the class's setup() method. This is especially useful for simple, sequential models.

Here's a minimal example, a single linear (dense) layer with one output unit and no activation function:

```
class LinearModel(nn.Module):
    @nn.compact
    def __call__(self, x):
        # Applies a single dense (fully connected) layer with 1 output neuron.
        # That is, it computes y = xW + b, where the output has dimension 1.
        return nn.Dense(features=1)(x)
```

We can instantiate the model like this:

```
model = LinearModel()
```

To initialize the model's parameters, use the `.init` method with a random key and a sample input. This allows Flax to infer the input and output shapes. Here, we pass in a dummy input with shape [1, 1]—one example (batch size of 1) with one input feature, matching the shape of our toy data:

```
rng = jax.random.PRNGKey(42)
variables = model.init(rng, jnp.ones([1, 1]))
```

This initializes the model's parameters. The result is a dictionary containing the weights and bias for the dense layer:

```
print_short_dict(variables)
```

Output:

```
{'params': {'Dense_0': {'kernel': Array([[-0.5220277]], dtype=float32), 'bias':
Array([0.], dtype=float32)}}}
```

Here:

- `kernel` is the learned weight matrix (shape [1, 1], since our input and output dimensions are both 1).

- `bias` is the learned bias term added after the matrix multiplication.

Note that while the Flax API has evolved over time, the core ideas remain stable: defining model layers, initializing parameters, and inferring shapes from inputs. Moreover, even if the syntax changes, these fundamentals will carry over to nearly any deep learning framework.

Why infer shapes from input? In Flax, the shapes of the model's weight and bias parameters are actually unknown until you run data through the model. That's because Flax follows a *functional* style: layers don't store input shape information when you define them. Instead, you provide a sample input during initialization, and Flax infers the necessary parameter shapes on the fly.

Other libraries, like PyTorch or Keras, use an object-oriented style, where layers often remember input shapes internally. This can make model construction feel more automatic, but Flax's approach gives you more control and makes model behavior easier to inspect and debug, especially when working with JAX's just-in-time (JIT) compilation.

Creating a training state

A *training state* in Flax is a container that packages together everything you need for training: the model's parameters, the optimizer, and the function used to apply the model. Let's build one:

```python
import optax
from flax.training import train_state

# Define an optimizer — here we use Adam with a learning rate of 1.0.
# (Note: in most real settings you'd use a smaller learning rate like 1e-3).
tx = optax.adam(1.0)

# Create the training state.
state = train_state.TrainState.create(
  apply_fn=model.apply,  # The model's forward pass function.
  params=variables["params"],  # The initialized model parameters.
  tx=tx,  # The optimizer.
)
```

The `TrainState` object is designed to make training in Flax cleaner and more manageable. It holds everything needed for model updates:

params
> The current model parameters.

tx
> The optimizer (in this case, Adam). "Tx" is short for transformation. In Optax (JAX's optimization library), optimizers are defined as transformations of gradients. For example, Adam transforms raw gradients using momentum and adaptive scaling.

apply_fn
> The function that runs the model's forward pass.

Importantly, the training state is immutable—rather than modifying it in place, each update returns a new `TrainState` with the updated parameters. This functional style is consistent with JAX's overall design and helps keep computations pure and traceable.

> Although `TrainState` is immutable and each update returns a new object, this doesn't lead to memory issues. JAX reuses memory efficiently, especially inside `@jit`-compiled functions.

Defining a loss function

The loss function measures how close the model's predictions are to the true targets. Here, we use MSE, which is common for regression tasks:

```
def calculate_loss(params, x, y):
    # Run a forward pass of the model to get predictions.
    predictions = model.apply({"params": params}, x)

    # Compute MSE loss.
    return jnp.mean((predictions - y) ** 2)
```

Note that `model.apply` works here because `model` was defined earlier and is available in the current scope (e.g., the same notebook or script). We don't need to pass it as an argument, because the function is still pure—all variable model state comes from the `params` we pass in.

Here's how you would call this function with your current model and data:

```
loss = calculate_loss(variables["params"], x_data, y_data)
print(f"Loss: {loss:.4f}")
```

Output:

```
Loss: 5.2768
```

This computes the current loss by applying the model to the data using the (random) parameters stored in `state`. Since the model hasn't learned from the data yet, the loss is expected to be relatively high. As training progresses, this number should steadily decrease—which would tell us that the model is improving at its task of predicting the targets from the inputs.

Defining the training step

The training step performs a forward pass, computes the loss and gradients, and updates the model parameters. We use `jax.jit` to compile the entire step for efficiency. While JAX can run on GPUs and TPUs without it, using `jit` ensures that the code is compiled into a single optimized graph—which runs significantly faster and takes full advantage of accelerator hardware.

```
@jax.jit
def train_step(state, x, y):
    # Compute the loss and its gradients with respect to the parameters.
    loss, grads = jax.value_and_grad(compute_loss)(state.params, x, y)

    # Apply gradient updates.
    new_state = state.apply_gradients(grads=grads)

    return new_state, loss
```

Often, you don't just want the gradients (to update the parameters). You also want to log the loss explicitly (e.g., to plot it over time). But computing gradients requires computing the loss anyway. Instead of doing this twice, `jax.value_and_grad` is a convenient utility that does both at once:

- It evaluates the function you give it (in this case, `calculate_loss`) to get the loss value.
- It computes the gradients of that loss with respect to the parameters.

This avoids redundant computation.

The result of our `train_step` function is a new `TrainState` with updated parameters, along with the current loss, which we can use to monitor training progress.

Note that often you will encounter the training step defined with a closure like so:

```
@jax.jit
def train_step(state, x, y):
  def calculate_loss(params):
    # state, x and y are not part of the function signature but are accessed.
    predictions = state.apply_fn({"params": params}, x)
    return jnp.mean((predictions - y) ** 2)

  loss, grads = jax.value_and_grad(calculate_loss)(state.params)
  state = state.apply_gradients(grads=grads)
  return state, loss
```

Here, `state`, `x`, and `y` are closed over—they're not part of the `calculate_loss` function's input signature, making the code more compact and easier to read and reason about.

Handling auxiliary outputs in the loss function

As a slight aside (but one that comes up often), sometimes we want to return extra information from the loss function—for example, predictions or metrics for logging—without affecting the gradient computation. JAX makes this easy with the `has_aux=True` flag. It tells `value_and_grad` to treat everything after the loss as "auxiliary" and to exclude it from gradient computation.

For example, let's modify our loss function to also return predictions and accommodate this by using `has_aux=True` in `jax.value_and_grad` in our `train_step`:

```
@jax.jit
def train_step(state, x, y):
  def calculate_loss(params):
    predictions = state.apply_fn({"params": params}, x)
    loss = jnp.mean((predictions - y) ** 2)
    return loss, predictions  # Return both loss and preds (aux info).
```

```
(loss, predictions), grads = jax.value_and_grad(calculate_loss, has_aux=True)(
    state.params
)
state = state.apply_gradients(grads=grads)
return state, (loss, predictions)
```

Without `has_aux=True`, JAX expects the loss function to return a single scalar. Returning anything else (like predictions) will raise an error. By setting `has_aux=True`, you're telling JAX: "only differentiate with respect to the loss; ignore any extra outputs, like predictions."

Defining the training loop

Now that all the components are in place, we can define the training loop that actually updates the model parameters based on the data.

In most machine learning workflows, training happens over either *steps* or *epochs*:

- A *step* refers to one update of the model using a single batch of data.
- An *epoch* is a full pass through the entire training dataset—typically consisting of many steps. In our toy example, we feed the entire dataset (100 input–output pairs) into the model at once, without batching. This means each step is equivalent to a full epoch. In more realistic settings, you'd typically break the data into batches, resulting in many steps per epoch.

Let's now train the model:

```
num_epochs = 150   # Number of full passes through the training data.

for epoch in range(num_epochs):
    state, (loss, _) = train_step(state, x_data, y_data)
    if epoch % 10 == 0:
        print(f"Epoch {epoch}, Loss: {loss:.4f}")
```

Output:

```
Epoch 0, Loss: 5.2768
Epoch 10, Loss: 0.9498
Epoch 20, Loss: 0.1091
Epoch 30, Loss: 0.0845
Epoch 40, Loss: 0.0283
Epoch 50, Loss: 0.0258
Epoch 60, Loss: 0.0106
Epoch 70, Loss: 0.0105
Epoch 80, Loss: 0.0106
Epoch 90, Loss: 0.0102
Epoch 100, Loss: 0.0101
Epoch 110, Loss: 0.0100
Epoch 120, Loss: 0.0100
```

```
Epoch 130, Loss: 0.0100
Epoch 140, Loss: 0.0100
```

The model converges quickly—its loss decreases rapidly to a stable, low value. After training, we can test how well the model has learned the underlying pattern by comparing its predictions to the true target values, as shown in Figure 1-2:

```python
# Generate test data (x values between 0 and 1).
x_test = jnp.linspace(0, 1, 10).reshape(-1, 1)
y_test = 2 * x_test + 1  # Ground truth: linear function without noise.

# Get model predictions.
y_pred = state.apply_fn({"params": state.params}, x_test)

plt.scatter(x_test, y_test, label="True values")
plt.plot(x_test, y_pred, color="red", label="Model predictions")
plt.xlabel("x")
plt.ylabel("y")
plt.legend()
plt.title("Linear Model Predictions vs. True Relationship")
plt.show()
```

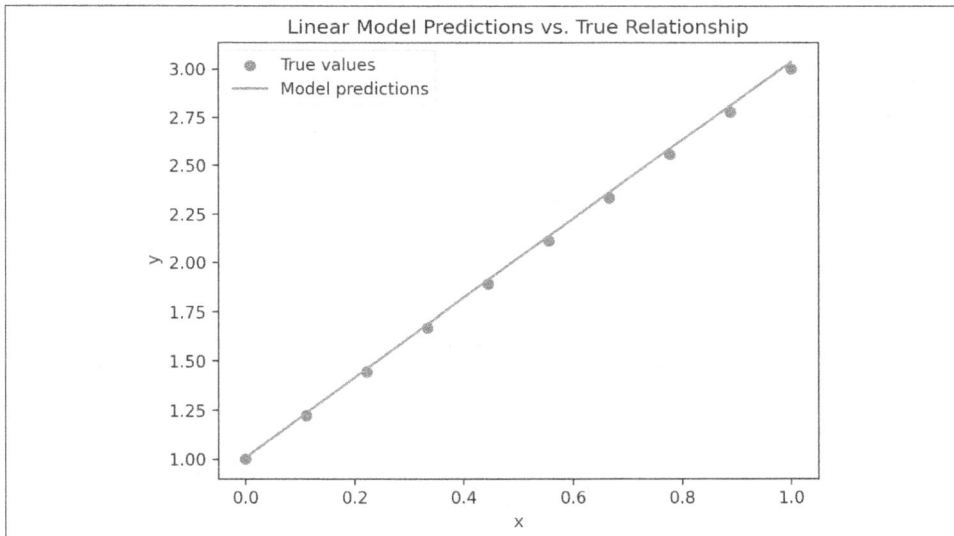

Figure 1-2. Scatterplot comparing the linear model predictions with the true values.

We can see that after training, the model has learned to approximate the true function y=2x+1 very closely. The predicted outputs are nearly identical to the expected values, which is exactly what we want in this toy regression task.

This may be a simple example, but it captures the core structure of almost every deep learning workflow: define a model, compute a loss, update parameters, and repeat. While real projects involve more complexity—batching, data pipelines, regularization, metrics, logging, and so on—the basic loop you've built here is the foundation of it all. You now have a solid mental model for how training works in JAX and Flax—and the foundation to build much more powerful systems.

> Where to go from here? Once you've built a working training loop (an important milestone), it's common to expand in several directions: add metrics that capture key aspects of model performance, split off a validation set to check generalization, and track progress over training time. These are all building blocks of real-world deep learning workflows, and we'll walk through them in upcoming chapters.

Machine Learning Tips

Here's a brief reminder of some machine learning concepts, focusing on what we use throughout the book. We explain these ideas in more detail as necessary when they come up.

Types of Tasks

In *classification*, the model predicts a label or a probability distribution over labels. There are three types of classification:

Binary classification
    Choose between two options; for example, whether a cell is healthy or not.

Multiclass classification
    Choose one label out of several; for example, based on patterns in a biological sample, a model might predict which part of the body it came from—like the brain, liver, or skin. Each sample belongs to exactly one class.

Multilabel classification
    Choose all labels that apply, not just one. For example, when analyzing a cell image, the model might predict which structures are visible—such as the nucleus, membrane, and mitochondria. In this case, multiple labels can be true for the same image.

Regression involves predicting a continuous value, such as the binding strength between two molecules.

We also use *representation learning*—learning useful embeddings or feature representations without direct supervision. These embeddings capture patterns or structure in the data, which can then be used for tasks like clustering, visualization, or as inputs to downstream models.

Types of Architectures

In this book, we primarily use the following model architectures:

Linear models and multilayer perceptrons (MLPs)
> These are fully connected networks that transform input vectors into output vectors through stacked dense layers. They are simple and widely used.

Convolutional neural networks (CNNs)
> CNNs apply spatial filters to model local structure, typically mapping images to images or feature maps. They are especially effective for image and sequence data.

Transformers
> Model long-range dependencies in sequences using attention. They operate on sets or sequences and are now state of the art in many areas of biology and language modeling.

Graph neural networks (GNNs)
> Operate on graph-structured data, passing messages between nodes and their neighbors. They are useful when data has relational or interaction-based structure.

Autoencoders
> Learn compact representations by encoding inputs into a latent space and reconstructing them. They are common in unsupervised learning and denoising tasks..

We'll explain each of these as they appear throughout the book.

Inductive Biases

Modern deep learning architectures are increasingly mature. Rather than trying to invent the next transformer, you'll often get further by composing existing building blocks in thoughtful ways, guided by domain knowledge.

Think about what *inductive biases* (assumptions about the structure of the data) make sense for your problem: Are nearby data points related (like the pixels in images)? Is there a sequential structure (like in DNA or text)? Do entities interact in a graph-like way? These structural assumptions help your model learn more efficiently by narrowing the search space of possible functions.

A common type of inductive bias is an *invariance*—where the model assumes that certain transformations shouldn't affect the output. For example:

- CNNs assume *translation invariance*: shifting an image slightly shouldn't change the prediction.

- GNNs assume *permutation invariance*: the order of nodes in a graph shouldn't affect the output.

- Transformers are *permutation invariant* by default: they ignore input order unless positional encodings are added. This gives them flexibility, but it also means they rely on you to encode any sequential structure.

Dataset Splits

Machine learning models are typically trained using three dataset partitions:

Training set
> Used to fit the model's parameters (its weights) by minimizing a loss function

Validation set
> Used to tune *hyperparameters* (discussed next) and evaluate performance during development

Test set
> Held out until the end to assess the model's final performance on truly unseen data, providing an estimate of how well it generalizes

This split structure is good practice because it helps ensure your model generalizes well to new data and gives you a reliable estimate of its real-world performance.

Hyperparameters

Hyperparameters control how a model is trained. They're set before training begins and are not updated by the optimizer. Common examples include:

Learning rate
> How quickly the model updates its weights

Batch size
> How many examples are processed together during one update

Model size
> The number of layers, hidden units, or attention heads

Regularization
> The dropout rate or weight decay, to prevent overfitting

We evaluate different hyperparameter settings using performance on the validation set. If we tune hyperparameters based on training performance alone, we risk *overfitting*—the model may simply memorize the training data, including noise or outliers.

Overfitting leads to poor performance on new data. In contrast, *generalization* means the model has learned patterns that apply beyond the training examples. This is a key requirement in biological applications, where test data can often come from different experiments or conditions.

Activation Functions

Activation functions introduce nonlinearity into neural networks, enabling them to model complex relationships between inputs and outputs.

As data flows through a model, it's transformed into intermediate tensors known as *activations*. These activations pass through a series of linear projections and nonlinear activation functions at each layer. The output of the final hidden layer is then often passed through a final activation function, which produces the model's ultimate prediction.

Choosing the right final activation is important—it should reflect the kind of data you're trying to predict. For example:

- Use `sigmoid` for binary classification, where outputs should lie between 0 and 1.
- Use `softmax` for multiclass classification, where outputs represent probabilities across categories.
- Use *no* final activation if your loss function expects raw logits (e.g., `sigmoid_cross_entropy` or `softmax_cross_entropy`). These loss functions apply the activation internally, so applying it yourself would be redundant or even harmful.
- Avoid ReLU or GELU as final activations when predicting real-valued outputs that can be negative—they will clip or distort negative values.
- For regression tasks, no activation is often best—or use tanh or sigmoid only if your target values are known to be bounded (e.g., between -1 and 1 or 0 and 1).

When in doubt, check the range of your target values, and choose an activation (or none) that allows the model to produce outputs in that range. The term *logits* is often used to describe the raw, unnormalized outputs of the final layer, especially before applying softmax or sigmoid. But it's a somewhat overloaded term that is sometimes used more loosely—for example, to describe intermediate values passed into a softmax, such as attention scores in a transformer.

The following are the most commonly used activation functions, with their shapes shown in Figure 1-3:

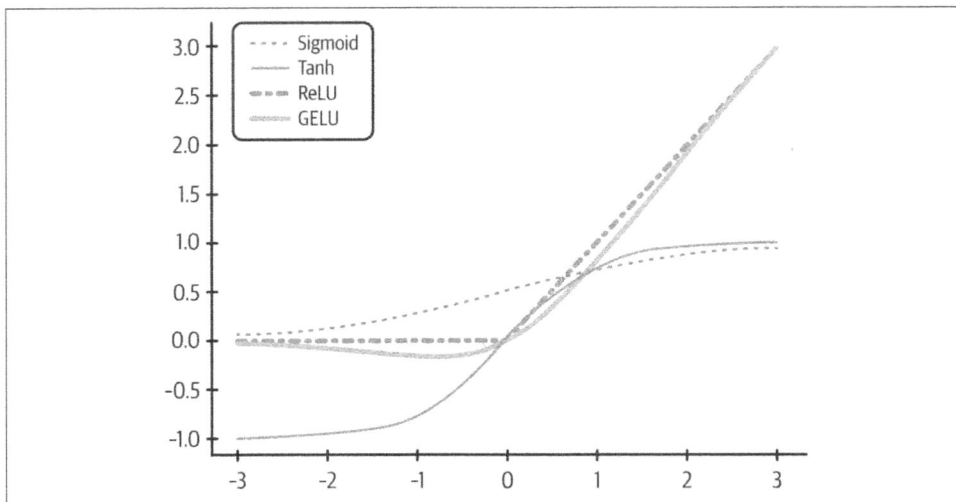

Figure 1-3. Common activation functions used in deep learning.

A bit more detail on these common activation functions:

ReLU (rectified linear unit)
  Outputs zero for negative inputs and the input itself for positive inputs. It is simple and effective, especially in deep networks.

GELU (Gaussian error linear unit)
  A smoother alternative to ReLU, often used in transformer models. It can yield slightly better results.

Sigmoid and tanh
  Older activation functions that squash values into fixed ranges. Sigmoid maps inputs to a range between 0 and 1, while tanh maps to a range between −1 and 1. They're useful in certain settings, such as output layers for binary classification, but can suffer from gradient issues in deeper models.

Softmax
  Converts a vector of values into a probability distribution that sums to 1. Used in the final layer of multiclass classification models. Not applied element-wise; it operates across the whole vector.

In this book, we'll usually stick to ReLU or GELU for hidden layers, and choose the final activation based on the prediction task.

Optimizers

Optimizers are algorithms that adjust a model's parameters (its weights and biases) to reduce the error (loss) during training. They do this using *gradient descent*, which computes how each parameter affects the loss and updates it to reduce the error.

We mostly use *Adam*, a widely used optimizer that adapts the learning rate for each parameter and combines ideas from momentum and RMSProp. It usually trains faster and more reliably than plain gradient descent, especially in noisy or sparse settings.

Initialization Strategy

Before training starts, we need to set initial values for model parameters. This step is more important than it sounds. Poor initialization can cause gradients to vanish or explode, making training unstable.

We typically use Xavier (Glorot) initialization, which is designed to keep the scale of activations and gradients roughly stable across layers. This helps training proceed smoothly. By default, Flax uses Xavier initialization for most layers like `Dense` and `Conv`, so you don't usually need to specify it manually.

Model Checkpointing

During training, it's often useful to periodically save your model parameters so that you can resume later or avoid losing progress. While production training pipelines often use robust checkpointing strategies (e.g., saving best-performing checkpoints, versioned histories, etc.), that level of complexity isn't necessary for the teaching examples in this book. Instead, we provide a lightweight utility that saves and restores just the most recent checkpoint. This is enough to pause and resume training or to store the final model output for later use. You'll see it used in several chapters to simplify experimentation and reduce boilerplate.

If you're building your own training loop for production or research, consider upgrading to a more complete checkpointing solution—for example, using Orbax (*https://oreil.ly/aazC8*), Flax's newer checkpointing system.

Early Stopping

In real-world training, it's often a good idea to stop training when your validation performance stops improving. This is known as *early stopping*, which helps prevent overfitting and saves compute.

In this book, we often show longer training runs to visualize how the loss evolves over time. But in practice, you'll likely want to use early stopping when training your own models.

Flax includes a simple utility for this:

```
from flax.training.early_stopping import EarlyStopping
```

You can track validation metrics and stop training when they stop improving for a certain number of steps (the `patience` parameter).

Selecting a Working Environment

Training large neural networks often benefits from *GPU acceleration*. There are a few ways to access a GPU, depending on your budget, goals, and technical comfort level.

One option is to use a local machine with a GPU, such as a gaming desktop or a workstation you've set up yourself. This gives you full control and avoids ongoing cloud costs, but it requires up-front hardware investment and setup. Another option is to rent GPU access from cloud providers like AWS, GCP, or Azure. These offer flexibility and scalability but can get expensive over time, especially if you're training large models.

For many beginners and small projects, Google Colab is a great place to start. It provides free, cloud-based Jupyter notebooks with GPU or TPU support and minimal setup.

In the following sections, we'll briefly walk through your options, from interactive notebooks to fully customized GPU development environments.

Selecting an Interactive Notebook

Jupyter notebooks are a popular and powerful tool for interactive coding, making them ideal for running the code examples in this book. They let you write code, execute it in cells, visualize results, and document your work, all in one place. This interactivity makes it easy to experiment, debug, and iterate quickly. Common notebook environments include JupyterLab, VSCode with Jupyter extensions, and Google Colab.

Google Colab in particular provides cloud-based notebooks with free access to GPUs and TPUs, making it a great option if you don't have a powerful local machine. It runs entirely in your browser and requires no setup beyond a Google account. You can install libraries with commands like `!pip install jax flax optax` and save notebooks directly to Google Drive. To enable hardware acceleration, navigate to the Runtime menu. From the drop-down menu, select "Change runtime type" and then select a GPU or TPU.

We provide all code in this book as Google Colab notebooks that you can open, run, and modify interactively.

Colab sessions can time out after periods of inactivity, so remember to save your work frequently.

While notebooks are excellent for exploration and prototyping, they can make version control, debugging, and maintaining complex projects more difficult. For longer-term work, setting up a dedicated development environment with GPU support offers more control, scalability, and reproducibility.

Structuring Your Code for Reuse and Debugging

Notebooks are great for interactive use, but as you begin building your own projects, it's worth thinking about how you will structure it early. Organize your code into clear, reusable Python modules: one for datasets, one for models, one for metrics and evaluation. Keeping these pieces separate makes your code easier to debug, scale, and reuse later.

This modular approach mirrors how this book is structured. Each subsection of each chapter generally focuses on a specific building block: dataset, model, training loop, and so on. As you follow along, you might try using this structure in your own code. Build clean components you can plug into future projects, and you'll move faster each time.

A good dataset class in particular will help you catch issues early. There are plenty of reasons you might get confused during a project. If your dataset and metrics are cleanly separated and easy to inspect, you can at least rule them out as the source of the problem.

Add frequent sanity checks. Can your model overfit a tiny dataset (like 10 examples)? Do loss and accuracy behave as expected when you shuffle labels?

Plot everything. Visualizing predictions, losses, or inputs will often reveal issues you wouldn't notice in the final numbers alone—like shifted targets, mislabeled data, empty tensors, or models that haven't learned anything at all.

Setting Up a GPU Development Environment

As your models grow more complex, you may want to move beyond notebooks into a more powerful GPU-enabled setup, either locally or in the cloud. A robust development environment can speed up experimentation, simplify debugging, and make your workflow more reproducible.

For local development, we recommend the following:

- Use Docker with NVIDIA Docker to create containerized environments that seamlessly access your GPU.
- Pair it with an editor like VSCode, which supports Docker and remote development for a smooth workflow.
- Use Git for version control to track changes, collaborate, and back up your work.

Mac users with Apple Silicon can try `jax-metal` (*https://oreil.ly/_FYSk*) to enable GPU acceleration via Apple's Metal backend. While it's improving rapidly, not all features are fully supported yet, so expect occasional compatibility issues.

> Paid cloud services like AWS, GCP, or Azure offer on-demand GPU instances if you don't have a local GPU. Alternatives like Paperspace, Lambda Labs, or RunPod can offer simpler setup and better value for smaller-scale projects.

Version Conflicts

Machine learning and scientific computing libraries evolve quickly—and not always in sync. You'll inevitably hit version conflicts between tools like NumPy, JAX, Flax, PyTorch, or Hugging Face Transformers. Strive for a balance: use versions recent enough to benefit from improvements, but not so new that they break compatibility with everything else.

Tools like `uv` (*https://oreil.ly/_aezy*)—a faster, more flexible alternative to `pip`—can help override incompatibilities and install packages even when their metadata says they're not compatible. It's not a permanent solution, but it can get you unstuck.

If you're working outside a notebook, always use a virtual environment (`venv`, `conda`, or `uv venv`) to keep dependencies isolated. Another option is to define your environment in a Docker container, which guarantees full reproducibility across machines. This setup is especially useful when working on remote GPU instances or in the cloud—something we'll explain later in the book.

When in doubt, check GitHub issues or forums—version mismatches are a common pain point for everyone, and at the very least, you can commiserate with others.

If you're working in a notebook, version issues can be especially tricky. Google Colab, for example, comes with many preinstalled packages—but these can be outdated or incompatible with the latest JAX/Flax stack. You can install or override versions directly in a cell using `!pip install`, but you may need to restart the runtime afterward for changes to take effect (Runtime > Restart runtime).

A Living Document

While this book endeavors to capture the state of the field at the time of writing, deep learning and biology remain fast-moving frontiers. A printed book is a static artifact; in contrast, frameworks evolve, APIs break, and new ideas emerge constantly.

We've done our best to future-proof examples and flag potential breaking changes (like Flax's transition from `linen` to `nnx`). Still, some discrepancies may arise over time. If you spot anything out of date—or have corrections, improvements, suggestions, or extensions—please let us know. The online repository can evolve even if the book is in a fixed state.

We also encourage you to look beyond these pages. Resources like D2L.ai, fast.ai, the JAX ecosystem, and preprints on bioRxiv and arXiv are excellent places to go deeper.

Most of all, experiment, build, break things, train bad models, and learn. The best way to grow in deep learning for biology is to get your hands dirty.

## CHAPTER 2

Learning the Language of Proteins

Life as we know it operates on *proteins*. The human genome holds about 20,000 *genes*, each made of DNA, that serve as blueprints for building different proteins. Some proteins have simple, well-understood functions—like collagen, which provides structural support and elasticity to tissues, or hemoglobin, which transports oxygen and carbon dioxide between the lungs and the rest of the body. Others have slightly more abstract roles: they act as messengers, modulators, or signal carriers, transmitting information within and between cells. For example, insulin is a protein hormone that signals cells to absorb sugar from the bloodstream.

We'll dive into how DNA and proteins work in more detail soon. But for now, imagine a protein as a blobby molecular machine bumping around in the crowded cell environment, occasionally making productive collisions. Its shape and movement may seem chaotic, but both have been fine-tuned by millions of years of evolution to carry out very specific molecular functions.

One key detail for this chapter: a protein can be represented as a sequence of its constituent building blocks, called *amino acids*. Just as English uses 26 letters to form words, proteins use an alphabet of 20 amino acids to form long chains with specific shapes and jobs. With that in mind, the goal of this chapter is simple: we'll train a model to predict a protein's function given its amino acid sequence. For example:

- Given the sequence of the COL1A1 collagen protein (`MFSFVDLR...`), we might predict its function is likely `structural` with probability 0.7, `enzymatic` with probability 0.01, and so on.

- Given the sequence of the INS insulin protein (`MALWMRLL...`), we might predict its function is likely *metabolic* with probability 0.6, *signaling* with probability 0.3, and so on.

To get hands-on with the material right away, open the companion Colab notebook and try running the code as you read the chapter. Exploring the examples interactively is one of the best ways to build intuition and make the ideas stick.

Biology Primer

We already highlighted that proteins are essential units of function within the cell, fulfilling a vast range of biological roles. A protein's function is very closely tied to its 3D structure, which in turn is determined by its primary amino acid sequence.

To recap the flow of information: a gene encodes the primary amino acid sequence of a protein. That sequence determines the protein's structure, and the structure governs its function.

Protein Structure

Protein structure is typically described in four hierarchical levels:

Primary structure
   The linear sequence of amino acids

Secondary structure
   Local folding into structural elements such as alpha helices and beta sheets

Tertiary structure
   The overall 3D shape formed by the complete amino acid chain

Quaternary structure
   The assembly of multiple protein subunits into a functional complex (not all proteins have this)

As an example, Figure 2-1 shows the structural organization levels of hemoglobin.

The human genetic code specifies 20 main amino acids. Each has a unique chemical structure, but they can be grouped by shared biochemical properties—such as hydrophobicity (how they interact with water), charge (positive, negative, or neutral), and polarity (how evenly electrical charge is distributed over the molecule).

Although biochemistry students are often expected to memorize all 20 amino acids, complete with names, structures, and single-letter codes (don't ask us how we know), it's more practical here to focus on their functional roles (summarized in Figure 2-2).

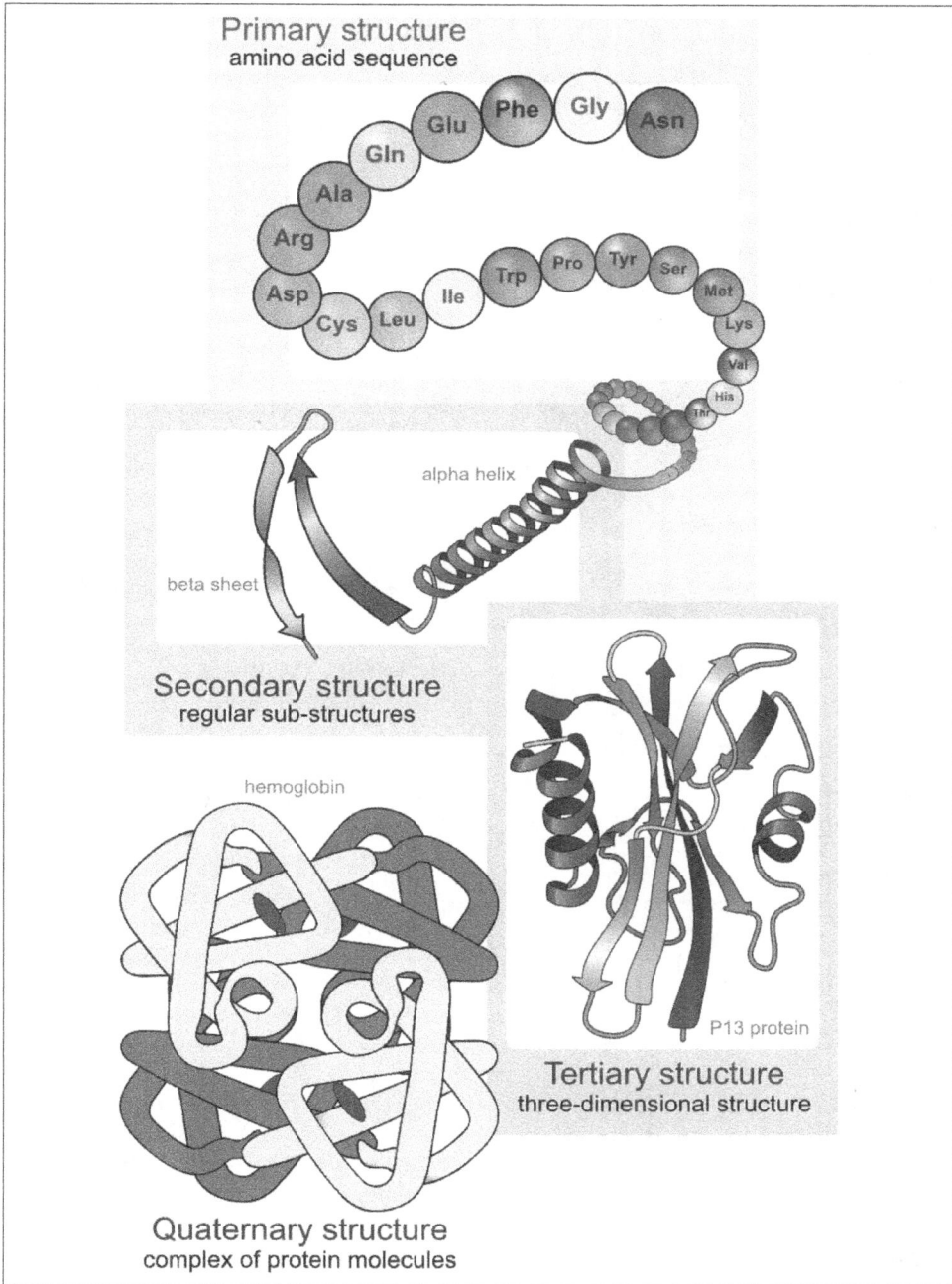

Figure 2-1. The four levels of protein structure, as illustrated by the hemoglobin protein.
Source: Wikipedia (https://oreil.ly/BD2Qa).

For instance, D (aspartic acid) and E (glutamic acid) are both negatively charged and often interchangeable without drastically altering a protein's function. But other amino acids play much more specific roles, and even a single substitution can dramatically alter how a protein folds or functions—sometimes with serious effects. In fact, many genetic diseases are caused by such point mutations. One famous example is sickle cell anemia, which is caused by a single-letter change in the gene for hemoglobin that replaces a hydrophilic amino acid (E) with a hydrophobic one (V), which ultimately leads to misshapen red blood cells.

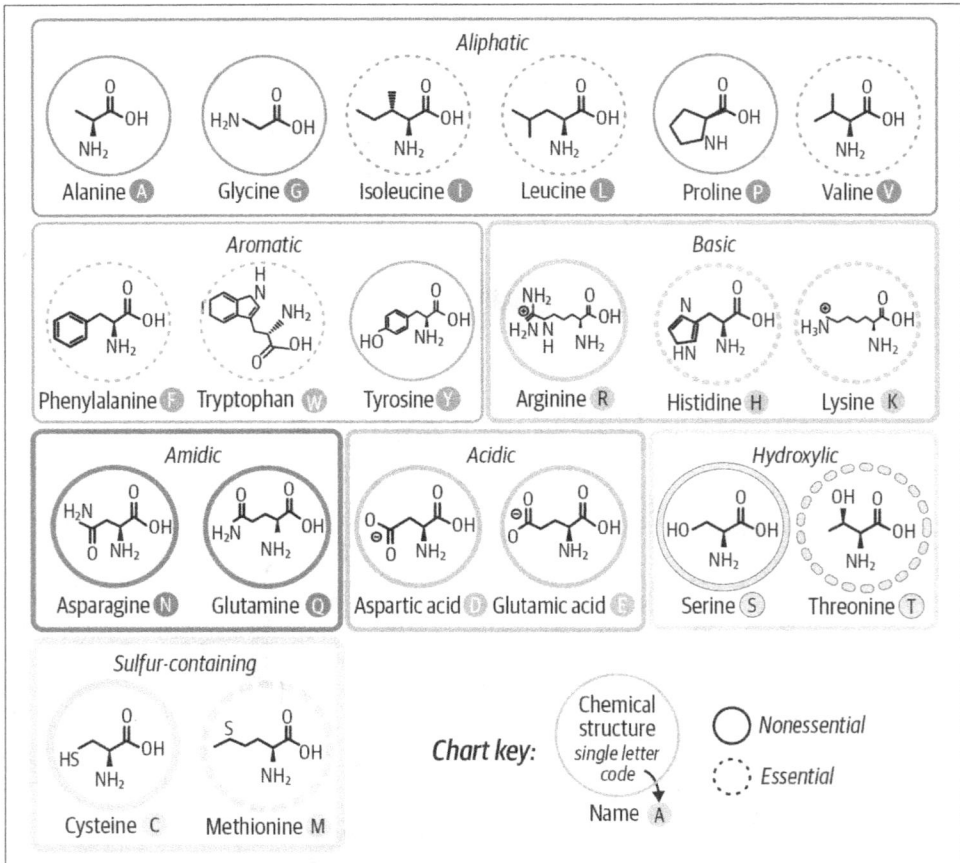

Figure 2-2. Chart showing the chemical structures of the 20 standard amino acids found in living organisms, grouped by biochemical similarity, color-coded by side-chain properties (e.g., acidic, basic, polar, nonpolar), and annotated with their names, one- and three-letter codes, and example DNA codons (the triplet DNA bases that code for that amino acid). Adapted from an infographic by Compound Interest (https://oreil.ly/o7Lyq).

With that introduction to protein structure, let's now look at function—what proteins actually do in the cell.

Protein Function

Proteins carry out nearly every task required for life: they catalyze chemical reactions, transmit signals, transport molecules, provide structural support, and regulate gene expression. Because of this diversity, systematically cataloging protein functions is a massive undertaking—and one of the most widely used frameworks for doing so is the *Gene Ontology* (GO) project.

The GO system organizes protein function into three broad categories, each capturing a different aspect of how proteins behave in the cell:

Biological process
> This contributes to—like cell division, response to stress, carbohydrate metabolism, or immune signaling.

Molecular function
> This describes the specific biochemical activity of the protein itself—such as binding to DNA or ATP (a molecule that stores and transfers energy in cells), acting as a kinase (an enzyme that attaches a small chemical tag called a phosphate group to other molecules to change their activity), or transporting ions across membranes.

Cellular component
> This indicates where in the cell the protein usually resides—such as the nucleus, mitochondria, or extracellular space. Although it's technically a location label and not a function *per se*, it often provides important clues about the protein's role (e.g., proteins in the mitochondria are probably involved in energy production). We'll return to this theme in Chapter 6.

Each protein can have multiple GO annotations across these categories. For example, a single protein might bind ATP (molecular function), drive muscle contraction (biological process), and localize to muscle fibers (cellular component). Some annotations are derived from direct experimental assays, while others are inferred computationally through similarity to known proteins. In this chapter, we'll work with a curated subset of high-confidence, experimentally validated GO annotations.

Predicting Protein Function

Why predict a protein's function from its sequence? This is actually a fundamental challenge in modern biology. Here are a few of the most common and impactful applications:

Biotechnology and protein engineering

If we can reliably predict function from sequence, we can begin to design new proteins with desired properties. This could be useful for designing enzymes for industrial chemistry, therapeutic proteins for medicine, or synthetic biology components.

Understanding disease mechanisms

Many diseases are caused by specific sequence changes (variants, or mutations) that disrupt protein function. A good predictive model can help identify how specific mutations alter function, offering insights into disease mechanisms and potential therapeutic targets.

Genome annotation

As we continue sequencing the genomes of new species, we're uncovering vast numbers of proteins whose functions remain unknown. For newly identified proteins—especially those that are distantly evolutionarily related to any known ones—computational prediction is essential for assigning functional hypotheses.

Metagenomics and microbiome analysis

When sequencing entire microbial communities, such as gut bacteria or ocean microbiota, many protein-coding genes have no close matches in existing databases. Predicting function from sequence helps uncover the roles of these unknown proteins, advancing our understanding of microbial ecosystems and their effects on hosts or the environment.

Although the task may sound somewhat straightforward—input a sequence, output a function—accurate protein function prediction is an extremely challenging problem. To succeed, a model must implicitly understand a range of highly complex biological principles: how amino acid sequence determines 3D structure (a Nobel Prize–winning machine learning problem in its own right), how structure enables function, and how these functions operate in the dynamic, crowded environment of the cell.

In this chapter, we won't aim for state-of-the-art performance. Instead, our goal is to build a simple working model and develop intuition for how protein sequences can be mapped to functional annotations. Along the way, we'll introduce several useful machine learning techniques—including using pretrained models to extract embeddings, visualizing those embeddings, and training lightweight classifiers on top of them—that will become recurring tools in later chapters.

Machine Learning Primer

We've briefly reviewed the biological background of proteins and how their function is encoded. Now, we'll turn to the machine learning techniques that allow us to learn from protein sequences in practice.

Large Language Models

It's hard these days to go anywhere without bumping into *large language models* (LLMs). Many recent breakthrough models in AI—such as ChatGPT, Gemini, Claude, and Llama—fall under this category. While these models involve immense engineering, the fundamental idea behind them is surprisingly simple: they're trained to predict the next token (e.g., a word or character) given the preceding context. There are slight variations—such as masked language models, which hide random tokens during training to encourage contextual reasoning—but the core principle remains the same.

One of the most surprising discoveries in modern AI has been that if you train a large enough model (in terms of the number of parameters) on enough data (in terms of total tokens), remarkable capabilities emerge without explicit supervision. These models can suddenly summarize text, translate between languages, and even generate creative writing like poems and stories—despite never being trained directly to do so.

This holds promise for biology. In many ways, biology is language-like: DNA and proteins are sequences built from discrete alphabets, with complex patterns and context-dependent "grammar." By training language models on massive corpora of biological sequences—using the same next-token prediction objective—we should be able to learn rich representations of biological information.

These learned representations can then be used for a wide range of downstream tasks, such as predicting a protein's function, inferring the effects of mutations, or identifying structural properties—all without needing to retrain a new model from scratch.

Later in this chapter, we will explore one of the most successful protein language models to date: ESM2.

Embeddings

One of the most powerful and versatile outputs of language models is their ability to generate *embeddings*. An embedding is a numerical vector—a list of floating-point numbers—that encodes the meaning or structure of an entity like a word, sentence, or protein sequence. For example, a protein might be represented by an embedding such as [0.1, -0.3, 1.3, 0.9, 0.2], which could capture aspects of its biochemical or structural properties in a compact numerical form.

Embeddings from language models are not just arbitrary numbers—they are structured so that similar inputs result in similar embeddings. Related words like lion, tiger, and panther cluster together in a linguistic "semantic space." Likewise, protein sequences with similar structure or function—such as collagen I and collagen II—will tend to have embeddings that are close together in what we might call a "protein space."

This idea generalizes to the concept of a *latent space*—a continuous, abstract space where similar entities are positioned close together based on learned patterns. In such spaces, we can perform powerful operations, such as interpolation, clustering, and generative design. For proteins, latent spaces can capture functional relationships that aren't apparent from sequence alone—for example, two proteins with very different sequences and evolutionary histories may have converged on similar functions and therefore appear close together in the latent space. These representations can also help predict new functions for uncharacterized proteins by comparing them to annotated neighbors in the space.

To identify proteins with similar structure or function, you can compare their embeddings using *cosine similarity*—a measure of how aligned two vectors are, regardless of their magnitude. This works even when sequences differ significantly at the amino acid level. By computing cosine similarities between a query protein and a set of known proteins, you can rank the closest matches in embedding space. These top hits often share functional roles, structural features, or evolutionary history.

Pretraining and Fine-tuning

Many machine learning tasks share underlying structure. Whether your goal is detecting hate speech, answering law school entrance questions, or writing poems about capybaras, your model first needs a strong foundation in how language works. Rather than training from scratch for every task, we typically start from a general-purpose model that's been pretrained on a huge, diverse dataset.

Pretraining gives a model broad knowledge and general capabilities. For a specific application, we often follow it with a smaller, focused training step called *fine-tuning*, where the model is trained further on a domain-specific dataset. This two-stage process is now standard in many areas of machine learning, especially as pretrained language models have become increasingly powerful.

In this first technical chapter of the book, we'll take a slightly different approach. Rather than fine-tuning the entire pretrained model, we'll treat it as a frozen feature extractor: we'll use its embeddings as input to a smaller classifier that we'll train from scratch. This strategy is efficient, requires little data, and still leverages the rich representations learned by the pretrained model. We'll explore full transfer learning with fine-tuning in later chapters.

Representations of Proteins and Protein LMs

Previously, we discussed what proteins are and how their structure is organized hierarchically—from a linear chain of amino acids, to local folding, to the final 3D form that enables their function. To make this less abstract, let's load up and visualize an example protein structure using the py3Dmol library:

```python
import py3Dmol
import requests


def fetch_protein_structure(pdb_id: str) -> str:
    """Grab a PDB protein structure from the RCSB Protein Data Bank."""
    url = f"https://files.rcsb.org/download/{pdb_id}.pdb"
    response = requests.get(url)
    return response.text



# The Protein Data Bank (PDB) is the main database of protein structures.
# Each structure has a unique 4-character PDB ID. Below are a few examples.
protein_to_pdb = {
    "insulin": "3I40",   # Human insulin - regulates glucose uptake.
    "collagen": "1BKV",   # Human collagen - provides structural support.
    "proteasome": "1YAR",   # Archaebacterial proteasome - degrades proteins.
}

protein = "collagen"   # @param ["insulin", "collagen", "proteasome"]
pdb_structure = fetch_protein_structure(pdb_id=protein_to_pdb[protein])

pdbview = py3Dmol.view(width=400, height=300)
pdbview.addModel(pdb_structure, "pdb")
pdbview.setStyle({"cartoon": {"color": "spectrum"}})
pdbview.zoomTo()
pdbview.show()
```

Running this code in our companion Colab notebook will display an interactive 3D rendering of your chosen protein. A screenshot of the visualization of collagen is shown in Figure 2-3.

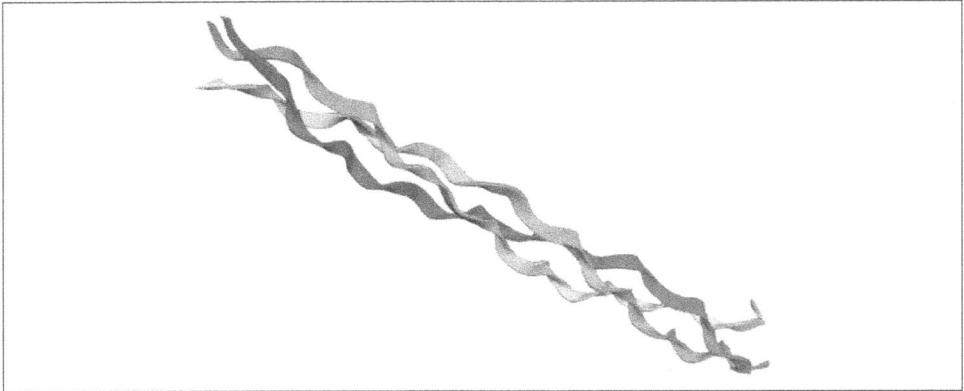

Figure 2-3. A 3D structure of the collagen protein rendered with py3Dmol. Collagen is a structural protein that forms triple-helical fibers, visible here as intertwined ribbonlike strands.

Try viewing the other examples, such as `insulin` and `proteasome`, to appreciate the incredible structural diversity of proteins. Their shapes often reflect their specialized roles. For example, the long, springy structure of collagen relates to its function as a flexible, supportive scaffold found throughout many tissues in the body.

Numerical Representation of a Protein

While 3D visualizations are useful for exploration, machine learning models require numerical input. To analyze or model proteins with machine learning techniques, we typically start from their 1D amino acid sequence.

Protein sequences for most known organisms can be retrieved from public databases such as Uniprot (*https://oreil.ly/9OqAK*). For example, here's the amino acid sequence of human insulin:

```
# Precursor insulin protein sequence (processed into two protein chains).
insulin_sequence = (
    "MALWMRLLPLLALLALWGPDPAAAFVNQHLCGSHLVEALYLVCGERGFFYTPKTRREAEDLQVGQVELGG"
    "GPGAGSLQPLALEGSLQKRGIVEQCCTSICSLYQLENYCN"
)
print(f"Length of the insulin protein precursor: {len(insulin_sequence)}.")
```

Output:

```
Length of the insulin protein precursor: 110.
```

This sequence representation is easy to store and manipulate, but it still needs to be converted to a numerical format before it can be used by machine learning models.

One-Hot Encoding of a Protein Sequence

The simplest way to convert a protein sequence into numerical form is with *one-hot encoding*. Here is how it works:

- There are 20 standard amino acids.

- Each amino acid is represented by a binary vector of length 20, where only one position is 1 (indicating the identity of that amino acid) and all other positions are 0.

- A protein sequence is then converted into a sequence of these one-hot vectors—one for each amino acid.

Let's walk through a toy example: encoding the short protein MALWN (the first five amino acids of the insulin precursor protein).

First, let's define the mapping between an amino acid letter code to an integer index:

```
from dlfb.utils.display import print_short_dict

amino_acids = [
  "R", "H", "K", "D", "E", "S", "T", "N", "Q", "G", "P", "C", "A", "V", "I",
  "L", "M", "F", "Y", "W",
]

amino_acid_to_index = {
  amino_acid: index for index, amino_acid in enumerate(amino_acids)
}

print_short_dict(amino_acid_to_index)
```

Output:

```
{'R': 0, 'H': 1, 'K': 2, 'D': 3, 'E': 4, 'S': 5, 'T': 6, 'N': 7, 'Q': 8, 'G': 9}
…(+10 more entries)
```

Given a protein sequence, we can convert it to a sequence of integers:

```
# Methionine, alanine, leucine, tryptophan, methionine.
tiny_protein = ["M", "A", "L", "W", "M"]

tiny_protein_indices = [
  amino_acid_to_index[amino_acid] for amino_acid in tiny_protein
]

tiny_protein_indices
```

Output:

```
[16, 12, 15, 19, 16]
```

And given a sequence of integers, we can convert it into a one-hot encoding (see Figure 2-4).

The protein's one-hot encoding:

	R	H	K	D	E	S	T	N	Q	G	P	C	A	V	I	L	M	F	Y	W
M	0	0	0	0	0	0	0	0	0	0	0	0	0	0	0	0	**1**	0	0	0
A	0	0	0	0	0	0	0	0	0	0	0	0	**1**	0	0	0	0	0	0	0
L	0	0	0	0	0	0	0	0	0	0	0	0	0	0	0	**1**	0	0	0	0
W	0	0	0	0	0	0	0	0	0	0	0	0	0	0	0	0	0	0	0	**1**
M	0	0	0	0	0	0	0	0	0	0	0	0	0	0	0	0	**1**	0	0	0

protein sequence: M A L W M

amino acid: R H K D E S T N Q G P C A V I L M F Y W

index: 0 1 2 3 4 5 6 7 8 9 10 11 12 13 14 15 16 17 18 19

Figure 2-4. One-hot encoding converts a protein's amino acid sequence into a binary matrix where each row corresponds to one amino acid and each column to a possible residue. Most values are zero, with a single "1" indicating the presence of a specific amino acid at each position.

In Figure 2-4, we see that:

- The resulting matrix has the shape [5, 20], where each of the five rows corresponds to one amino acid in the sequence, and each column represents one of the 20 standard amino acids.
- Each row contains all zeros except for a single 1 in the position corresponding to that amino acid's identity, preserving its categorical nature without implying any numerical ordering or similarity.

Why not just skip the one-hot encoding step and use amino acid indices directly?

The issue is that numeric indices (like 3 versus 17) imply an artificial order and relative similarity, even though amino acids are categorical entities without meaningful numerical relationships.

One-hot encoding avoids this by assigning each amino acid a distinct binary vector—ensuring that the model treats them as equally separate and avoids inferring nonexistent patterns from arbitrary index values.

In code, we can use the handy `jax.nn.one_hot` utility from the JAX library to get this embedding:

```
import jax

one_hot_encoded_sequence = jax.nn.one_hot(
    x=tiny_protein_indices, num_classes=len(amino_acids)
)

print(one_hot_encoded_sequence)
```

Output:

```
[[0. 0. 0. 0. 0. 0. 0. 0. 0. 0. 0. 0. 0. 0. 0. 0. 1. 0. 0. 0.]
 [0. 0. 0. 0. 0. 0. 0. 0. 0. 0. 0. 0. 1. 0. 0. 0. 0. 0. 0. 0.]
 [0. 0. 0. 0. 0. 0. 0. 0. 0. 0. 0. 0. 0. 0. 1. 0. 0. 0. 0. 0.]
 [0. 0. 0. 0. 0. 0. 0. 0. 0. 0. 0. 0. 0. 0. 0. 0. 0. 0. 0. 1.]
 [0. 0. 0. 0. 0. 0. 0. 0. 0. 0. 0. 0. 0. 0. 0. 0. 1. 0. 0. 0.]]
```

We can visualize the resulting one-hot encoding matrix as a heatmap as in Figure 2-5 (essentially re-creating the earlier Figure 2-4):

```
import seaborn as sns

fig = sns.heatmap(
    one_hot_encoded_sequence, square=True, cbar=False, cmap="inferno"
)
fig.set(xlabel="Amino Acid Index", ylabel="Protein Sequence");
```

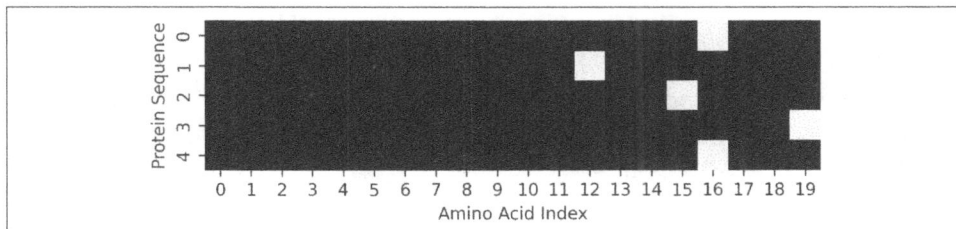

Figure 2-5. One-hot encoded representation of a toy protein sequence (MALWM), visualized with a heatmap. This binary matrix encodes the identity of each residue without implying any similarity between them.

Now that we've constructed a basic numerical representation of a protein, we're ready to move beyond this simplistic format and explore *learned embeddings*—dense vector representations that encode much more biological meaning about each amino acid.

Learned Embeddings of Amino Acids

In the rest of this chapter, we'll use a pretrained protein language model called ESM2 (*https://oreil.ly/iZmXA*), released by Meta in 2023 (ESM stands for *evolutionary scale modeling*). These models are hosted on the Hugging Face platform (*https://hugging*

face.co). If you haven't encountered it yet, Hugging Face is a fantastic resource with thousands of pretrained models (*https://huggingface.co/models*) ready for you to use and explore.

We'll explore how the ESM2 model works in more detail shortly, but first, let's examine how it represents individual amino acids. We'll access the model using the Hugging Face transformers library. ESM2 is based on the *transformer* neural network architecture introduced in 2017 (*https://oreil.ly/XPvFW*), which has become the standard for modeling sequences like text and proteins.

> Ideally, we'd load the ESM2 model using JAX/Flax, but it's only officially available in PyTorch at the moment. In practice, being comfortable with multiple deep learning frameworks is useful—so here we'll use PyTorch to load the model and extract embeddings, which we'll then process and build on top of using JAX.
>
> The rest of the book will use JAX/Flax exclusively, but this brief mixing of frameworks is a good example of how flexible real-world workflows can be.

```
from transformers import AutoTokenizer, EsmModel

# Model checkpoint name taken from this GitHub README:
# https://github.com/facebookresearch/esm#available-models-and-datasets-
model_checkpoint = "facebook/esm2_t33_650M_UR50D"
tokenizer = AutoTokenizer.from_pretrained(model_checkpoint)
model = EsmModel.from_pretrained(model_checkpoint)
```

We can check the model's token-to-index mapping:

```
vocab_to_index = tokenizer.get_vocab()
print_short_dict(vocab_to_index)
```

Output:

```
{'<cls>': 0, '<pad>': 1, '<eos>': 2, '<unk>': 3, 'L': 4, 'A': 5, 'G': 6, 'V': 7,
'S': 8, 'E': 9}
…(+23 more entries)
```

This is similar to the manual amino acid indexing we did earlier, but it includes special tokens like <unk> for unknown residues, <eos> for end-of-sequence, and rare amino acids like U (selenocysteine) and O (pyrrolysine).

Let's use the ESM2 tokenizer to encode our tiny protein sequence:

```
tokenized_tiny_protein = tokenizer("MALWM")["input_ids"]
tokenized_tiny_protein
```

Output:

```
[0, 20, 5, 4, 22, 20, 2]
```

If desired, we can drop the special start (`<cls>`) and end (`<eos>`) tokens:

```
tokenized_tiny_protein[1:-1]
```

Output:

```
[20, 5, 4, 22, 20]
```

Now we'll extract the learned token embeddings from the model using `model.get_input_embeddings()`:

```
token_embeddings = model.get_input_embeddings().weight.detach().numpy()
token_embeddings.shape
```

Output:

```
(33, 1280)
```

Each of the 33 possible tokens is embedded into a 1,280-dimensional space. While humans can't visualize such high-dimensional spaces directly, we can apply dimensionality reduction techniques like t-SNE or UMAP to project the embeddings down to two dimensions. This allows us to inspect how the model organizes different tokens in a more interpretable form:

```
import pandas as pd
from sklearn.manifold import TSNE

tsne = TSNE(n_components=2, random_state=42)
embeddings_tsne = tsne.fit_transform(token_embeddings)
embeddings_tsne_df = pd.DataFrame(
    embeddings_tsne, columns=["first_dim", "second_dim"]
)
embeddings_tsne_df.shape
```

Output:

```
(33, 2)
```

We can see that the t-SNE–transformed array has shape (`33, 2`), meaning that each of the 33 tokens has been projected into a 2D space. Figure 2-6 shows a scatterplot of these points, giving us a visual sense of how the model organizes token embeddings:

```
fig = sns.scatterplot(
    data=embeddings_tsne_df, x="first_dim", y="second_dim", s=50
)
fig.set_xlabel("First Dimension")
fig.set_ylabel("Second Dimension");
```

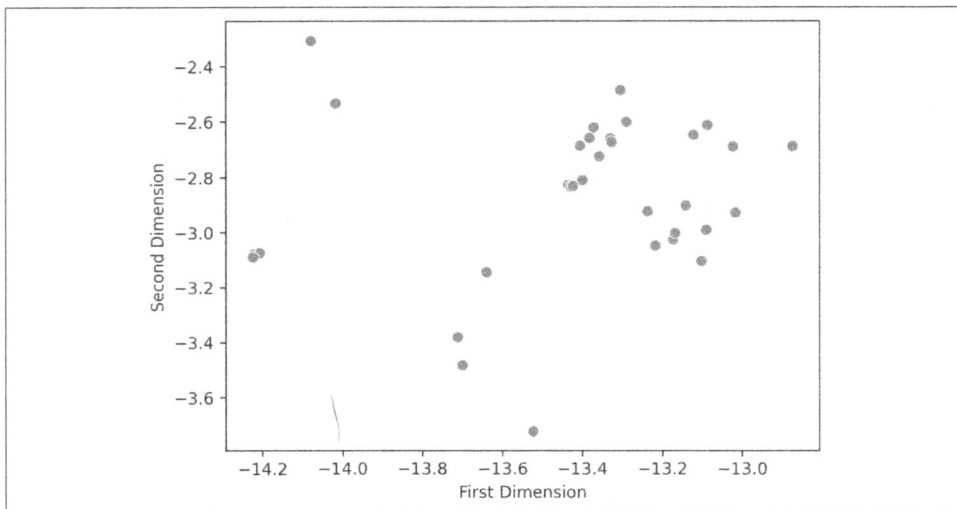

Figure 2-6. A 2D t-SNE projection of the learned token embeddings from the ESM2 model. Even without labels, clusters begin to emerge—hinting that the model has learned to organize tokens in a meaningful way.

To sanity-check whether similar types of tokens cluster in the 2D embedding space, we can label each token using known amino acid properties (like those shown earlier in the chapter) and replot the t-SNE projection in Figure 2-7:

```python
from adjustText import adjust_text

embeddings_tsne_df["token"] = list(vocab_to_index.keys())

token_annotation = {
  "hydrophobic": ["A", "F", "I", "L", "M", "V", "W", "Y"],
  "polar uncharged": ["N", "Q", "S", "T"],
  "negatively charged": ["D", "E"],
  "positively charged": ["H", "K", "R"],
  "special amino acid": ["B", "C", "G", "O", "P", "U", "X", "Z"],
  "special token": [
    "-",
    ".",
    "<cls>",
    "<eos>",
    "<mask>",
    "<null_1>",
    "<pad>",
    "<unk>",
  ],
}

embeddings_tsne_df["label"] = embeddings_tsne_df["token"].map(
  {t: label for label, tokens in token_annotation.items() for t in tokens}
```

```
)

fig = sns.scatterplot(
  data=embeddings_tsne_df,
  x="first_dim",
  y="second_dim",
  hue="label",
  style="label",
  s=50,
)
fig.set_xlabel("First Dimension")
fig.set_ylabel("Second Dimension")
texts = [
  fig.text(point["first_dim"], point["second_dim"], point["token"])
  for _, point in embeddings_tsne_df.iterrows()
]
adjust_text(
  texts, expand=(1.5, 1.5), arrowprops=dict(arrowstyle="->", color="grey")
);
```

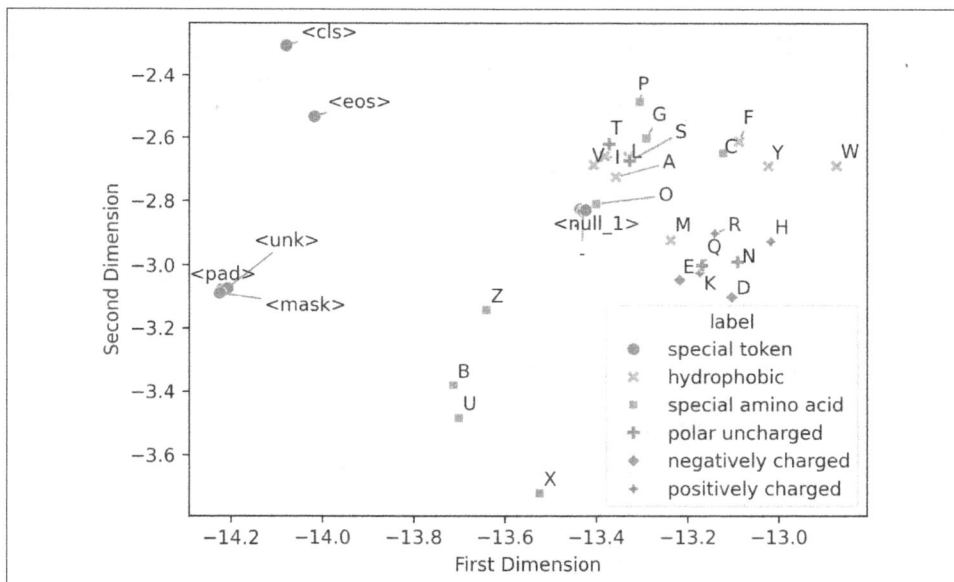

Figure 2-7. Coloring the t-SNE projection by amino acid properties reveals clear clusters of amino acids with similar biochemical roles that tend to group together in embedding space, reflecting the model's ability to capture meaningful biological structure. Technical non–amino acid tokens also group together in this latent space.

Tokens with similar biochemical properties tend to cluster together. For instance, hydrophobic amino acids like F, Y, and W group in the upper right, while special-purpose tokens such as <cls> and <eos> appear together on the left side of the plot.

This structure suggests that the model has learned meaningful distinctions among amino acids based on the roles they play within protein sequences.

Now that we've explored what these token embeddings look like, let's dive into how the ESM2 model actually works—and how it learns such representations in the first place.

The ESM2 Protein Language Model

Now that you're more familiar with token embeddings, let's discuss how the ESM2 model actually works. ESM2 is a *masked language model* (MLM), which means it was trained by repeatedly masking a random subset of amino acids in each protein sequence and asking the model to predict them. In the case of ESM2, a randomly selected 15% of the amino acids in each sequence were masked during training. Figure 2-8 illustrates this visually, comparing it to masked language modeling in natural language tasks:

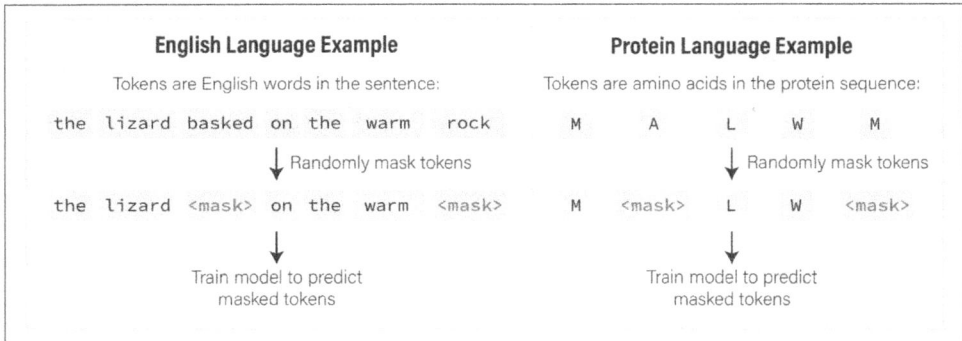

Figure 2-8. A comparison between masked language modeling in natural and protein language models. In natural language, models are trained to predict missing words (or sometimes subwords) from surrounding context. Protein language models use the same principle: randomly masking amino acids in a sequence and training the model to predict them from the surrounding context.

Let's try masking one amino acid in the insulin protein sequence and see whether the model can predict it:

```
insulin_sequence = (
    "MALWMRLLPLLALLALWGPDPAAAFVNQHLCGSHLVEALYLVCGERGFFYTPKTRREAEDLQVGQVELGG"
    "GPGAGSLQPLALEGSLQKRGIVEQCCTSICSLYQLENYCN"
)

masked_insulin_sequence = (
    # Let's mask the `L` amino acid in the 29th position (0-based indexing):
    #       ...LALLALWGPDPAAAFVNQH  L  CGSHLVEALYLVCGERGFF...
    "MALWMRLLPLLALLALWGPDPAAAFVNQH<mask>CGSHLVEALYLVCGERGFFYTPKTRREAEDLQVGQVELGG"
```

```
    "GPGAGSLQPLALEGSLQKRGIVEQCCTSICSLYQLENYCN"
)

# Tokenize the masked insulin sequence.
masked_inputs = tokenizer(masked_insulin_sequence)["input_ids"]

# Check that we indeed have a <mask> token in the place that we expect it. Note
# that the tokenizer adds a <cls> token to the start of the sequence, so we in
# fact expect the <mask> token at position 30 (not 29).
assert masked_inputs[30] == vocab_to_index["<mask>"]
```

The <mask> token tells the model to predict the amino acid at that position. To do this, we load the full language model, EsmForMaskedLM, which includes the language prediction head.

> To accelerate inference, we'll use a smaller ESM2 model variant (150M parameters with 640-dimensional embeddings) rather than the large, 650M model with 1,280-dimensional embeddings used earlier. This is a good reminder that many models on Hugging Face come in different sizes, and swapping between them is often as simple as changing a model checkpoint.
>
> Of course, there's a trade-off—smaller models may capture less information and typically perform worse on complex tasks. Still, they're great for rapid prototyping and exploring model behavior.

We load up the model:

```
from transformers import EsmForMaskedLM

# Model checkpoint name taken from this GitHub README:
# https://github.com/facebookresearch/esm#available-models-and-datasets-
model_checkpoint = "facebook/esm2_t30_150M_UR50D"
tokenizer = AutoTokenizer.from_pretrained(model_checkpoint)
masked_lm_model = EsmForMaskedLM.from_pretrained(model_checkpoint)
```

And we'll run it to get predictions for the masked token. We see that the model correctly predicts the token L (leucine) with very high probability in Figure 2-9.

```
import matplotlib.pyplot as plt

model_outputs = masked_lm_model(
    **tokenizer(text=masked_insulin_sequence, return_tensors="pt")
)
model_preds = model_outputs.logits

# Index into the predictions at the <mask> position.
mask_preds = model_preds[0, 30].detach().numpy()

# Apply softmax to convert the model's predicted logits to probabilities.
mask_probs = jax.nn.softmax(mask_preds)
```

```
# Visualize the predicted probability of each token.
letters = list(vocab_to_index.keys())
fig, ax = plt.subplots(figsize=(6, 4))
plt.bar(letters, mask_probs, color="grey")
plt.xticks(rotation=90)
plt.title("Model Probabilities for the Masked Amino Acid");
```

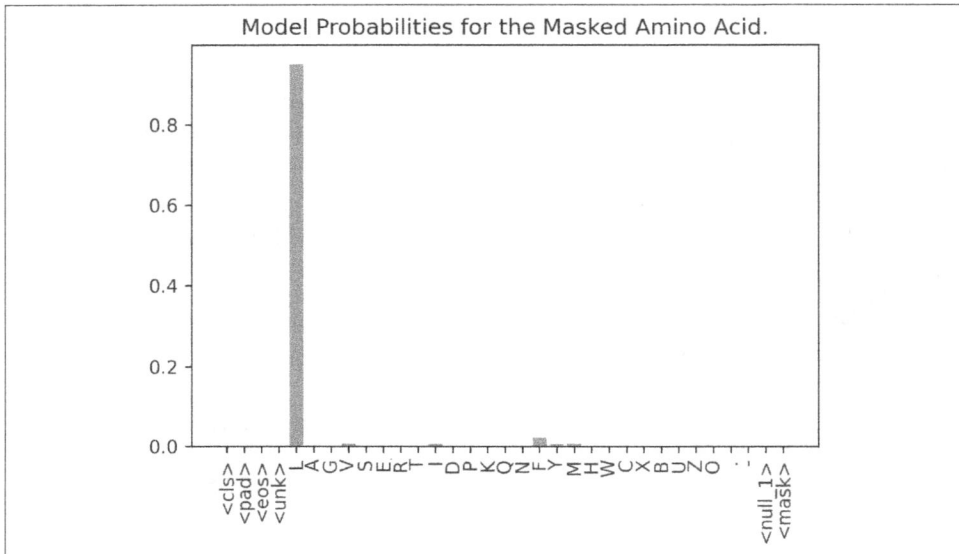

Figure 2-9. Model prediction for a masked leucine (L) in the insulin sequence. The model confidently predicts the correct amino acid (L) with high probability, showing that it has learned common sequence patterns in proteins.

Let's rewrite this code as a more general form as `MaskPredictor`, with methods that mask a sequence, make a prediction, and plot the predictions:

```
class MaskPredictor:
    """Predict masked amino acids using a protein language model."""

    def __init__(self, tokenizer: PreTrainedTokenizer, model: PreTrainedModel):
        """Initialize with a tokenizer and pretrained model."""
        self.tokenizer = tokenizer
        self.model = model

    def plot_predictions(self, sequence: str, mask_index: int) -> Figure:
        """Plot predicted probabilities for the masked amino acid."""
        mask_probs = self.predict(sequence, mask_index)
        fig, _ = plt.subplots(figsize=(6, 4))
        plt.bar(list(self.tokenizer.get_vocab().keys()), mask_probs, color="grey")
        plt.xticks(rotation=90)
        plt.title(
```

```
        "Model Probabilities for the Masked Amino Acid\n"
        f"at Index={mask_index} (True Amino Acid = {sequence[mask_index]})."
    )
    return fig

def predict(self, sequence: str, mask_index: int) -> jax.Array:
    """Return model probabilities for masked amino acid at a position."""
    masked_sequence = self.mask_sequence(sequence, mask_index)
    masked_inputs = self.tokenizer(masked_sequence, return_tensors="pt")
    model_outputs = self.model(**masked_inputs)
    mask_preds = model_outputs.logits[0, mask_index + 1].detach().numpy()
    mask_probs = jax.nn.softmax(mask_preds)
    return mask_probs

@staticmethod
def mask_sequence(sequence: str, mask_index: int) -> str:
    """Insert mask token at specified index in the input sequence."""
    if mask_index < 0 or mask_index > len(sequence):
        raise ValueError("Mask index outside of sequence range.")
    return f"{sequence[0:mask_index]}<mask>{sequence[(mask_index + 1):]}"
```

Let's try it on a different position—index 26, where the correct amino acid is N (asparagine). The result is shown in Figure 2-10:

```
MaskPredictor(tokenizer, model=masked_lm_model).plot_predictions(
    sequence=insulin_sequence, mask_index=26
);
```

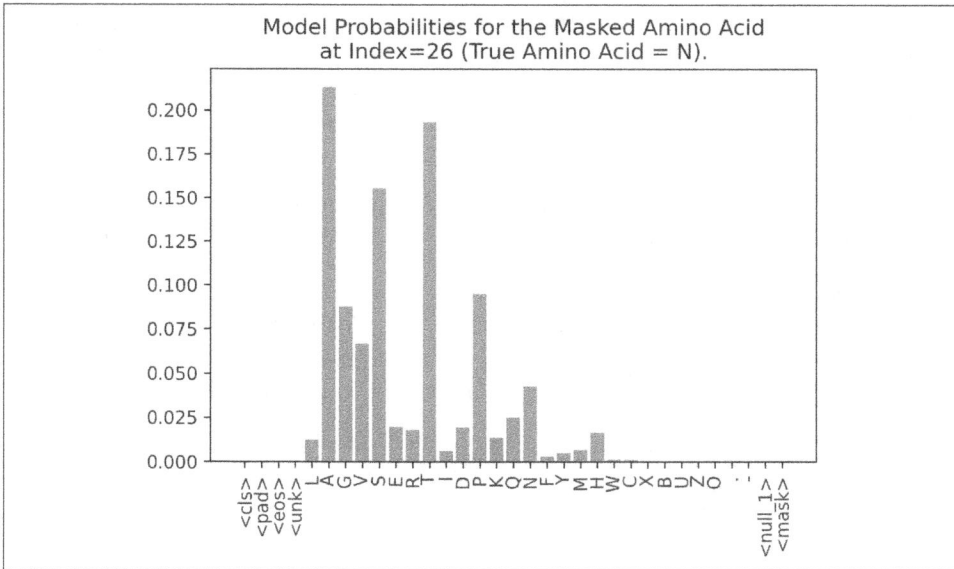

Figure 2-10. Model prediction for a masked asparagine (N) in the insulin sequence. Here, the model is more uncertain—it assigns moderate probability to several possible amino acids, indicating that this position is harder to predict based on surrounding context.

In this case, the model doesn't strongly prefer any one amino acid. It assigns moderate probability to several, including A, T, and S (with the model assigning the true amino acid N a fairly low probability). This uncertainty could reflect the biochemical flexibility of that position—some regions of proteins can tolerate different residues due to redundancy, structural flexibility, or lack of strict functional constraints. These are often called "permissive" positions and are common in disordered (unstructured) or surface regions of proteins.

This example illustrates that the model has learned and understands the probabilistic grammar of proteins. The next question is: how can we leverage this understanding to represent an entire protein, and not just one amino acid at a time?

Strategies for Extracting an Embedding for an Entire Protein

So far, we've explored how the ESM2 model represents individual amino acids. But many downstream tasks—like predicting protein function—require a fixed-length representation for the entire protein sequence. How can we convert a variable-length sequence of amino acids into a single embedding vector that captures the protein's overall structure and meaning?

Several strategies are commonly used:

Concatenation of amino acid embeddings
One simple approach is to loop through each amino acid in a sequence, extract its embedding, and concatenate them into one long vector. For example, if a protein has length 10 and each amino acid has a 640-dimensional embedding, this yields a protein embedding of length 10 × 640 = 6400. While this preserves fine-grained information of each amino acid, it has several drawbacks:

Variable length
Different proteins will yield different-length embeddings, which complicates model input formatting.

Scalability
Long proteins produce huge embeddings. For example, titin—the longest known human protein at ~34,000 amino acids—would produce an embedding with over *43 million* values. That's unwieldy for most models.

Limited modeling
This approach treats amino acids independently, ignoring the contextual relationships that are central to protein function.

Averaging of amino acid embeddings
A more compact approach is to average the token embeddings across the sequence. Using the same example of a length-10 protein with 640-dim

embeddings, we take the mean across all 10 embeddings to produce a final 640-dimensional vector.

- This has the advantage of producing fixed-size vectors, regardless of protein length.
- It's efficient and sometimes used, but also crude—averaging discards ordering and interaction information. It's like summarizing a novel by averaging all its word vectors: some meaning survives, but the nuance is lost.

Using the model's contextual sequence embeddings

A more principled approach is to extract the hidden representations for the entire sequence directly from the language model. Since ESM2 is trained to predict masked tokens based on their surrounding context, its internal layers encode rich, contextualized embeddings for every amino acid in the sequence.

- Concretely, we can pass a protein sequence through ESM2 and extract the final hidden layer activations, resulting in a tensor of shape (L', D), where L' is the number of output tokens (which may differ from the input length L), and D is the model's hidden size (e.g., 640).
- We then apply mean pooling across the sequence length to produce a fixed-length embedding of shape (D,). While averaging may seem simplistic, it often works surprisingly well—because the model has already integrated contextual information into each token's representation using self-attention, the pooled vector still captures meaningful dependencies across the sequence.

This final approach is the most common and powerful in practice—and it's the one we'll explore in the next section.

Extracellular Versus Membrane Protein Embeddings

We'll introduce the GO dataset properly in the next section on protein function prediction. For now, let's use it to associate each UniProt protein accession and sequence with its known cellular location:

```python
import pandas as pd

from dlfb.utils.context import assets

protein_df = pd.read_csv(assets("proteins/datasets/sequence_df_cco.csv"))
protein_df = protein_df[~protein_df["term"].isin(["GO:0005575", "GO:0110165"])]
num_proteins = protein_df["EntryID"].nunique()
print(protein_df)
```

Output:

```
        EntryID            Sequence  taxonomyID         term  aspect  Length
0        095231  MRLSSSPPRGPQQLSS...        9606   GO:0005622     CCO     258
1        095231  MRLSSSPPRGPQQLSS...        9606   GO:0031981     CCO     258
2        095231  MRLSSSPPRGPQQLSS...        9606   GO:0043229     CCO     258
...         ...                 ...         ...          ...     ...     ...
337551   E7ER32  MPPLKSPAAFHEQRRS...        9606   GO:0031974     CCO     798
337552   E7ER32  MPPLKSPAAFHEQRRS...        9606   GO:0005634     CCO     798
337553   E7ER32  MPPLKSPAAFHEQRRS...        9606   GO:0005654     CCO     798

[294731 rows x 6 columns]
```

For each protein sequence identified by an `EntryID`, the `term` column provides its GO annotation for cellular localization.

Let's focus on two specific locations:

`extracellular` *(GO:0005576)*
    Proteins secreted outside the cell, often involved in signaling, immune response, or structural roles

`membrane` *(GO:0016020)*
    Proteins embedded in or associated with cell membranes, frequently functioning in transport, signaling, or cell–cell interaction

---

The Connection Between Cell Location and Sequence Features

We're filtering proteins based on their cellular location, but the model we're using is trained purely on sequence. So what's the connection?

The key point is that certain types of proteins have characteristic sequence features that correlate with where they function in the cell. For example, membrane proteins often contain stretches of amino acids that anchor them into the cell's outer membrane. These regions tend to be water repellent (hydrophobic), helping them interact with the oily membrane environment.

By contrast, extracellular proteins—those sent outside the cell—typically include short signal sequences that direct their export. They also often form stable structures through chemical bridges called disulfide bonds and may include regions that facilitate binding to other molecules.

These structural features are encoded in the amino acid sequence and should, in theory, be picked up by pretrained language models like ESM2—even though the model was never trained on location labels. In this section, we're essentially testing whether such structural signals are reflected in the learned embeddings.

We'll filter the dataset to proteins annotated with only one of these two locations:

```
# Filter protein dataframe to proteins with a single location.
num_locations = protein_df.groupby("EntryID")["term"].nunique()
proteins_one_location = num_locations[num_locations == 1].index
protein_df = protein_df[protein_df["EntryID"].isin(proteins_one_location)]

go_function_examples = {
  "extracellular": "GO:0005576",
  "membrane": "GO:0016020",
}

sequences_by_function = {}

min_length = 100
max_length = 500  # Cap sequence length for speed and memory.
num_samples = 20

for function, go_term in go_function_examples.items():
  proteins_with_function = protein_df[
    (protein_df["term"] == go_term)
    & (protein_df["Length"] >= min_length)
    & (protein_df["Length"] <= max_length)
  ]
  print(
    f"Found {len(proteins_with_function)} human proteins\n"
    f"with the molecular function '{function}' ({go_term}),\n"
    f"and {min_length}<=length<={max_length}.\n"
    f"Sampling {num_samples} proteins at random.\n"
  )
  sequences = list(
    proteins_with_function.sample(num_samples, random_state=42)["Sequence"]
  )
  sequences_by_function[function] = sequences
```

Output:

```
Found 164 human proteins
with the molecular function 'extracellular' (GO:0005576),
and 100<=length<=500.
Sampling 20 proteins at random.

Found 65 human proteins
with the molecular function 'membrane' (GO:0016020),
and 100<=length<=500.
Sampling 20 proteins at random.
```

We'll now extract embeddings from these sequences. The function `get_mean_embed dings` computes the mean hidden state across each sequence, summarizing the model's representation of protein sequences:

```
def get_mean_embeddings(
  sequences: list[str],
  tokenizer: PreTrainedTokenizer,
  model: PreTrainedModel,
```

```
    device: torch.device | None = None,
) -> np.ndarray:
    """Compute mean embedding for each sequence using a protein LM."""
    if not device:
        device = get_device()

    # Tokenize input sequences and pad them to equal length.
    model_inputs = tokenizer(sequences, padding=True, return_tensors="pt")

    # Move tokenized inputs to the target device (CPU or GPU).
    model_inputs = {k: v.to(device) for k, v in model_inputs.items()}

    # Move model to the target device and set it to evaluation mode.
    model = model.to(device)
    model.eval()

    # Forward pass without gradient tracking to obtain embeddings.
    with torch.no_grad():
        outputs = model(**model_inputs)
        mean_embeddings = outputs.last_hidden_state.mean(dim=1)

    return mean_embeddings.detach().cpu().numpy()
```

Now we'll extract embeddings using a smaller ESM2 model, which produces 320-dimensional representations and requires significantly less memory than larger variants:

```
model_checkpoint = "facebook/esm2_t6_8M_UR50D"
tokenizer = AutoTokenizer.from_pretrained(model_checkpoint)
model = EsmModel.from_pretrained(model_checkpoint)
```

We then calculate the embeddings:

```
# Compute mean protein embeddings for each location.
protein_embeddings = {
    loc: get_mean_embeddings(sequences_by_function[loc], tokenizer, model)
    for loc in ["extracellular", "membrane"]
}

# Reformat data.
labels, embeddings = [], []
for location, embedding in protein_embeddings.items():
    labels.extend([location] * embedding.shape[0])
    embeddings.append(embedding)
    print(f"{location}: {embedding.shape}")
```

Output:

```
extracellular: (20, 320)
membrane: (20, 320)
```

Each set of 20 sampled proteins is now represented as a (20, 320) embedding matrix. This means that for each sequence—regardless of its original length—we obtain a fixed-size vector of 320 dimensions. These vectors correspond to the mean of the final hidden layer activations across all tokens in the sequence, and should capture some information about the overall protein structure.

To visualize how these embeddings might relate to protein localization, we project them into two dimensions using t-SNE, a common method for visualizing high-dimensional data. Figure 2-11 shows that the extracellular and membrane proteins tend to form distinct clusters in this space.

```python
import numpy as np
import seaborn as sns
from sklearn.manifold import TSNE

embeddings_tsne = TSNE(n_components=2, random_state=42).fit_transform(
    np.vstack(embeddings)
)
embeddings_tsne_df = pd.DataFrame(
    {
        "first_dimension": embeddings_tsne[:, 0],
        "second_dimension": embeddings_tsne[:, 1],
        "location": np.array(labels),
    }
)

fig = sns.scatterplot(
    data=embeddings_tsne_df,
    x="first_dimension",
    y="second_dimension",
    hue="location",
    style="location",
    s=50,
    alpha=0.7,
)
plt.title("tSNE of Protein Embeddings")
fig.set_xlabel("First Dimension")
fig.set_ylabel("Second Dimension");
```

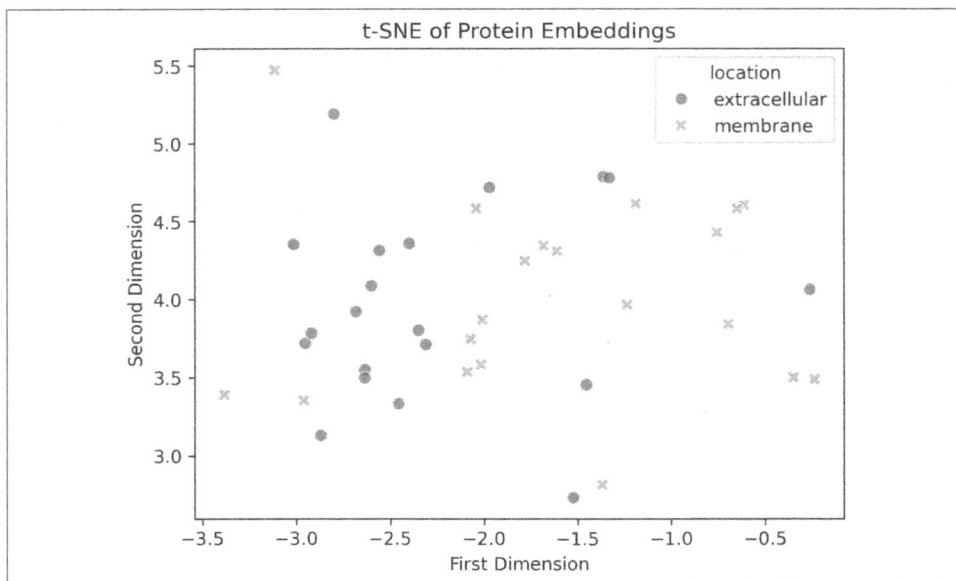

Figure 2-11. Two-dimensional t-SNE projection of the 320-dimensional embeddings from a small ESM2 model. Even with this lightweight model, we observe a tendency for extracellular and membrane proteins to form separate clusters, suggesting that the embeddings contain information relevant to cellular localization.

While the separation isn't perfect, there's a clear trend: extracellular proteins tend to cluster in a different region of embedding space than membrane proteins. It's quite striking that the model picks up on this purely from sequence. This suggests that the learned embeddings reflect biologically meaningful patterns—even without any explicit supervision for cellular location.

With this initial exploration complete, we now turn to the central machine learning task of this chapter: predicting protein function. Let's begin by preparing the dataset.

Preparing the Data

Many machine learning books and blog posts jump straight into the exciting parts—training and evaluating models—as soon as possible. But in practice, training is often a small fraction of the overall workflow. A significant portion of time is spent understanding, cleaning, and structuring the data. And when things go wrong with a model, the root cause is often found in the data. So, rather than handing you a polished CSV from the ether, we'll walk through the data preparation process step-by-step—starting from real-world resources and working through the steps needed to turn them into something a model can use.

Our goal is to fine-tune a model to predict protein function from sequence, which means assembling a dataset of (protein_sequence, protein_function) pairs. Fortunately, biologists have developed systematic frameworks for defining protein functions, and curated datasets already exist. One of the most widely used resources is the CAFA (Critical Assessment of Functional Annotation) (*https://oreil.ly/87EN_*) challenge, a community-driven competition where teams build models to predict protein function. We'll use CAFA data as our raw material, but we'll still need to process and structure it ourselves.

> If you're familiar with AlphaFold and protein structure prediction, you may have heard of the similarly named CASP (Critical Assessment of Structure Prediction), which plays a similar role in the protein structure community. Public benchmarks like these have been instrumental in driving progress across a wide range of computational biology problems.

Let's now explore the CAFA dataset.

Loading the CAFA3 Data

There have been several rounds of CAFA, but the CAFA3 dataset is the most recent publicly available one. We first downloaded the "CAFA3 Targets" and "CAFA3 Training Data" files from the CAFA website (*https://oreil.ly/87EN_*). Let's start by loading the label file, which tells us the functional annotations for each protein:

```
labels = pd.read_csv(
    assets("proteins/datasets/train_terms.tsv.zip"), sep="\t", compression="infer"
)
print(labels)
```

Output:

```
              EntryID         term aspect
0           A0A009IHW8  GO:0008152    BPO
1           A0A009IHW8  GO:0034655    BPO
2           A0A009IHW8  GO:0072523    BPO
...                ...         ...    ...
5363860         X5M5N0  GO:0005515    MFO
5363861         X5M5N0  GO:0005488    MFO
5363862         X5M5N0  GO:0003674    MFO

[5363863 rows x 3 columns]
```

This dataframe contains three columns:

EntryID
    The UniProt ID of the protein

term
: A GO accession code describing a specific protein function

aspect
: The GO category the function belongs to; one of three types of function described in the introduction: biological process (BPO), molecular function (MFO), and cellular component (CCO)

The `term` column contains only GO accession codes. To make these more interpretable, we'd ideally like to know their corresponding human-readable descriptions. This information isn't included directly in the CAFA files, but it is available via the Gene Ontology downloads page (*https://oreil.ly/uNhm2*). The ontology is stored in graph format as a `.obo` file, and we can use the obonet Python library to parse it. Here's how we retrieve the term descriptions:

```python
import obonet

def get_go_term_descriptions(store_path: str) -> pd.DataFrame:
    """Return GO term to description mapping, downloading if needed."""
    if not os.path.exists(store_path):
        url = "https://current.geneontology.org/ontology/go-basic.obo"
        graph = obonet.read_obo(url)

        # Extract GO term IDs and names from the graph nodes.
        id_to_name = {id: data.get("name") for id, data in graph.nodes(data=True)}
        go_term_descriptions = pd.DataFrame(
            zip(id_to_name.keys(), id_to_name.values()),
            columns=["term", "description"],
        )
        go_term_descriptions.to_csv(store_path, index=False)

    else:
        go_term_descriptions = pd.read_csv(store_path)
    return go_term_descriptions
```

The function will load the annotations from a local file if it already exists, or download and cache them if not:

```python
go_term_descriptions = get_go_term_descriptions(
    store_path=assets("proteins/datasets/go_term_descriptions.csv")
)
print(go_term_descriptions)
```

Output:

```
              term            description
0       GO:0000001  mitochondrion in...
1       GO:0000002  mitochondrial ge...
2       GO:0000006  high-affinity zi...
...            ...                    ...
40211   GO:2001315  UDP-4-deoxy-4-fo...
```

```
40212  GO:2001316  kojic acid metab...
40213  GO:2001317  kojic acid biosy...

[40214 rows x 2 columns]
```

We can then merge the human-readable term descriptions back onto the labels dataframe:

```
labels = labels.merge(go_term_descriptions, on="term")
labels
```

Output:

```
           EntryID        term aspect            description
0        A0A009IHW8  GO:0008152    BPO      metabolic process
1        A0A009IHW8  GO:0034655    BPO  nucleobase-conta...
2        A0A009IHW8  GO:0072523    BPO  purine-containin...
...             ...         ...    ...                    ...
4933955      X5M5N0  GO:0005515    MFO        protein binding
4933956      X5M5N0  GO:0005488    MFO                binding
4933957      X5M5N0  GO:0003674    MFO      molecular_function

[4933958 rows x 4 columns]
```

In this chapter, we'll focus specifically on molecular functions (MFO)—that is, what a protein does at the biochemical level. Later, you may want to extend this chapter's approach to include the other two GO categories.

Let's take a look at which molecular functions are most commonly annotated in the dataset:

```
labels = labels[labels["aspect"] == "MFO"]
print(labels["description"].value_counts())
```

Output:

```
description
molecular_function                        78637
binding                                   57380
protein binding                           47987
                                            ...
kaempferide 7-O-methyltransferase activity    1
protopine 6-monooxygenase activity            1
costunolide 3beta-hydroxylase activity        1
Name: count, Length: 6973, dtype: int64
```

We can already see that the distribution of function annotations is highly skewed. Some terms—like molecular_function, binding, and protein binding—appear tens of thousands of times, while others occur only once. Labels like molecular_function are arguably overly generic and provide little meaningful information, making them unhelpful for machine learning. We'll filter these out in a later step.

Next, let's load the protein sequences associated with each protein ID. This information is stored in the file *train_sequences.fasta*, a standard format for representing biological sequences such as proteins and DNA. We can use BioPython's `SeqIO` module to parse the *.fasta* file into a format we can work with.

> A quick aside: no one starts out knowing what BioPython's `SeqIO` module is, or how *.fasta* files work, or what GO annotations mean —and that's completely normal. Working at the intersection of biology and machine learning means constantly encountering new tools and terminology. Frequent looking up of new terms and tools is not just OK, it's expected.

We'll convert the *.fasta* sequences into a pandas dataframe to make them easier to manipulate:

```python
from Bio import SeqIO

sequences_file = assets("proteins/datasets/train_sequences.fasta")
fasta_sequences = SeqIO.parse(open(sequences_file), "fasta")

data = []
for fasta in fasta_sequences:
  data.append(
    {
      "EntryID": fasta.id,
      "Sequence": str(fasta.seq),
      "Length": len(fasta.seq),
    }
  )
sequence_df = pd.DataFrame(data)
print(sequence_df)
```

Output:

```
            EntryID           Sequence  Length
0            P20536  MNSVTVSHAPYTITYH...     218
1            O73864  MTEYRNFLLLFITSLS...     354
2            O95231  MRLSSSPPRGPQQLSS...     258
...             ...                ...     ...
142243       Q5RGB0  MADKGPILTSVIIFYL...     448
142244  A0A2R8QMZ5  MGRKKIQITRIMDERN...     459
142245  A0A8I6GHU0  HCISSLKLTAFFKRSF...     138

[142246 rows x 3 columns]
```

We've also computed the length of each sequence, since protein lengths can vary widely and this information will be useful later when filtering data.

One important detail: the CAFA dataset includes proteins from many different organisms. To isolate human proteins, we'll use the associated taxonomy file provided in the download:

```
taxonomy_file = assets("proteins/datasets/train_taxonomy.tsv.zip")
taxonomy = pd.read_csv(taxonomy_file, sep="\t", compression="infer")
print(taxonomy)
```

Output:

```
        EntryID  taxonomyID
0        Q8IXT2        9606
1        Q04418      559292
2        A8DYA3        7227
...         ...         ...
142243  A0A2R8QBB1      7955
142244   P0CT72      284812
142245   Q9NZ43        9606

[142246 rows x 2 columns]
```

This file contains a taxonomy ID (taxonomyID) for each protein, based on NCBI's organism classification system. We'll merge this onto our sequence dataframe and keep only proteins with taxonomyID == 9606, which corresponds to *Homo sapiens*:

```
sequence_df = sequence_df.merge(taxonomy, on="EntryID")
sequence_df = sequence_df[sequence_df["taxonomyID"] == 9606]
```

Now let's get an overview of the number of unique proteins and molecular function terms in our filtered dataset:

```
sequence_df = sequence_df.merge(labels, on="EntryID")
print(
  f'Dataset contains {sequence_df["EntryID"].nunique()} human proteins '
  f'with {sequence_df["term"].nunique()} molecular functions.'
)
```

Output:

```
Dataset contains 16336 human proteins with 4101 molecular functions.
```

Let's also take a look at the resulting sequence_df after merging in the function labels:

```
print(sequence_df)
```

Output:

```
        EntryID          Sequence  Length  taxonomyID       term  aspect  \
0        O95231  MRLSSSPPRGPQQLSS...     258        9606  GO:0003676     MFO
1        O95231  MRLSSSPPRGPQQLSS...     258        9606  GO:1990837     MFO
2        O95231  MRLSSSPPRGPQQLSS...     258        9606  GO:0001216     MFO
...         ...               ...     ...         ...         ...     ...
152523   Q86TI6  MGAAAVRWHLCVLLAL...     347        9606  GO:0005515     MFO
152524   Q86TI6  MGAAAVRWHLCVLLAL...     347        9606  GO:0005488     MFO
152525   Q86TI6  MGAAAVRWHLCVLLAL...     347        9606  GO:0003674     MFO
```

```
              description
0        nucleic acid bin...
1        sequence-specifi...
2        DNA-binding tran...
...                    ...
152523      protein binding
152524              binding
152525    molecular_function

[152526 rows x 7 columns]
```

From this table, we can already see that many proteins are associated with multiple molecular functions. To quantify this, we examine the distribution of the number of functions per protein in Figure 2-12:

```python
sequence_df.groupby("EntryID")["term"].nunique().plot.hist(
    bins=100, figsize=(5, 3), color="grey", log=True
)
plt.xlabel("Number of Molecular Function Annotations per Protein")
plt.ylabel("Frequency (log scale)")
plt.title("Distribution of Function Counts per Protein")
plt.tight_layout()
```

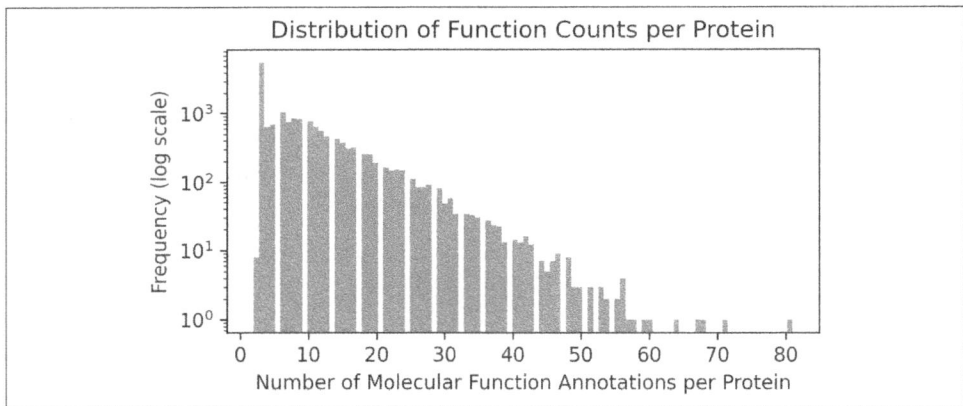

Figure 2-12. Distribution of the number of molecular functions annotated per protein. The y-axis is shown on a logarithmic scale to make rare cases more visible. While most proteins have fewer than 20 annotated functions, a small number of proteins are associated with more than 50 distinct molecular roles.

This pattern reflects a complex biological reality: while many proteins carry out a single, well-defined function, others are involved in a wide variety of molecular roles. For example, some proteins act as enzymes, bind to other molecules, and participate in multiple pathways. From a machine learning perspective, this means our model must be able to assign multiple function labels to a single protein and also cope with the fact that some labels are much rarer than others.

Let's now take a closer look at the most frequent molecular function labels. Some terms are so broad and universally assigned that they offer little meaningful insight. For example, molecular function applies to nearly all proteins, binding covers 93%, and protein binding appears in 89% of cases. These labels will tend to dominate the loss during training and can cause the model to fixate on predicting them at the expense of more meaningful functions. As a dataset preprocessing step, we'll explicitly remove these overly generic terms:

```
uninteresting_functions = [
  "GO:0003674",  # "molecular function". Applies to 100% of proteins.
  "GO:0005488",  # "binding". Applies to 93% of proteins.
  "GO:0005515",  # "protein binding". Applies to 89% of proteins.
]

sequence_df = sequence_df[~sequence_df["term"].isin(uninteresting_functions)]
sequence_df.shape
```

Output:

```
(106501, 7)
```

On the opposite end of the spectrum, some molecular functions are extremely rare—for example, GO:0099609 (microtubule lateral binding) appears only once. To learn meaningful associations, our model needs enough training examples per function. So we'll filter out the rarest labels and keep only those that appear in at least 50 proteins:

```
common_functions = (
  sequence_df["term"]
  .value_counts()[sequence_df["term"].value_counts() >= 50]
  .index
)

sequence_df = sequence_df[sequence_df["term"].isin(common_functions)]
sequence_df["term"].value_counts()
```

Output:

```
term
GO:0003824    3875
GO:1901363    2943
GO:0003676    2469
              ...
GO:0031490      51
GO:0019003      50
GO:0015179      50
Name: count, Length: 303, dtype: int64
```

This gives us a cleaner set of function labels that are more amenable to learning.

Thresholds used during data processing—like how many times a label must appear to be included—are somewhat arbitrary, but they can significantly affect model performance. These decisions are effectively hyperparameters and should be tuned based on the specific task, dataset size, and model capacity.

Now we'll reshape the dataframe so that each row corresponds to one protein, and each column corresponds to a molecular function label. We'll use the `pivot` function in pandas to create this multilabel format:

```python
sequence_df = (
    sequence_df[["EntryID", "Sequence", "Length", "term"]]
    .assign(value=1)
    .pivot(
        index=["EntryID", "Sequence", "Length"], columns="term", values="value"
    )
    .fillna(0)
    .astype(int)
    .reset_index()
)
print(sequence_df)
```

Output:

term	EntryID	Sequence	Length	GO:0000166	GO:0000287	...	\
0	A0A024R6B2	MIASCLCYLLLPATRL...	670	0	0	...	
1	A0A087WUI6	MSRKISKESKKVNISS...	698	0	0	...	
2	A0A087X1C5	MGLEALVPLAMIVAIF...	515	0	0	...	
...	...	...	...	...	...	...	...
10706	Q9Y6Z7	MNGFASLLRRNQFILL...	277	0	0	...	
10707	X5D778	MPKGGCPKAPQQEELP...	421	0	0	...	
10708	X5D7E3	MLDLTSRGQVGTSRRM...	237	0	0	...	

term	GO:1901702	GO:1901981	GO:1902936	GO:1990782	GO:1990837
0	0	0	0	0	0
1	0	0	0	0	0
2	0	0	0	0	0
...	...	...	...	...	...
10706	0	0	0	0	0
10707	0	0	0	0	0
10708	0	0	0	0	0

```
[10709 rows x 306 columns]
```

Great—this dataset is now in a format that's almost ready for machine learning. Before we move on, let's run a few final sanity checks.

First, how many unique proteins do we have?

```python
sequence_df["EntryID"].nunique()
```

Output:

```
10709
```

This number is in the right ballpark. There are roughly 21,000 protein-coding genes in the human genome, and since we applied several filtering steps, we expect a somewhat smaller number. It's always worth keeping rough order-of-magnitude expectations in mind—if we saw 1,000 or 1,000,000 here, we'd suspect something was off.

Next, let's check whether any protein sequences are duplicated:

```
sequence_df["Sequence"].nunique()
```

Output:

```
10698
```

It seems that a few protein sequences are repeated. For example, the entries P0DP23, P0DP24, and P0DP25 all share the same sequence:

```
print(sequence_df[sequence_df["EntryID"].isin(["P0DP23", "P0DP24", "P0DP25"])])
```

Output:

```
term EntryID          Sequence  Length  GO:0000166  GO:0000287 ... \
1945  P0DP23  MADQLTEEQIAEFKEA...     149           0           0 ...
1946  P0DP24  MADQLTEEQIAEFKEA...     149           0           0 ...
1947  P0DP25  MADQLTEEQIAEFKEA...     149           0           0 ...

term  GO:1901702  GO:1901981  GO:1902936  GO:1990782  GO:1990837
1945           0           0           0           0           0
1946           0           0           0           0           0
1947           0           0           0           0           0

[3 rows x 306 columns]
```

These seem to be legitimate biological duplicates—proteins with different Uniprot identifiers but identical sequences—so we'll keep them in the dataset.

At this point, we have a final dataset linking 10,709 human proteins to one or more of 303 molecular functions.

Since our simple mean embedding approach can be quite memory intensive, we'll filter the dataset to include only proteins with a maximum length of 500 amino acids. This helps avoid out-of-memory errors during model inference and training:

```
print(sequence_df.shape)
sequence_df = sequence_df[sequence_df["Length"] <= 500]
print(sequence_df.shape)
```

Output:

```
(10709, 306)
(5957, 306)
```

This roughly halves the dataset, which is perfectly fine for initial prototyping. You can always remove this constraint later if time and memory allow.

Now that we have a clean and compact dataset, let's process it further for compatibility with machine learning.

Splitting the Dataset into Subsets

We will split our dataset into three distinct subsets:

Training set
Used to fit the model. The model sees this data during training and uses it to learn patterns.

Validation set
Used to evaluate the model's performance during development. We use this to tune hyperparameters and compare model variants.

Test set
Used only once, for final evaluation. Crucially, we avoid using this data to guide model design decisions. It serves as our best estimate of how well the model would generalize to completely unseen data.

We'll split the proteins by their `EntryID`, ensuring that each protein appears in only one subset:

```python
from sklearn.model_selection import train_test_split

# 60% of the proteins will go into the training set.
train_sequence_ids, valid_test_sequence_ids = train_test_split(
    list(set(sequence_df["EntryID"])), test_size=0.40, random_state=42
)

# Split the remaining 40% evenly between validation and test sets.
valid_sequence_ids, test_sequence_ids = train_test_split(
    valid_test_sequence_ids, test_size=0.50, random_state=42
)
```

Now we'll extract the rows for each split from our dataframe `sequence_df`:

```python
sequence_splits = {
    "train": sequence_df[sequence_df["EntryID"].isin(train_sequence_ids)],
    "valid": sequence_df[sequence_df["EntryID"].isin(valid_sequence_ids)],
    "test": sequence_df[sequence_df["EntryID"].isin(test_sequence_ids)],
}
```

```
for split, df in sequence_splits.items():
    print(f"{split} has {len(df)} entries.")
```

Output:

```
train has 3574 entries.
valid has 1191 entries.
test has 1192 entries.
```

This gives us clean, nonoverlapping training, validation, and test sets—each containing a subset of proteins we'll use throughout model development and evaluation.

Converting Protein Sequences into Their Mean Embeddings

We will now convert the sequences from each dataset split into their corresponding mean embeddings, just as we did earlier. Since this step can be time-consuming—especially with larger models—it's worth thinking about how to do it efficiently. Using a GPU can significantly speed up computation, but we can also avoid repeating work by computing the embeddings only once, storing them to disk, and loading them later.

To make this process more convenient, we'll use a pair of helper functions to store and load sequence embeddings:

```python
def store_sequence_embeddings(
    sequence_df: pd.DataFrame,
    store_prefix: str,
    tokenizer: PreTrainedTokenizer,
    model: PreTrainedModel,
    batch_size: int = 64,
    force: bool = False,
) -> None:
    """Extract and store mean embeddings for each protein sequence."""
    model_name = str(model.name_or_path).replace("/", "_")
    store_file = f"{store_prefix}_{model_name}.feather"

    if not os.path.exists(store_file) or force:
        device = get_device()

        # Iterate through protein dataframe in batches, extracting embeddings.
        n_batches = ceil(sequence_df.shape[0] / batch_size)
        batches: list[np.ndarray] = []
        for i in range(n_batches):
            batch_seqs = list(
                sequence_df["Sequence"][i * batch_size : (i + 1) * batch_size]
            )
            batches.extend(get_mean_embeddings(batch_seqs, tokenizer, model, device))

        # Store each of the embedding values in a separate column in the dataframe.
        embeddings = pd.DataFrame(np.vstack(batches))
        embeddings.columns = [f"ME:{int(i)+1}" for i in range(embeddings.shape[1])]
        df = pd.concat([sequence_df.reset_index(drop=True), embeddings], axis=1)
```

```
    df.to_feather(store_file)


def load_sequence_embeddings(
    store_file_prefix: str, model_checkpoint: str
) -> pd.DataFrame:
    """Load stored embedding DataFrame from disk."""
    model_name = model_checkpoint.replace("/", "_")
    store_file = f"{store_file_prefix}_{model_name}.feather"
    return pd.read_feather(store_file)
```

Let's use the more powerful (but computationally expensive) ESM2 model with 640-dimensional embeddings and store the embeddings for each split using the `store_sequence_embeddings` function:

```
model_checkpoint = "facebook/esm2_t30_150M_UR50D"
tokenizer = AutoTokenizer.from_pretrained(model_checkpoint)
model = EsmModel.from_pretrained(model_checkpoint)

for split, df in sequence_splits.items():
    store_sequence_embeddings(
        sequence_df=df,
        store_prefix=assets(f"proteins/datasets/protein_dataset_{split}"),
        tokenizer=tokenizer,
        model=model,
    )
```

Once the embeddings are stored, we can load them back into memory whenever needed. Here's a glimpse of the resulting training dataset that the model will learn from:

```
train_df = load_sequence_embeddings(
    assets("proteins/datasets/protein_dataset_train"),
    model_checkpoint=model_checkpoint,
)

print(train_df)
```

Output:

```
        EntryID           Sequence  Length  GO:0000166  GO:0000287  ...  \
0     A0A0C4DG62  MAHVGSRKRSRSRSRS...     218           0           0  ...
1     A0A1B0GTB2  MVITSENDEDRGGQEK...      48           0           0  ...
2         A0AVI4  MDSPEVTFTLAYLVFA...     362           0           0  ...
...          ...                 ...     ...         ...         ...  ... ...
3571      Q9Y6W5  MPLVTRNIEPRHLCRQ...     498           0           0  ...
3572      Q9Y6W6  MPPSPLDDRVVVALSR...     482           0           0  ...
3573      Q9Y6Y9  MLPFLFFSTLFSSIFT...     160           0           0  ...

        ME:636    ME:637    ME:638    ME:639    ME:640
0     0.062926  0.040286  0.030008 -0.033614  0.023891
1     0.129815 -0.044294  0.023842 -0.020635  0.125583
2     0.153848 -0.075747  0.024440 -0.123321  0.020945
```

```
 ...        ...        ...        ...        ...        ...
3571 -0.001535 -0.084161 -0.014317 -0.141801 -0.040719
3572  0.120192 -0.086032 -0.016481 -0.108710 -0.077937
3573  0.114847 -0.028570  0.084638  0.038610  0.087047

[3574 rows x 946 columns]
```

You'll notice a series of columns labeled ME:1 through ME:640. These represent the mean-pooled hidden states from the final layer of the ESM2 model—effectively a fixed-length numerical summary of each protein sequence. These embeddings capture biochemical and structural information learned during pretraining and will serve as the input features for our classifier.

This dataframe becomes the input to a convert_to_tfds function, which we've defined to make it easier to prepare the datasets for each split:

```python
import tensorflow as tf

def convert_to_tfds(
    df: pd.DataFrame,
    embeddings_prefix: str = "ME:",
    target_prefix: str = "GO:",
    is_training: bool = False,
    shuffle_buffer: int = 50,
) -> tf.data.Dataset:
    """Convert embedding DataFrame into a TensorFlow dataset."""
    dataset = tf.data.Dataset.from_tensor_slices(
        {
            "embedding": df.filter(regex=f"^{embeddings_prefix}").to_numpy(),
            "target": df.filter(regex=f"^{target_prefix}").to_numpy(),
        }
    )
    if is_training:
        dataset = dataset.shuffle(shuffle_buffer).repeat()
    return dataset
```

Let's now use our convert_to_tfds function to build a TensorFlow-compatible dataset from the training DataFrame:

```python
train_ds = convert_to_tfds(train_df, is_training=True)
```

Fetching a batch of data from these datasets is straightforward. We just batch the dataset, convert it to a NumPy iterator, and retrieve a batch by calling next:

```python
batch_size = 32

batch = next(train_ds.batch(batch_size).as_numpy_iterator())
batch["embedding"].shape, batch["target"].shape
```

Output:

```
((32, 640), (32, 303))
```

These shapes confirm that each input is a 640-dimensional embedding vector (from the ESM2 model), and each target is a 303-dimensional binary vector representing the presence or absence of each molecular function label.

> Because the training dataset includes `.repeat()`, it yields batches indefinitely by looping over the data. This is useful for training, where we want to cycle through the dataset multiple times. In contrast, the validation and test datasets are not repeated—so their batches will eventually be exhausted, which is exactly what we want during evaluation, where each example should be seen only once.

To streamline the dataset setup, we've wrapped the entire pipeline into a single helper function, `build_dataset`:

```python
def build_dataset(
    store_file_prefix: str, model_checkpoint: str
) -> dict[str, tf.data.Dataset]:
    """Build train/valid/test TensorFlow datasets from stored embeddings."""
    dataset_splits = {}

    for split in ["train", "valid", "test"]:
        dataset_splits[split] = convert_to_tfds(
            df=load_sequence_embeddings(
                store_file_prefix=f"{store_file_prefix}_{split}",
                model_checkpoint=model_checkpoint,
            ),
            is_training=(split == "train"),
        )
    return dataset_splits
```

This function loads the saved mean embeddings from disk for all three splits and constructs `tf.data.Dataset` objects that are ready for training:

```python
dataset_splits = build_dataset(
    assets("proteins/datasets/protein_dataset"), model_checkpoint=model_checkpoint
)
```

With this, we now have our data fully preprocessed and ready to use in training a model.

Training the Model

We will now train a simple Flax (*https://oreil.ly/MjH5C*) linear model on top of the mean protein embeddings. Recall that each protein sequence has a variable length, but we've already transformed them into fixed-size embeddings. Our goal is to predict which of the 303 possible molecular functions each protein performs. This is a *multilabel classification* problem, meaning each protein may be associated with several function labels simultaneously.

In this setup, we'll train a lightweight MLP (multilayer perceptron)—a stack of dense layers with nonlinearities. Importantly, we are not fine-tuning the original ESM2 model: it remains frozen, and our model simply learns on top of its embeddings.

Here's the model code:

```python
import flax.linen as nn
from flax.training import train_state


class Model(nn.Module):
    """Simple MLP for protein function prediction."""

    num_targets: int
    dim: int = 256

    @nn.compact
    def __call__(self, x):
        """Apply MLP layers to input features."""
        x = nn.Sequential(
          [
            nn.Dense(self.dim * 2),
            jax.nn.gelu,
            nn.Dense(self.dim),
            jax.nn.gelu,
            nn.Dense(self.num_targets),
          ]
        )(x)
        return x

    def create_train_state(self, rng: jax.Array, dummy_input, tx) -> TrainState:
        """Initialize model parameters and return a training state."""
        variables = self.init(rng, dummy_input)
        return TrainState.create(
          apply_fn=self.apply, params=variables["params"], tx=tx
        )
```

Some notes on this very lightweight model:

- It uses `nn.Sequential` to stack layers, which keeps the definition clean and readable for this simple model.

- We use a GELU (Gaussian Error Linear Unit) activation function, which is a smooth, nonlinear alternative to ReLU.

- The final layer is an `nn.Dense` layer projecting to the number of function labels (`num_targets`). It returns logits, not probabilities—so we'll apply a suitable activation (like sigmoid) inside the loss function to convert these logits into predicted probabilities.

- This model is frozen on top of the ESM2 embeddings—meaning it does not update the transformer weights. It learns only to map fixed embeddings to functional labels. This is efficient and interpretable, and it reduces memory usage during training.

You may also have noticed that we attached a convenience function, `create_train_state`, to the model class for creating a training state. This encapsulates model initialization, parameter registration, and optimizer setup into a single `TrainState` object. It's particularly useful because it allows us to construct the training state right when everything needed—the model, dummy input for shape inference, and optimizer config—is readily available.

Let's instantiate the model with the correct number of output targets, based on how many GO term columns we have in the training dataframe:

```
targets = list(train_df.columns[train_df.columns.str.contains("GO:")])

mlp = Model(num_targets=len(targets))
```

This model is now ready to be trained to predict which molecular functions a protein is involved in, using the precomputed embeddings as input.

Defining the Training Loop

With the model and dataset ready, we can now define a function to perform a single training step. This step includes:

- A forward pass through the model
- Computing the loss
- Calculating gradients
- Updating the model parameters using those gradients

Here's how we implement it:

```
@jax.jit
def train_step(state, batch):
    """Run a single training step and update model parameters."""

    def calculate_loss(params):
        """Compute sigmoid cross-entropy loss from logits."""
        logits = state.apply_fn({"params": params}, x=batch["embedding"])
        loss = optax.sigmoid_binary_cross_entropy(logits, batch["target"]).mean()
        return loss

    grad_fn = jax.value_and_grad(calculate_loss, has_aux=False)
    loss, grads = grad_fn(state.params)
    state = state.apply_gradients(grads=grads)
    return state, loss
```

In this setup:

- We use a sigmoid activation and binary cross-entropy loss, appropriate for multilabel classification. The logits go through a sigmoid activation, not softmax—because we want independent yes/no predictions for each possible protein function. Remember that each protein could have many functions at once.
- `@jax.jit` compiles the training step for better performance.

Next, let's implement some metrics to evaluate how well the model is doing beyond the loss alone, using tools from `sklearn`:

```python
import sklearn

def compute_metrics(
    targets: np.ndarray, probs: np.ndarray, thresh=0.5
) -> dict[str, float]:
    """Compute accuracy, recall, precision, auPRC, and auROC."""
    if np.sum(targets) == 0:
        return {
            m: 0.0 for m in ["accuracy", "recall", "precision", "auprc", "auroc"]
        }
    return {
        "accuracy": metrics.accuracy_score(targets, probs >= thresh),
        "recall": metrics.recall_score(targets, probs >= thresh).item(),
        "precision": metrics.precision_score(
            targets,
            probs >= thresh,
            zero_division=0.0,
        ).item(),
        "auprc": metrics.average_precision_score(targets, probs).item(),
        "auroc": metrics.roc_auc_score(targets, probs).item(),
    }
```

We'll track the following evaluation metrics for each function label:

Accuracy

The fraction of correct predictions across all labels. In multilabel classification with imbalanced data (like this), accuracy can be misleading—most labels are zero, so a model that always predicts "no function" would appear accurate. Still, it's an intuitive metric and we'll include it for now.

Recall

The proportion of actual function labels the model correctly predicted (i.e., true positives/all actual positives). High recall means the model doesn't miss many true functions.

Precision

The proportion of predicted function labels that are correct (i.e., true positives/all predicted positives). High precision means the model avoids false alarms.

Area under the precision-recall curve (auPRC)

Summarizes the tradeoff between precision and recall at different thresholds. Particularly useful in highly imbalanced settings like this one.

Area under the receiver operating characteristic curve (auROC)

Measures the model's ability to distinguish positive from negative examples across all thresholds. While it's a standard metric of discrimination ability, it can sometimes be misleading in highly imbalanced datasets, as it gives equal weight to both classes.

In a multilabel setting, we calculate these metrics for each protein function (i.e., per target/label), then average them to get a holistic view of model performance.

We apply these metrics calculations during the evaluation step `eval_step`:

```python
def eval_step(state, batch) -> dict[str, float]:
    """Run evaluation step and return mean metrics over targets."""
    logits = state.apply_fn({"params": state.params}, x=batch["embedding"])
    loss = optax.sigmoid_binary_cross_entropy(logits, batch["target"]).mean()
    target_metrics = calculate_per_target_metrics(logits, batch["target"])
    metrics = {
        "loss": loss.item(),
        **pd.DataFrame(target_metrics).mean(axis=0).to_dict(),
    }
    return metrics


def calculate_per_target_metrics(logits, targets):
    """Compute metrics for each target in a multi-label batch."""
    probs = jax.nn.sigmoid(logits)
    target_metrics = []
    for target, prob in zip(targets, probs):
        target_metrics.append(compute_metrics(target, prob))
    return target_metrics
```

The evaluation computes metrics per protein in the batch. For each protein, we:

- Apply sigmoid to its 303 logits to get function probabilities.

- Threshold those probabilities (e.g., at 0.5) to get binary predictions.

- Compare these to the true function labels to compute metrics like accuracy, precision, recall, auPRC, and auROC.

We repeat this for every protein in the batch and then average the resulting metrics across proteins. This tells us how well the model predicts sets of functions per protein. It does not report performance per GO term. If we wanted per-function metrics (e.g., how well the model predicts GO:0003677), we'd need to compute metrics column-wise instead.

In the next chunk of code, everything comes together into a `train` function, and variations of this basic setup will be repeated in every chapter. We have the training loop where we first initialize our model training state and then loop over the dataset in batches to train the model and evaluate it every so often:

```python
def train(
    state: TrainState,
    dataset_splits: dict[str, tf.data.Dataset],
    batch_size: int,
    num_steps: int = 300,
    eval_every: int = 30,
):
    """Train model using batched TF datasets and track performance metrics."""
    # Create containers to handle calculated during training and evaluation.
    train_metrics, valid_metrics = [], []

    # Create batched dataset to pluck batches from for each step.
    train_batches = (
        dataset_splits["train"]
        .batch(batch_size, drop_remainder=True)
        .as_numpy_iterator()
    )

    steps = tqdm(range(num_steps))  # Steps with progress bar.
    for step in steps:
        steps.set_description(f"Step {step + 1}")

        # Get batch of training data, convert into a JAX array, and train.
        state, loss = train_step(state, next(train_batches))
        train_metrics.append({"step": step, "loss": loss.item()})

        if step % eval_every == 0:
            # For all the evaluation batches, calculate metrics.
            eval_metrics = []
            for eval_batch in (
                dataset_splits["valid"].batch(batch_size=batch_size).as_numpy_iterator()
            ):
                eval_metrics.append(eval_step(state, eval_batch))
            valid_metrics.append(
                {"step": step, **pd.DataFrame(eval_metrics).mean(axis=0).to_dict()}
            )

    return state, {"train": train_metrics, "valid": valid_metrics}
```

A few notes on this training loop:

Efficient batch sampling
>Training data is streamed via `.as_numpy_iterator()`, and the `.repeat()` in the dataset ensures infinite looping over the data.

Regular evaluation
>Every `eval_every` step, the model is evaluated on the full validation set to monitor progress using metrics we defined previously, like auPRC and auROC.

Metric aggregation
>Validation metrics are computed batch-wise and then averaged across all batches using `pd.DataFrame(...).mean(axis=0)`. This gives a stable estimate of performance across the entire validation set.

Let's now train the model. But first, a quick trick: to avoid unnecessarily repeating training from scratch every time you rerun your code cell, we use the `@restorable` decorator. This lightweight utility checks whether a trained model already exists at a specified path. If it does, it:

- Skips retraining
- Restores the model into a valid `TrainState`
- Returns the model along with any saved metrics

This makes your workflow much faster and more reproducible, especially during iterative development and debugging. Let's take a look at how this is used:

```python
import optax

from dlfb.utils.restore import restorable

# Initiate training state with dummy data from a single batch.
rng = jax.random.PRNGKey(42)
rng, rng_init = jax.random.split(key=rng, num=2)

state, metrics = restorable(train)(
    state=mlp.create_train_state(
        rng=rng_init, dummy_input=batch["embedding"], tx=optax.adam(0.001)
    ),
    dataset_splits=dataset_splits,
    batch_size=32,
    num_steps=300,
    eval_every=30,
    store_path=assets("proteins/models/mlp"),
)
```

Some additional parameters worth mentioning are the optimizer (here, optax.adam) and the total number of training steps (num_steps). Given that we have 2,100 training examples and a batch size of 32, it will take about 66 steps for the model to see the entire training set once. Setting num_steps=300 means the model will see each training data point several times.

Having trained the model with the previous train call, we can now evaluate its training dynamics and performance on the validation set, as shown in Figure 2-13:

```python
import matplotlib.pyplot as plt
import seaborn as sns

from dlfb.utils.metric_plots import DEFAULT_SPLIT_COLORS

fig, ax = plt.subplots(nrows=1, ncols=2, figsize=(9, 4))

# Plot training loss curve.
learning_data = pd.concat(
    pd.DataFrame(metrics[split]).melt("step").assign(split=split)
    for split in ["train", "valid"]
)

sns.lineplot(
    ax=ax[0],
    x="step",
    y="value",
    hue="split",
    data=learning_data[learning_data["variable"] == "loss"],
    palette=DEFAULT_SPLIT_COLORS,
)
ax[0].set_title("Loss over training steps.")

# Plot validation metrics curves.
sns.lineplot(
    ax=ax[1],
    x="step",
    y="value",
    hue="variable",
    style="variable",
    data=learning_data[learning_data["variable"] != "loss"],
    palette="Set2",
)
plt.legend(loc="center left", bbox_to_anchor=(1, 0.5))
ax[1].set_title("Validation metrics over training steps.");
```

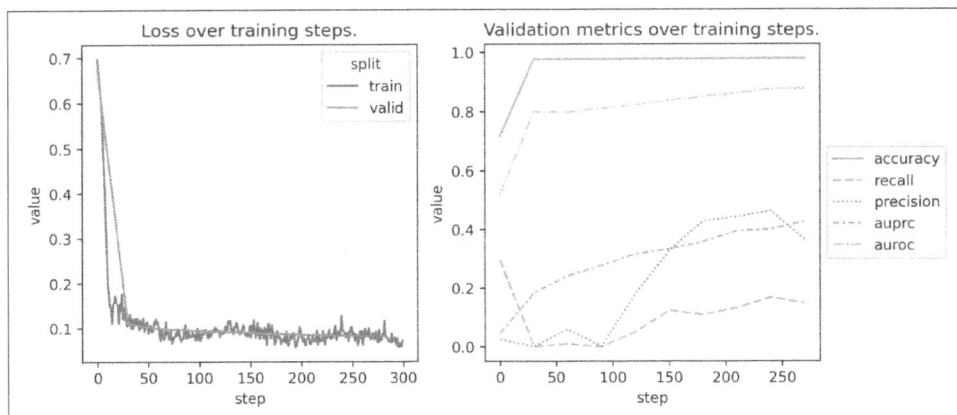

Figure 2-13. Training and evaluation of the MLP model over 300 steps. On the left, loss curves for the training and validation splits show rapid convergence, with stability reached after ~30 steps. On the right, auPRC, precision, and recall improve gradually. Accuracy and auROC metrics are very high due to class imbalance and are not very informative for this problem.

In the left panel, we observe that both training and validation loss drop sharply within the first ~30 steps and then stabilize. This is a typical learning curve, indicating rapid convergence without substantial instability (e.g., no major spikes or divergence). It suggests that the model—a shallow MLP operating on top of frozen pretrained embeddings—quickly captures the low-hanging signal in the data.

In the right panel, we track several evaluation metrics over time:

- Accuracy and auROC start high and remain flat, but these can be misleading in imbalanced, multilabel settings like this one. Since most function labels are negative (i.e., a protein lacks the majority of all possible functions), a model that mostly predicts zeros can still achieve a high score on these metrics. For that reason, we don't put much weight on these metrics in this context.

- auPRC steadily improves and does not fully plateau, suggesting the model continues to learn subtle distinctions and could potentially benefit from further training (i.e., by increasing num_steps).

- Precision improves more quickly than recall, indicating the model becomes increasingly confident in its predictions but still fails to capture some true positives.

Together, these trends indicate that while most of the learning happens early on, there may still be headroom—particularly in recall and auPRC—if training were extended further or if a more powerful architecture were used.

It can be slightly tedious to manually log metrics inside every training loop and then hook up custom plotting code to visualize them. To streamline this, later chapters introduce a MetricsLogger (for capturing values) and MetricsPlotter (for rendering them).

Beyond that, many modern machine learning workflows use hosted (or self-hosted) dashboards to automatically collect, store, and display metrics in real time. These tools help monitor experiments, compare training runs, and share results across teams. We encourage you to check them out. Popular options include:

- TensorBoard (*https://oreil.ly/tSPIP*)
- Weights & Biases (W&B) (*https://oreil.ly/Loybs*)
- MLflow (*https://oreil.ly/A1faJ*)

It's great to see the model training successfully and loss and metrics curves trending in the right direction—but that's just the beginning. The real insight comes from analyzing the model's predictions, understanding where it performs well, and identifying its limitations.

Examining the Model Predictions

With a trained model in hand, it's time to explore its strengths and weaknesses. We'll start by generating predictions for the entire validation set and storing them in a dataframe for easier inspection:

```
valid_df = load_sequence_embeddings(
    store_file_prefix=f"{assets('proteins/datasets/protein_dataset')}_valid",
    model_checkpoint=model_checkpoint,
)

# Use batch size of 1 to avoid dropping the remainder.
valid_probs = []
for valid_batch in dataset_splits["valid"].batch(1).as_numpy_iterator():
    logits = state.apply_fn({"params": state.params}, x=valid_batch["embedding"])
    valid_probs.extend(jax.nn.sigmoid(logits))

valid_true_df = valid_df[["EntryID"] + targets].set_index("EntryID")
valid_prob_df = pd.DataFrame(
    np.stack(valid_probs), columns=targets, index=valid_true_df.index
)
```

To get a high-level sense of how the model is performing, we can visualize the full prediction matrix as a heatmap. In Figure 2-14, we plot two side-by-side heatmaps: one showing the true protein-function annotations (left) and the other showing the model's predicted probabilities (right). Each column corresponds to a protein function, and each row to a protein:

```
fig, ax = plt.subplots(nrows=1, ncols=2, figsize=(11, 4))

sns.heatmap(
    ax=ax[0],
    data=valid_true_df,
    yticklabels=False,
    xticklabels=False,
    cmap="flare",
)
ax[0].set_title("True functional annotations by protein.")
ax[0].set_xlabel("Functional category")

sns.heatmap(
    ax=ax[1],
    data=valid_prob_df,
    yticklabels=False,
    xticklabels=False,
    cmap="flare",
)
ax[1].set_title("Predicted functional annotations by protein.")
ax[1].set_xlabel("Functional category");
```

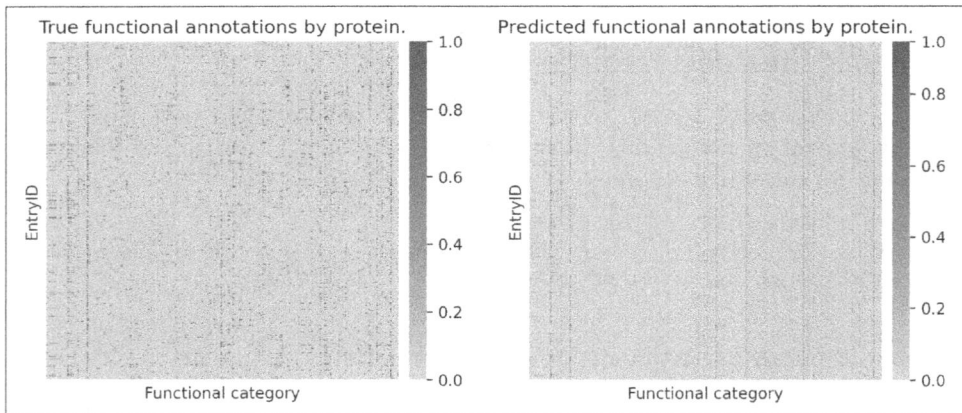

Figure 2-14. Heatmap overview of protein function prediction. The left panel shows the ground truth functional annotations for each protein in the validation set, while the right panel shows the model's predicted probabilities. Both matrices are sparse, with vertical bands reflecting common function labels.

This visualization is quite zoomed out and high level, but it helps build intuition about overall model behavior:

- Some protein functions appear frequently in the dataset (visible as vertical stripes), and the model tends to predict these relatively well.

- Rare functions are harder to capture—the model often misses them entirely, leading to sparse or empty columns in the predicted heatmap.

- A few functions are over-predicted, visible as faint vertical lines across many proteins, suggesting the model is overly confident for those categories.
- Many cells in the predicted matrix show intermediate color tones, which reflect more uncertain probabilities (not a confident near-0 or near-1).

We'll now shift from this qualitative view to a quantitative one by evaluating model performance on each protein function individually:

```
metrics_by_function = {}
for function in targets:
  metrics_by_function[function] = compute_metrics(
    valid_true_df[function].values, valid_prob_df[function].values
  )

overview_valid = (
  pd.DataFrame(metrics_by_function)
  .T.merge(go_term_descriptions, left_index=True, right_on="term")
  .set_index("term")
  .sort_values("auprc", ascending=False)
)
print(overview_valid)
```

Output:

```
            accuracy    recall  precision     auprc     auroc  \
term
GO:0004930  0.958858  0.000000   0.000000  0.948591  0.982272
GO:0004888  0.945424  0.177215   1.000000  0.849885  0.968354
GO:0003824  0.848027  0.731591   0.819149  0.849362  0.909372
...              ...       ...        ...       ...       ...
GO:0003774  0.000000  0.000000   0.000000  0.000000  0.000000
GO:0051015  0.000000  0.000000   0.000000  0.000000  0.000000
GO:1902936  0.000000  0.000000   0.000000  0.000000  0.000000

                      description
term
GO:0004930  G protein-couple...
GO:0004888  transmembrane si...
GO:0003824    catalytic activity
...                          ...
GO:0003774  cytoskeletal mot...
GO:0051015  actin filament b...
GO:1902936  phosphatidylinos...

[303 rows x 6 columns]
```

This analysis reveals substantial variation in model performance across protein functions. For instance, the model performs well on functions like GO:0004930 (G protein–coupled receptor activity), but it struggles with others, such as GO:0003774 (cytoskeletal motor activity). However, interpreting these results requires caution: some metrics may be based on very few validation examples, and performance is naturally limited for functions that are underrepresented during training. A high score on a frequent function may simply reflect ample training data, while low scores on rare functions may be expected.

Thresholded and Continuous Evaluation Metrics

Our evaluation metrics fall into two categories: thresholded and continuous.

- *Precision* and *recall* are computed from binary predictions—i.e., after applying a fixed threshold (typically >0.5) to the model's output probabilities.
- *auPRC* (area under the precision–recall curve) and *auROC* (area under the receiver operating characteristic curve) are *threshold independent*. They assess how well the model ranks positive examples above negatives across all possible thresholds.

Although a bit counterintuitive, it's entirely possible for precision and recall to be 0 while auPRC and auROC remain high. This happens when the model assigns higher probabilities to the correct labels, but those probabilities never exceed the decision threshold. In such cases, thresholded metrics show failure, while ranking-based metrics still reflect meaningful signal.

If we wanted to address this issue with the current thresholded metrics, we could lower the decision threshold—for example, to 0.2 or 0.3—to encourage more positive predictions. The threshold can be tuned automatically using metrics like the F1 score (the harmonic mean of precision and recall).

Let's take a closer look at whether there's a relationship between how often a protein function appears in the training data and how well the model learns to predict it in the validation set:

```python
# Compute number of occurrences of each function in the training set.
overview_valid = overview_valid.merge(
    pd.DataFrame(train_df[targets].sum(), columns=["train_n"]),
    left_index=True,
    right_index=True,
)
print(overview_valid)
```

Output:

	accuracy	recall	precision	auprc	auroc	\
GO:0004930	0.958858	0.000000	0.000000	0.948591	0.982272	
GO:0004888	0.945424	0.177215	1.000000	0.849885	0.968354	
GO:0003824	0.848027	0.731591	0.819149	0.849362	0.909372	
...	...	...	...	...	...	
GO:0003774	0.000000	0.000000	0.000000	0.000000	0.000000	
GO:0051015	0.000000	0.000000	0.000000	0.000000	0.000000	

```
GO:1902936    0.000000    0.000000    0.000000    0.000000    0.000000

                    description    train_n
GO:0004930    G protein-couple...      138
GO:0004888    transmembrane si...      228
GO:0003824     catalytic activity     1210
...                       ...          ...
GO:0003774    cytoskeletal mot...        5
GO:0051015    actin filament b...       17
GO:1902936    phosphatidylinos...       18

[303 rows x 7 columns]
```

At a glance, it seems that functions with higher predictive performance (e.g., higher auPRC) also tend to have more training examples. In Figure 2-15, we visualize this relationship more clearly with a scatterplot:

```
fig = sns.scatterplot(
    x="train_n", y="auprc", data=overview_valid, alpha=0.5, s=50, color="grey"
)
fig.set_xlabel("# Train instances")
fig.set_ylabel("Validation auPRC");
```

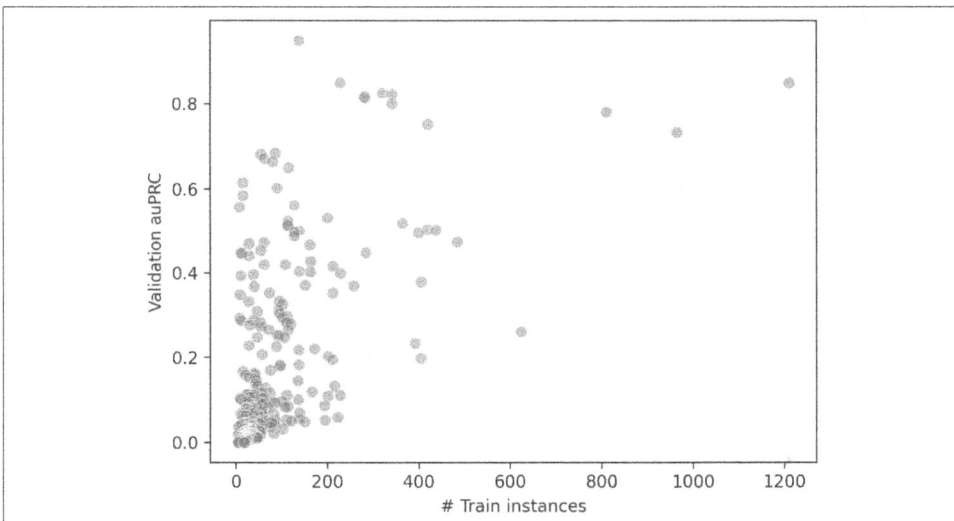

Figure 2-15. Relationship between training frequency and predictive performance (auPRC) across protein functions. Commonly observed functions in the training set tend to be predicted more accurately by the model.

This plot shows a clear trend: protein functions that occur more frequently in the training set tend to be predicted more accurately by the model on the validation set (as measured by auPRC). This aligns with expectations—machine learning models usually perform better on well-represented classes. It also highlights the challenge of class imbalance: rare functions are often poorly predicted, not necessarily due to biological complexity but because the model has limited data to learn from.

But how do we know whether a specific auPRC score is actually good? An auPRC value of, say, 0.8 for a certain protein function might sound promising—but is that better than chance? Is it meaningful? To interpret these scores, we need something to compare them against.

Evaluating Model Usefulness

To ground our evaluation, we'll compare our model against two simple baselines:

Coin flip
For each protein function, randomly predict 0 or 1 with equal probability. This gives us a baseline for total ignorance.

Proportional guessing
Predict 1 for each function with probability equal to its frequency in the training set. This reflects prior class distribution knowledge, but without any learning.

These baselines help contextualize the model's performance. If our trained model doesn't outperform these simple heuristics, it's a sign that it may not have learned meaningful structure from the data.

Here are implementations for the baselines:

```python
def make_coin_flip_predictions(
    valid_true_df: pd.DataFrame, targets: list[str]
) -> pd.DataFrame:
    """Make random coin flip predictions for each protein function."""
    predictions = np.random.choice([0.0, 1.0], size=valid_true_df.shape)
    return pd.DataFrame(predictions, columns=targets, index=valid_true_df.index)


def make_proportional_predictions(
    valid_true_df: pd.DataFrame, train_df: pd.DataFrame, targets: list[str]
) -> pd.DataFrame:
    """Make random protein function predictions proportional to frequency."""
    percent_1_train = dict(train_df[targets].mean())
    proportional_preds = []
    for target_column in targets:
        prob_1 = percent_1_train[target_column]
        prob_0 = 1 - prob_1
        proportional_preds.append(
            np.random.choice([0.0, 1.0], size=len(valid_true_df), p=[prob_0, prob_1])
        )
    return pd.DataFrame(
        np.stack(proportional_preds).T, columns=targets, index=valid_true_df.index
    )
```

These baselines should give us simple but informative reference points. Let's now apply these prediction methods, alongside our trained model:

```
prediction_methods = {
  "coin_flip_baseline": make_coin_flip_predictions(valid_true_df, targets),
  "proportional_guess_baseline": make_proportional_predictions(
    valid_true_df, train_df, targets
  ),
  "model": valid_prob_df,
}
```

Now let's evaluate the baselines in exactly the same way as our model—by computing per-protein metrics and averaging them:

```
metrics_by_method = {}
for method, preds_df in prediction_methods.items():
  metrics_by_method[method] = pd.DataFrame(
    [
      compute_metrics(valid_true_df.iloc[i], preds_df.iloc[i])
      for i in range(len(valid_true_df))
    ]
  ).mean()

print(pd.DataFrame(metrics_by_method))
```

Output:

	coin_flip_baseline	proportional_guess_baseline	model
accuracy	0.500916	0.956447	0.978569
recall	0.499229	0.093555	0.128532
precision	0.023883	0.079994	0.424301
auprc	0.025307	0.039701	0.412350
auroc	0.500027	0.535605	0.882679

Our model clearly outperforms both baselines across all metrics—especially in precision, auPRC, and auROC. This is expected, as the trained model leverages actual sequence features to make more informed predictions. As noted earlier, accuracy is not a reliable metric in this setting, and even simple proportional guessing achieves a deceptively high accuracy due to class imbalance.

Most of the model's performance gains come from a large increase in precision, while the improvement in recall is more modest. This means the model is good at correctly identifying positive cases when it makes a prediction, but it tends to miss many true positives—it's cautious and biased toward predicting "no function."

This highlights a key trade-off: the model is conservative but accurate. Depending on your application, you may want to tune this behavior—for example, by lowering the decision threshold to improve recall, as discussed earlier.

Next, we'll break down the model's strengths and weaknesses by individual protein function and compare performance against both baselines. This allows us to see which specific functions the model predicts well—and where it struggles:

```
auprc_by_function = {}

for method, preds_df in prediction_methods.items():
  metrics_by_function = {}

  for function in targets:
    metrics_by_function[function] = compute_metrics(
      valid_true_df[function], preds_df[function]
    )

  auprc_by_function[method] = (
    pd.DataFrame(metrics_by_function)
    .T.merge(go_term_descriptions, left_index=True, right_on="term")
    .set_index("term")
    .sort_values("auprc", ascending=False)
  )["auprc"].to_dict()
```

In Figure 2-16, we visualize the function-level auPRC scores as a bar plot to highlight which functional categories the model handles best:

```
best_performing = (
  pd.DataFrame(auprc_by_function)
  .merge(go_term_descriptions, left_index=True, right_on="term")
  .set_index("term")
  .sort_values("model", ascending=False)
  .head(20)
  .melt("description")
)

fig, ax = plt.subplots(figsize=(8, 5))
sns.barplot(
  x="description",
  y="value",
  hue="variable",
  data=best_performing,
)
ax.set_title("The model's 20 best performing protein functions")
ax.set_ylabel("Validation auPRC")
plt.xticks(rotation=90);
```

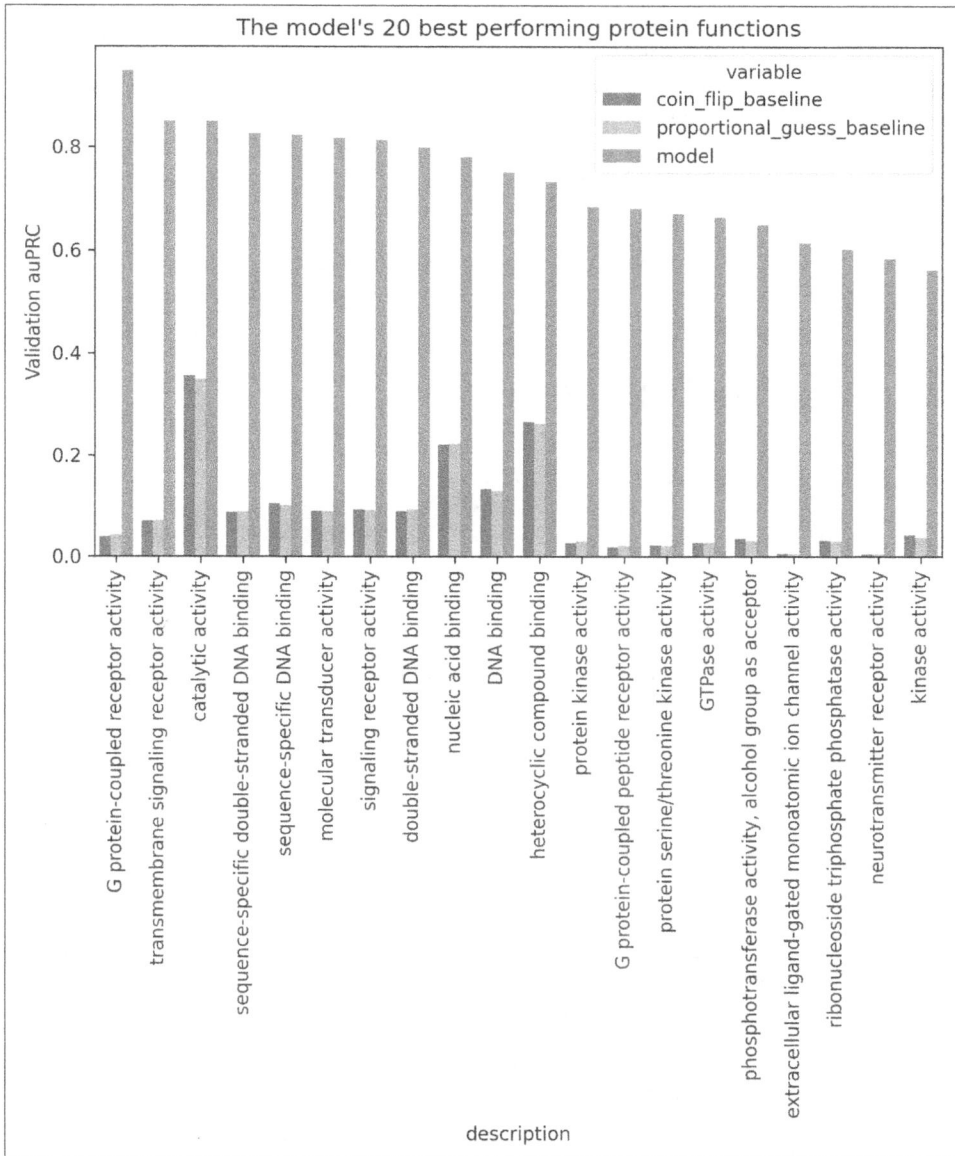

Figure 2-16. Top 20 protein functions ranked by model auPRC on the validation set. Bars show the auPRC achieved by the model, compared against two simple baselines (i.e., coin flips and proportional guessing).

Many of the top-performing protein functions in the plot are related to membrane or signaling roles (e.g., GPCR activity, kinase activity, transmembrane receptor activity). One possible reason is that these functions often involve well-conserved sequence features—such as transmembrane helices or catalytic domains—that may be easier for models to learn. While speculative, this aligns with the idea that functions tied to strong structural or biochemical motifs may produce clearer sequence-level signals than more context-dependent roles.

Together, these results suggest that the model is capable of detecting meaningful biological signal for certain classes of protein function—and that it significantly outperforms simple baselines.

Conducting a Final Check on the Test Set

Take a look at the next section for ideas on how to extend and improve this model. Once you are satisfied with your exploration, we can move on to the final step of this project: making the final predictions on the test set. Remember not to touch the test set until the last stage of your project.

> Be sure not to touch the test set until you've fully finalized your model—including all hyperparameters, architectures, and training choices. Evaluating on the test set repeatedly can lead to overly optimistic results and undermine the validity of your findings.

We'll make predictions on the test set of proteins in the same way we did for the validation set:

```
eval_metrics = []

for split in ["valid", "test"]:
  split_metrics = []

  for eval_batch in dataset_splits[split].batch(32).as_numpy_iterator():
    split_metrics.append(eval_step(state, eval_batch))

  eval_metrics.append(
    {"split": split, **pd.DataFrame(split_metrics).mean(axis=0).to_dict()}
  )
print(pd.DataFrame(eval_metrics))
```

Output:

	split	loss	accuracy	recall	precision	auprc	auroc
0	valid	0.080156	0.978457	0.126869	0.418515	0.411870	0.880883
1	test	0.080675	0.978032	0.125820	0.435193	0.410439	0.879234

The test set metrics closely mirror those observed on the validation set, which is good. In many workflows, test performance is slightly lower due to repeated use of the validation set during development—potentially leading to mild overfitting. However, in this case, we haven't done extensive tuning, so the gap is minimal. Because the test set was held out throughout, its results provide a more reliable estimate of how the model will generalize to truly unseen data. These are the metrics we would report externally.

Improvements and Extensions

The model we've built demonstrates that protein function can be predicted from sequence using pretrained embeddings and a lightweight classifier. However, many directions remain to improve, interpret, and extend this work. We split these ideas into two broad categories: analysis-driven insights and machine learning improvements.

But before diving into technical upgrades, it's worth stepping back to revisit the bigger picture:

Why are you doing this?
> Who will use the model, and what do they actually need? Can you share this prototype with users now to gather early feedback?

When are you done?
> Is the current model already good enough? What specific improvements would meaningfully increase its utility? What benchmarks exist for this or similar tasks?

What matters most?
> Is performance across all functions equally important, or do you care about a specific class (e.g., enzymes versus nonenzymes)? Focusing your optimization accordingly can save time.

Do you need interpretability?
> For some applications, understanding why a model makes a prediction may matter more than maximizing performance.

Ideally, you'll have thought about some of these questions before starting the modeling—but revisiting them now can help guide your next steps.

Biological and Analytical Exploration

Even with a fixed model, we can learn a lot more by probing its behavior and comparing it to biological expectations:

Threshold tuning

Our results showed that the model has high auPRC but low recall at a default probability threshold of 0.5. You could optimize this threshold (e.g., per protein function or globally) using a metric like F1 score to find a better trade-off between precision and recall.

Species generalization

The current dataset is human only, but this might be unnecessarily limited. Try including protein-function pairs from other species to see if performance improves.

Function-specific performance drivers

Why does the model do well on some functions (e.g., GPCR activity) but poorly on others (e.g., growth factor activity)? You could investigate whether function prevalence, sequence length, or other properties correlate with performance.

Examine protein multifunctionality

Does the model struggle more with proteins that have many functions? Group proteins by number of annotated functions and plot performance (e.g., auPRC) to see if there's a trend.

False positives that might be real

Find proteins where the model confidently predicts a function that isn't labeled. Could the model be correct and the annotation missing? How might you follow this up?

Machine Learning Improvements

From a machine learning perspective, here are a few directions you could explore:

Tune the MLP

Our model is a small MLP on top of frozen embeddings. Try adding more layers, dropout, or batch normalization to increase capacity while controlling overfitting.

Alternative input encodings

We used the mean-pooled embedding, which loses sequence order information. Try attention pooling or a small 1D CNN or transformer on top of the token-level embeddings.

Feature engineering

You could augment the input to include protein length, species (if you extend beyond human), or even simple statistics like embedding norms. These additional features might help the model distinguish protein types more effectively.

Train a per-function head:
> Instead of predicting all functions jointly, try training separate models (or heads) for each function. This can help when tasks are highly imbalanced or unrelated. Alternatively, you could cluster GO functions into a few categories and train one model per cluster.

Predict function hierarchically
> Rather than treating each function independently, you could use the GO hierarchy to add structure to predictions—for example, predicting broad function categories first and then refining to more specific ones.

Try alternative base models
> You could plug in other protein language models from Hugging Face or explore combining embeddings from multiple models by concatenating them.

Unfreeze the language model
> The ESM2 embeddings are pretrained on a generic task. Fine-tuning the language model directly for protein function classification may boost performance, though it requires more compute and a more involved training setup.

> While it's tempting to chase performance gains through increasingly complex models, always align your efforts with the actual goals of your project. Improving interpretability or expanding biological coverage may be more valuable than inching up another point on a leaderboard.

Summary

In this chapter, we took our first hands-on step into the world of deep learning for biology. Starting with a dataset of human proteins, we explored how to extract meaningful representations using a pretrained protein language model, trained a simple classifier to predict protein function, and evaluated its performance using quantitative metrics.

Along the way, we encountered practical challenges typical of biological modeling: getting comfortable with a new modeling setup, dealing with imbalanced label distributions, and carefully interpreting evaluation metrics.

In the next chapter, we'll build on these foundations by shifting our focus from proteins to DNA. You'll define convolutional neural networks from scratch in Flax and train them end to end to model regulatory sequences, predict functional elements, and discover motif patterns directly from genomic data.

## CHAPTER 3

Learning the Logic of DNA

In this chapter, we'll build a deep learning model to predict whether a DNA sequence is bound by a class of proteins called *transcription factors* (TFs). Transcription factors play a central role in gene regulation: they bind to specific DNA sequences and influence whether nearby genes are turned on or off. By recognizing these sequence patterns, we can begin to decode the regulatory logic embedded in the genome.

Unlike the previous chapter—where we used an off-the-shelf protein model from Hugging Face—here we'll start defining and training our own models from scratch. This gives us more control and helps us better understand how deep learning works on biological data. We'll explore both convolutional and transformer-based architectures and introduce interpretation techniques to help us understand how our models make predictions.

We will tackle this problem in stages, gradually increasing the complexity:

1. Start simple

First, we'll train a basic convolutional network to predict whether a DNA sequence binds a single transcription factor called CTCF. Its binding behavior is relatively easy to predict, making it a great first target. We'll build the full pipeline: loading data, training the model, and checking whether it captures meaningful biological signals.

2. Increase complexity

Next, we'll scale up to predicting whether a sequence binds any of 10 different TFs. We'll introduce regularization and normalization, improve our evaluation metrics, and begin inspecting individual predictions. We'll also use mutation experiments and input gradients to highlight which parts of the sequence the model relies on—offering a first step toward interpretability.

3. Incorporate advanced techniques

Finally, we'll try adding transformer layers to explore whether they improve performance and continue dissecting model behavior to understand how different architectural choices influence learning.

This staged approach—building up from simple to more complex—is one we recommend in general. It helps keep models interpretable, makes debugging easier, and builds confidence along the way.

Before diving in, we'll do a quick refresher on the biological and machine learning concepts that underpin this chapter.

> To get the most out of this chapter, open the companion Colab notebook and run the code cells as you follow along. Executing the code interactively will deepen your understanding and give you space to experiment with the concepts in real time.

Biology Primer

It's astonishing that all the instructions for building an entire human body are encoded in the DNA of a single cell. Every human starts as one tiny cell—about 100 micrometers wide—with its DNA tightly packed into a nucleus just 6 micrometers across. This DNA acts as the blueprint for processes like cell division and differentiation, eventually giving rise to the diverse tissues and cell types that make up an entire human body.

> A few terminology clarifications: The *genome* refers to the complete set of DNA in an organism, including all of its genes and other genetic material. While *genetics* typically studies individual genes or small gene sets, *genomics* takes a broader view, often analyzing entire genomes across individuals or even species.

The human genome is vast—more than 3 billion base pairs long—and carries with it billions of years of evolutionary history. But what is this molecule, really, and how does it encode biological function?

What Exactly Is DNA?

DNA is the molecule of inheritance—the fundamental code of life. Its double-helix structure was first revealed in 1953 by Watson, Crick, and Franklin, marking a pivotal moment in the biological sciences. Nearly half a century later, the first complete draft of the human genome was published in 2001,[1] laying the foundation for modern genetics and genomics. But these milestones are relatively recent, and while we now know a great deal about *what* is in the genome, we still understand surprisingly little about *how* it actually works.

We do know that DNA is built from four chemical letters, or nucleotide bases: A (adenine), C (cytosine), G (guanine), and T (thymine). These bases form long sequences that carry genetic instructions. The full human genome contains around 3.2 billion of these letters, packed into 23 pairs of chromosomes. To fit inside the tiny nucleus of a cell, this DNA wraps around proteins and coils into compact, highly organized structures known as *chromatin*, as shown in Figure 3-1.

Despite decades of research, the genome remains full of unanswered questions. Only about 2% of it directly codes for proteins—what is the rest doing? How can all the cells in your body share the same DNA, yet behave so differently? What controls when a gene is used, and how do changes in the environment or during development affect this process?

These mysteries lie at the heart of gene regulation—and increasingly, deep learning is helping us explore them.

---

1 International Human Genome Sequencing Consortium, "Initial Sequencing and Analysis of the Human Genome." Nature 409 (2001): 860–921.

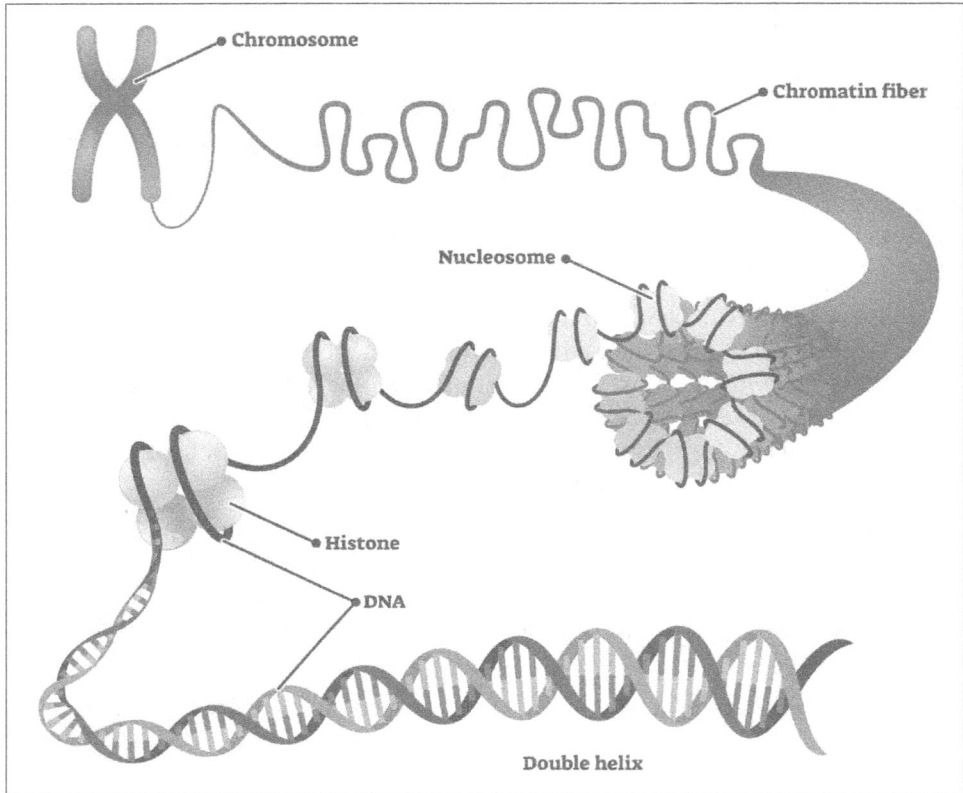

Figure 3-1. DNA is packed into the nucleus in multiple layers of structure. Starting with the double helix, it first wraps around histone proteins to form nucleosomes (beads on a string), which fold into chromatin fibers, and ultimately into chromosomes. Source: National Institute of Environmental Health Sciences (https://oreil.ly/h8M44).

Coding and Noncoding Regions

The human genome contains around 20,000 protein-coding genes. These make up the *coding* regions—stretches of DNA that are *transcribed* into RNA and then *translated* into proteins. Each protein carries out specific tasks, from building cellular structures to catalyzing chemical reactions. Together, they perform most of the cell's essential functions.

Yet, protein-coding genes account for only about 2% of the genome. The remaining 98% is noncoding DNA. While it doesn't produce proteins, noncoding DNA plays critical regulatory roles, helping to control when and where genes are used. In fact, most genetic variants associated with human disease fall in noncoding regions—though we still have limited understanding of how most of them exert their effects.

Some noncoding DNA produces RNAs that regulate gene expression, while other regions help organize the 3D structure of the genome or serve as docking sites for regulatory proteins. One especially important category of noncoding region is transcription factor binding sites. These are short DNA sequences where *transcription factors* (TFs) attach to help regulate gene activity—and they're the central focus of this chapter.

How Transcription Factors Orchestrate Gene Activity

TFs are proteins that control which genes are used, when, and in what context. They do this by binding to short, specific DNA sequences called *motifs*, often located near genes. By binding these motifs, TFs can activate or repress transcription. You can think of them as conductors of a genomic orchestra—directing the performance by determining which regions get "played" and when.

TFs are involved in nearly every biological process, from guiding development to coordinating how cells respond to stress or infection. Humans have around 1,600 different TFs, each of which has evolved to recognize specific DNA motifs. These motifs are short sequence patterns—typically 6 to 15 base pairs long—that TFs preferentially bind to. For example, the well-studied transcription factor CTCF binds a core motif with the central pattern CCCTC. TFs don't bind randomly across the genome—they search for these preferred sequences.

These motifs form specific 3D shapes in the DNA helix, and the protein-binding domains of TFs are shaped to complement them—much like a key fitting into a lock. In reality, the interaction is often more flexible and dynamic than the analogy suggests, but the idea of a physical match still holds.

To make this interaction more concrete, Figure 3-2 shows crystallographic structures of different TFs bound to DNA.

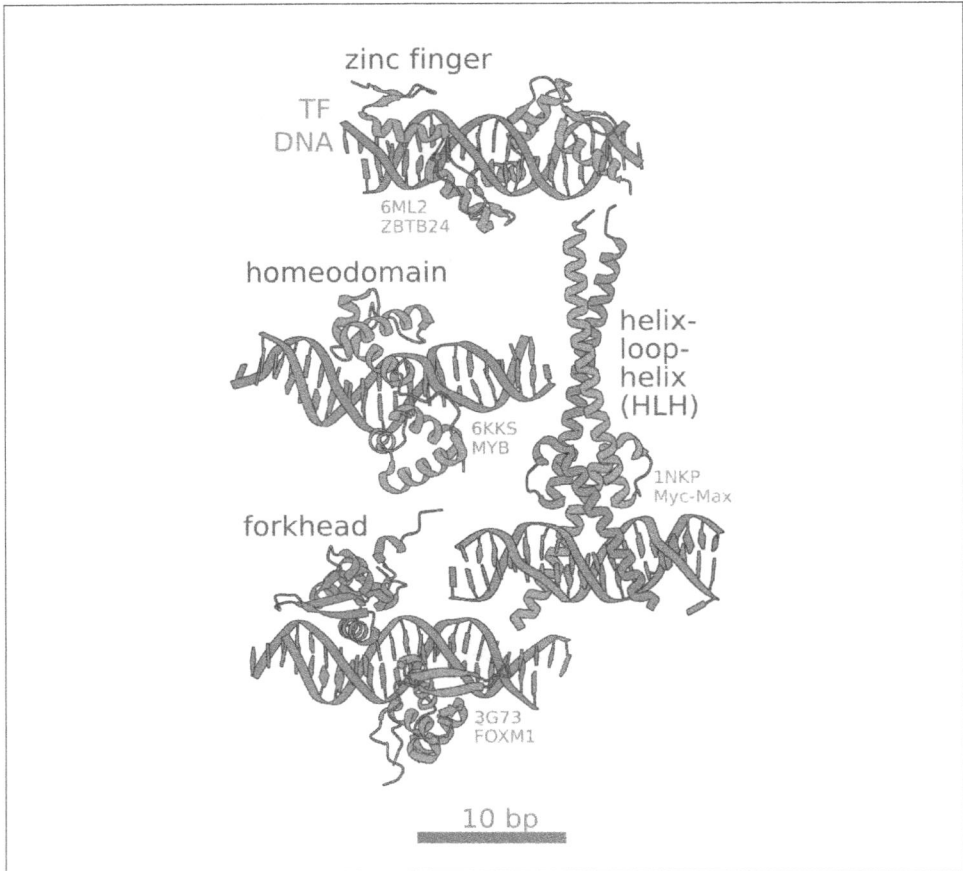

Figure 3-2. Crystallographic structures showing how different types of TF binding domains interact with DNA. Each gray structure represents a different TF binding domain—zinc finger, homeodomain, helix-loop-helix, and forkhead—bound to a double-stranded DNA molecule. These protein segments recognize specific DNA motifs by the physical shape those sequences form. Like a key fitting into a lock, the structure of the protein complements the shape of the DNA at its binding site. Shown here are actual resolved structures from the Protein Data Bank (PDB: 6ML2, 6KKS, 1NKP, 3G73). Source: Wikipedia (https://oreil.ly/Nd4Tv).

However, not every matching motif is actually bound in a living cell. In fact, the genome contains far more motif matches than actual binding events. That's because binding depends on many additional factors:

Chromatin accessibility
> DNA that's tightly packed into chromatin is harder for proteins to access. TFs are more likely to bind in regions of open chromatin.

DNA methylation
> Certain TFs, like CTCF, are sensitive to methylation (a certain chemical modification of DNA bases) at their binding sites, which can block binding even if the motif is present.

Cellular signals
> Signals inside the cell—such as hormones or stress responses—can activate or deactivate a TF's ability to bind.

Other proteins
> Helper or blocking proteins in the local environment can facilitate or inhibit binding.

Cooperative binding
> Many TFs work in complexes or recruit others to stabilize binding and control gene activity.

And perhaps most importantly, real cells are dynamic. Molecules move, concentrations change, and TF binding events happen on short timescales. Most genomic datasets, by contrast, are static snapshots—freeze-frames of a constantly changing scene. That's worth keeping in mind when interpreting binding data in this chapter.

Measuring Where Transcription Factors Bind

In deep learning, we're incredibly reliant on experimental data—we need something to learn from. When it comes to studying TFs, that data typically comes from laboratory experiments that measure where in the genome a particular TF binds.

The cornerstone wet lab method is ChIP-seq (*https://oreil.ly/IovSO*), or chromatin immunoprecipitation followed by sequencing. TFs don't permanently stick to DNA—they bind and unbind constantly. ChIP-seq captures a snapshot of this dynamic process by chemically cross-linking (gluing) proteins to DNA, essentially freezing them in place. The DNA fragments bound by a specific TF can then be isolated, sequenced, and mapped back to the genome to determine where the protein was bound.

ChIP-seq data is typically visualized as peaks over the DNA sequence—regions of the genome where TF binding was enriched. The height of a peak reflects how strong or frequent the binding was at that site, while a flat zero signal means no detectable binding occurred there.

To simplify the data for modeling, ChIP-seq peaks can be binarized: instead of retaining the full quantitative signal, we apply a threshold and record only whether the TF was bound in a given region. This reduces the task to a binary classification problem—does the TF bind to this DNA sequence or not—which is the setup we'll use throughout this chapter.

Machine Learning Primer

With the biology background knowledge in place, we now review some foundational machine learning concepts that we'll use in this chapter. If you're already familiar with deep learning, feel free to skim this section as a refresher. We'll briefly cover *convolutional* and *transformer*-based architectures, how they can be applied to biological sequence data like DNA, and what kinds of insights they can provide.

Convolutional Neural Networks

Convolutional neural networks (CNNs) are one of the most widely used deep learning architectures. Their core strength lies in their ability to automatically learn useful patterns from raw, grid-like input data—whether that's pixels in an image or bases in a DNA sequence—without the need for hand-engineered features.

CNNs were originally developed for image recognition tasks. In images, low-level features like edges and textures appear in small local patches, while higher-level concepts like shapes or objects are formed by combining these local features. CNNs mirror this structure by using small, learnable *filters* that slide across the input, extracting local patterns at each position. As we stack more layers, the model combines local features into more abstract and global representations.

This ability to model hierarchical structure turns out to be useful far beyond images. Whenever there are meaningful local patterns in data—like substructures in molecules, motifs in DNA, or phonemes in speech—CNNs often perform well.

We cover CNNs in more detail in Chapter 5, where we build a skin cancer classifier and explore common model design patterns. For now, we'll provide a short overview of the key components you'll need for this chapter.

Let's briefly walk through the key components of a typical CNN:

Convolutional layers
> These are the heart of the CNN. A convolutional layer contains multiple filters (also called *kernels*)—small weight matrices that slide across the input and compute dot products. Each filter acts like a pattern detector, lighting up when it

finds a good match in the input. The result of applying a convolution is a *feature map* showing where the pattern occurred.

Pooling layers

These downsample the feature maps to reduce dimensionality and computation. *Max pooling*, for example, keeps only the strongest signal in each region, helping the model focus on the most salient features.

Normalization

Layers like *batch normalization* rescale activations to make training more stable and efficient. They reduce internal *covariate shift*—the tendency for activations to drift during training—and often speed up convergence.

Fully connected layers

These sit at the end of the network and use the features extracted by earlier layers to make final predictions—such as whether a DNA sequence is bound by a transcription factor.

One key property of CNNs is that they are *translation equivariant*, meaning that a pattern can be recognized regardless of where it appears in the input. This is especially useful for DNA: a binding motif is still a binding motif whether it's at position 10 or position 90 of the sequence.

Convolutions for DNA Sequences

Although CNNs were originally developed for images (which are 2D grids of pixels), the architecture can easily be adapted to 1D data like DNA sequences. In genomics, DNA is commonly represented as a one-hot encoded matrix, where each base (A, C, G, T) is turned into a binary vector. A sequence of 100 bases would thus become a 100×4 matrix.

We then apply *1D convolutions*—filters that slide across the sequence in one dimension, looking for patterns in short windows of DNA bases. These filters often end up learning to identify the presence of DNA motifs: the short sequence patterns that have biological meaning we mentioned before, such as transcription factor binding sites. For example:

- Shallow layers might learn to detect low-level DNA features such as simple GC-rich or AT-rich regions in DNA.
- Mid-level filters may identify known TF motifs.
- Deeper layers might learn higher-order combinations—such as co-occurring motifs or long-range dependencies.

Importantly, the model learns all of this automatically from labeled data. It doesn't need to be told what motif to look for—it discovers useful patterns by optimizing for the task at hand, such as predicting TF binding.

> CNNs have become a standard architecture in sequence-based biology tasks because they offer a good balance between power, speed, and interpretability. Compared to other deep learning architectures, CNNs are relatively lightweight, easy to train, and often easier to interpret: you can visualize which motifs a filter has learned and where they appear in a sequence.

Their main limitation, however, is that they operate on fixed windows of sequence and struggle to model interactions between distant bases. For problems involving long-range dependencies—relationships between elements far apart in a sequence—we often turn to a different class of models.

Transformers

While CNNs are excellent at detecting local patterns, transformers are particularly powerful for modeling relationships across long distances in a sequence. Their core mechanism—*self-attention*—allows the model to dynamically determine which parts of the input are relevant for predicting any given output, regardless of how far apart those parts are.

Transformer-based models have revolutionized deep learning since their debut in 2017.[2] Originally designed for natural language processing tasks like translation and summarization, transformers have since become the state of the art in a wide range of domains, including genomics and protein modeling. Earlier architectures like RNNs and CNNs can also handle sequences, but transformers have largely surpassed them for tasks that require understanding global sequence context.

The key innovation is *self-attention*: a mechanism that lets each token in the sequence "attend to" every other token, computing how much influence they should have. This is useful for:

- Language: Where a word's meaning depends on faraway context
- Genomics: Where a regulatory DNA motif located thousands of bases away might affect gene activity

---

2 Vaswani, Ashish, Noam Shazeer, Niki Parmar, Jakob Uszkoreit, Llion Jones, Aidan N. Gomez, Lukasz Kaiser, and Illia Polosukhin. 2017. "Attention Is All You Need" (*https://oreil.ly/-Qa6G*). *arXiv (Cornell University)* 30 (June): 5998–6008.

This flexibility allows transformers to learn arbitrary and complex dependencies—something CNNs struggle with. The main drawback of transformers is scalability: self-attention requires computations that grow quadratically with sequence length. For very long inputs like whole-genome sequences, this can become a bottleneck. However, many efficient transformer variants (e.g., Linformer, Longformer, Performer) have been developed to partially address this weakness.

> If you'd like to dive deeper into transformers, we recommend the excellent blog post "The Illustrated Transformer" (*https://oreil.ly/WxzGl*) by Jay Alammar or the 3Blue1Brown video (*https://oreil.ly/dLW40*) on attention. We'll only touch on the basics here.

The foundation of transformers lies in their ability to assign *attention*—a mechanism that lets each position in a sequence dynamically focus on other positions. This is what enables them to model long-range dependencies.

Attention

At a high level, attention is a process that enriches the embedding of each token by incorporating information from every other token in the sequence. This makes each token more context aware, as illustrated in Figure 3-3.

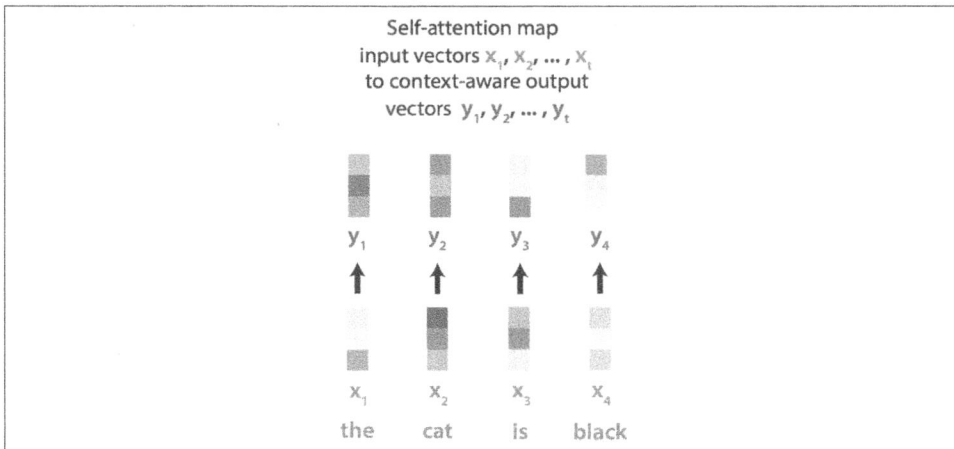

Figure 3-3. High-level visualization of the attention mechanism transforming input token embeddings into context-aware output embeddings. Each of the four tokens in the sequence ("the cat is black") attends to every other token, allowing the model to capture relationships and dependencies across the entire sequence. The output is also four tokens, each enriched with contextual information from the others.

Briefly, here's how it works. The model first creates three versions of the input embeddings—queries (Q), keys (K), and values (V)—via learned linear transformations. Each query is compared to all keys using a dot product, producing a score that reflects how relevant one token is to another. These scores are then normalized via a softmax function, producing the attention weights. Finally, each token's new embedding is computed as a weighted sum of the value vectors, with the attention weights determining how much each value contributes.

Let's walk through a simple example. Suppose we have the input sequence the, cat, is, black. In a transformer model, each token doesn't just pass through the network on its own; it decides how much attention to pay to every other token. For example, when processing the word cat, the model might assign a high attention weight to the, recognizing that articles and nouns are often linked. This helps the model understand grammatical relationships and contextual meaning.

In genomics, attention can serve a similar purpose. Imagine a model processing a DNA sequence to predict TF binding. An attention mechanism allows the model to ask: how relevant is this motif to another element upstream or downstream—perhaps thousands of bases away? Just as a word's meaning is shaped by the words around it, the function of a sequence motif may depend on other elements scattered across the genome.

Once attention has enriched the token embeddings with contextual information, the result is passed through a *feedforward network*—typically a small, multilayer perceptron applied independently to each token. This network introduces nonlinearity and helps the model capture more complex patterns. The output is then passed through *residual connections* (which help with gradient flow) and *layer normalization* (which stabilizes training).

All together, this sequence—attention, feedforward layers, residual connections, and normalization—forms one transformer block. A full transformer model is typically built by stacking many of these blocks, allowing information to flow and be refined across layers. As tokens pass through successive layers, their representations become increasingly rich, capturing everything from local patterns to global structure.

Query, Key, and Value Intuition

You might wonder: why bother transforming the input into queries, keys, and values at all? The key idea is that Q, K, and V aren't just redundant copies of the original token embeddings. They're learned projections that allow the model to look at the same sequence from different perspectives:

- The *query* is like a lens that each token uses to express what it wants to pay attention to. In our sentence example, the word black might use its query to find the noun it modifies—cat. In DNA, a regulatory region might "look for" compatible regulatory motifs elsewhere in the sequence. The query says, "I'm looking for X."

- The *key* is how each token presents itself to others: it encodes what kind of information it offers. Continuing the analogy, each word or DNA element "advertises" what kind of content it has, saying: "I contain Y."

- The *value* is the actual content that gets passed along if the query decides the key is relevant. In other words, the query compares itself to all keys to compute attention weights and then uses those weights to pull a mixture of values from across the sequence.

This separation of roles allows the model to reason more flexibly. Instead of treating all parts of the sequence as equally relevant, each token decides what matters to it right now, based on its query and the other tokens' keys. The values then supply the useful content.

> This design makes attention work like a smart lookup system: tokens advertise what they contain (keys), queries look for matches, and then the actual content (values) is pulled in weighted proportion to how well the key matched the query.

Importantly, the Q, K, and V projections are all learned from data. They start out random, and the model figures out—through training—how best to shape these representations for the task at hand, whether that's learning grammar, predicting regulatory activity, or modeling protein interactions.

Multiheaded Attention

Multiheaded attention (MHA) enhances the attention mechanism by running several attention operations—called *heads*—in parallel. Each head learns to focus on different parts or patterns in the input sequence. For instance, one head might focus on local motifs, while another might detect longer-range interactions or subtle contextual cues.

By combining these multiple perspectives, MHA allows the model to capture a richer and more diverse set of relationships within the data, beyond what a single attention operation could learn.

> While often the different heads learn somewhat redundant patterns, this parallel structure increases the model's expressive power and flexibility. The outputs from all heads are concatenated and linearly transformed to produce the final representation, which then feeds into subsequent model layers.

In essence, MHA lets transformers attend to multiple types of interactions simultaneously, which is particularly useful for complex biological sequences where various signals and dependencies exist at different scales.

Representing Positional Information

One final point: basic self-attention is *position invariant*, meaning it does not inherently capture the order or position of tokens in a sequence. To address this, transformer models include positional encodings or other mechanisms that inject information about the relative or absolute positions of tokens, enabling the model to understand sequence order. In the original transformer paper, sinusoidal functions of the token index were added to the token embeddings to provide this positional information.

With this brief overview of transformers complete, let's move on to some model interpretation techniques we'll apply in this chapter.

Model Interpretation

A common criticism of deep learning models is that they act as a *black box*—they may produce accurate predictions, but it's often unclear how exactly those predictions are made or what internal reasoning the model uses. While deep learning models are generally less interpretable than simpler methods like linear models or decision trees, there are several techniques to probe their inner workings. These techniques fall under the umbrella of *model interpretation*.

Model interpretation for deep learning is a vast and active research area, so here we provide a brief overview of the most commonly used techniques in the DNA modeling space:

Mutagenesis
To understand which input features the model relies on, we systematically alter (or *mutate*) parts of the input and observe how the model's predictions change. For example, when predicting gene expression from DNA sequence, we can shuffle, replace, or delete certain bases and see if the prediction shifts significantly. Large changes indicate that the mutated region is important for the model.

- Pro: This is direct and intuitive. It provides rich, localized insights.
- Con: It is computationally expensive, since each mutation requires a separate forward pass through the model.

Input gradients

A faster but approximate method involves computing the gradient of the model's output with respect to each input feature. This gradient shows how sensitive the prediction is to small changes in each input element.

- Pro: It is efficient, as it requires only one backward pass to generate an importance map.
- Con: It can be noisy and less precise, making it harder to distinguish signal from noise.

Attention mechanisms

For models that include attention, we can inspect the attention weights to see where the model is focusing when making predictions.

- Pro: This provides a naturally interpretable visualization of model focus and interactions.
- Con: Attention scales quadratically with input sequence length, meaning that in practice, we can't use attention to model very long DNA strings without first condensing them down in some way.

There are many extensions and refinements of these methods, and model interpretation is increasingly standard in deep learning biology research papers.

Next, we'll dive deeper into the two interpretation approaches that we will implement in this chapter: *in silico* mutagenesis and input gradients.

In Silico Saturation Mutagenesis

In silico saturation mutagenesis (ISM) may sound complex, but it essentially involves systematically making every possible alteration (or mutation) to a biological sequence —such as DNA or protein—and generating a separate model prediction for each mutated sequence. Because this requires many forward passes through the model, it is computationally expensive. However, the detailed insights it provides into how each possible variation affects the output often justify the cost.

> Terminology breakdown: it's called *mutagenesis* because we induce mutations, *saturation* because every possible mutation is tested, and *in silico* since these predictions are done computationally rather than experimentally in the lab.

Figure 3-4 shows an example plot we will generate later in this chapter.

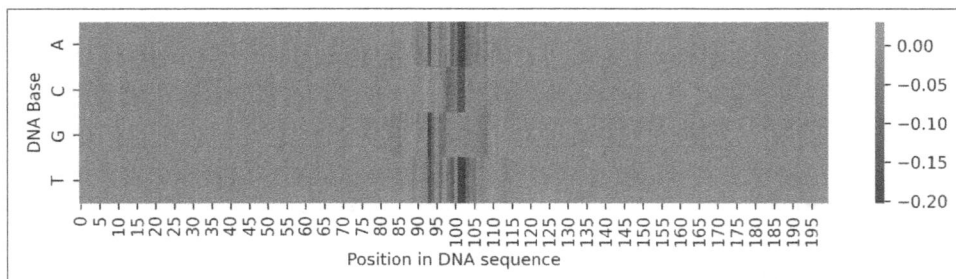

Figure 3-4. Example of an in silico saturation mutagenesis plot showing predicted probabilities of transcription factor binding across a 200-base DNA sequence. The x-axis represents the DNA sequence positions, while the y-axis shows the three possible mutations at each position (including the mutation to the original base, which has no effect and is set to zero). In this example, most mutation effects are negative (darker color), indicating that changing bases generally reduces the predicted binding probability. This approach is computationally expensive due to the large number of forward passes required to generate this output matrix (here, 3 mutations × 200 positions = 600 model predictions).

ISM plots help to quickly highlight which parts of the sequence are most important for the model's predictions. Because they visualize the most salient or impactful input regions, they are often referred to as *saliency maps*. In this example, the plot suggests that mutating any of the central bases likely disrupts a motif that the protein binds to, as these mutations generally lead to lower predicted binding probabilities (indicated by the negative values).

Input Gradients

Input gradients provide a faster way to generate a saliency map, summarizing which parts of the sequence most influence the model's predictions. Conceptually, they are the derivatives of the model's output with respect to its input features. If you've trained neural networks before, you've already encountered gradients—typically computed with respect to model parameters to guide weight updates.

Input gradients follow the same principle, but they shift the focus from parameters to the input itself. By calculating how the output changes in response to tiny perturbations at each input position, we can assess the model's sensitivity. For DNA sequences, this means identifying which bases have the greatest influence on predictions like TF binding. A large gradient at a given base suggests that altering it would significantly impact the model's output—signaling that the base is important.

Concretely, for a TF binding prediction:

- A *large negative gradient* at a base suggests that changing it would significantly lower the binding probability, perhaps disrupting the motif the TF needs to fit properly.

- A *large positive gradient* suggests that changing it would significantly increase the binding probability, maybe by strengthening an existing motif or creating a new one, thus improving the physical binding affinity between the DNA and the TF.

> What does "making a small change to the input" mean when the input is a one-hot encoded DNA sequence, rather than a continuous scalar value? Each DNA sequence, each base (A, T, C, G) is generally represented as a binary vector (e.g., [1, 0, 0, 0] for A). During gradient calculation, however, we treat these vectors as if they could vary continuously—allowing fractional values like [0.9, 0.1, 0, 0]. While such fractional bases aren't biologically meaningful, this mathematical abstraction lets us compute gradients and gain insights into which positions are influential for predictions.

You can think of input gradients as a faster but more approximate alternative to in silico saturation mutagenesis. Input gradients provide a general idea of important regions, whereas saturation mutagenesis directly tests every possible mutation's effect but is computationally expensive.

This concludes the biology and machine learning primers. Now let's dive into exploring and modeling the data to predict transcription factor binding in DNA sequences.

Building a Simple Prototype

The modeling task we'll tackle in this chapter is a *binary classification problem*: given a 200-base DNA sequence, can we predict whether it binds to a specific TF called CTCF? Among the 1,500+ TFs in humans, CTCF stands out for its role in organizing the genome's 3D architecture—folding DNA into loops and domains that regulate gene activity.

CTCF is also a great first target because its binding behavior is relatively easy to model. It recognizes a well-characterized, highly conserved motif with strong sequence specificity, meaning its binding sites are more predictable than those of many other TFs. That makes CTCF an ideal entry point for building and interpreting sequence-based models of TF binding. Later, we will expand our scope to predict the binding of 10 different TFs.

As with other chapters in this book, we'll begin by exploring the dataset, building a simple prototype model, and then iteratively extending and improving it.

Building a Dataset

The dataset we will use looks like Figure 3-5.

Input	Output
200 base long DNA sequence	Binary label of whether the DNA binds the transcription factor
...ACGTACACAGTACCGGATTCA...	1
...GTCGACGTATATGCGTAATCG...	0
...AAACTCGCGGGCATACGCGAT...	0
...GGTCTATATTTATTTCGATCA...	1
...TTCGAGATTCCGATATGATCG...	0
...	...

Figure 3-5. The input dataset consists of DNA sequences, each 200 bases long, with an associated binary label indicating whether the protein CTCF binds to it.

This task is inspired by one of the evaluation challenges presented in a recent 2024 paper preprint,[3] which sourced the dataset from a 2023 genomics interpretation study.[4]

Loading the Labeled Sequences

We start by examining the training dataset:

```
import pandas as pd

from dlfb.utils.context import assets

train_df = pd.read_csv(assets("dna/datasets/CTCF_train_sequences.csv"))
train_df
```

Output:

```
               sequence  label  transcription_factor  subset
0       TACCACATGAGTTCTC...      1                  CTCF   train
1       CATCAACACTCGTGCG...      0                  CTCF   train
2       GCACACAGCGCAGGAA...      1                  CTCF   train
...                   ...    ...                   ...     ...
61080   CCTCCCTCCCATCCCC...      1                  CTCF   train
```

---

3 Tang, Z., Somia, N., Yu, Y., & Koo, P. K. (2024). Evaluating the representational power of pre-trained DNA language models for regulatory genomics (*https://doi.org/10.1101/2024.02.29.582810*). bioRxiv (Cold Spring Harbor Laboratory).

4 Majdandzic, A., Rajesh, C., & Koo, P. K. (2023). Correcting gradient-based interpretations of deep neural networks for genomics (*https://doi.org/10.1186/s13059-023-02956-3*). Genome Biology, 24(1).

```
61081   CAGGAATGCACCGGAA...        0                    CTCF   train
61082   AAAACAGAAACTGAAA...        0                    CTCF   train

[61083 rows x 4 columns]
```

The two classes appear fairly balanced (equally represented) in the training dataset, so we won't have to do any rebalancing by downsampling the majority class here:

```
train_df["label"].value_counts()
```

Output:

```
label
1    30545
0    30538
Name: count, dtype: int64
```

To use the DNA sequences numerically, we need to convert them into one-hot format. The function dna_to_one_hot performs this mapping:

```python
import numpy as np

def dna_to_one_hot(dna_sequence: str) -> np.ndarray:
    """Convert DNA into a one-hot encoded format with channel ordering ACGT."""
    base_to_one_hot = {
      "A": (1, 0, 0, 0),
      "C": (0, 1, 0, 0),
      "G": (0, 0, 1, 0),
      "T": (0, 0, 0, 1),
      "N": (1, 1, 1, 1),  # N represents any unknown or ambiguous base.
    }
    one_hot_encoded = np.array([base_to_one_hot[base] for base in dna_sequence])
    return one_hot_encoded
```

Let's see what the one-hot encoding looks like on a sample DNA sequence, "AAACGT":

```python
dna_to_one_hot("AAACGT")
```

Output:

```
array([[1, 0, 0, 0],
       [1, 0, 0, 0],
       [1, 0, 0, 0],
       [0, 1, 0, 0],
       [0, 0, 1, 0],
       [0, 0, 0, 1]])
```

We can apply this converter to the entire training dataset to generate the numerical training data x_train as follows:

```python
x_train = np.array([dna_to_one_hot(seq) for seq in train_df["sequence"]])
y_train = train_df["label"].values[:, None]
```

Here, y_train contains the binary target labels: 0 means the sequence does not bind CTCF, and 1 means the sequence does bind CTCF.

The dataset loading code for this problem is fairly straightforward, but we can wrap it into a convenient function called load_dataset for cleaner use:

```python
def load_dataset(sequence_db) -> dict[str, np.ndarray]:
    """Load sequences and labels from a CSV into numpy arrays."""
    df = pd.read_csv(sequence_db)
    return {
        "labels": df["label"].to_numpy()[:, None],
        "sequences": np.array([dna_to_one_hot(seq) for seq in df["sequence"]]),
    }
```

Converting the Data to a TensorFlow Dataset

Next, we convert the training data into a TensorFlow dataset. This format makes it easy to efficiently iterate over batches during model training, especially when shuffling and repeating the data for multiple epochs:

```python
def convert_to_tfds(
    dataset, batch_size: int | None = None, is_training: bool = False
):
    """Convert DNA sequences and labels to a TensorFlow dataset."""
    ds = tf.data.Dataset.from_tensor_slices(dataset)
    if is_training:
        ds = ds.shuffle(buffer_size=len(dataset["sequences"]))
        ds = ds.repeat()
    batch_size = batch_size or len(dataset["labels"])
    ds = ds.batch(batch_size)
    ds = ds.prefetch(tf.data.experimental.AUTOTUNE)
    return ds

batch_size = 32

train_ds = convert_to_tfds(
    load_dataset(assets("dna/datasets/CTCF_train_sequences.csv")),
    batch_size=batch_size,
    is_training=True,
)
```

In the preceding code, we created the training dataset train_ds with batching, shuffling, and repetition enabled. Let's sanity-check by pulling one batch from the dataset and inspecting its shape and contents:

```python
batch = next(train_ds.as_numpy_iterator())
print(f'Batch sequence shape: {batch["sequences"].shape}')
print(f'Batch sequence instances: {batch["sequences"][:3,:3,]}...')
print(f'Batch labels shape: {batch["labels"].shape}')
print(f'Batch labels instances: {batch["labels"][:3,]}...')
```

Output:

```
Batch sequence shape: (32, 200, 4)
Batch sequence instances: [[[0 1 0 0]
  [0 0 1 0]
  [0 0 1 0]]

 [[1 0 0 0]
  [1 0 0 0]
  [0 1 0 0]]

 [[1 0 0 0]
  [0 1 0 0]
  [0 1 0 0]]]...
Batch labels shape: (32, 1)
Batch labels instances: [[0]
 [1]
 [0]]...
```

This looks sensible. We see the data has shape (32, 200, 4), indicating a batch size of 32, sequence length of 200, and 4 channels per DNA base as expected. The labels have shape (32, 1) since each label is a simple binary 0 or 1.

For validation, since the dataset is smaller, we can structure it as one single big batch:

```
valid_ds = load_dataset(assets("dna/datasets/CTCF_valid_sequences.csv"))
```

With our data now in the correct format, we are ready to build and train our first simple model.

Defining a Simple Convolutional Model

Next, we define a simple CNN composed of two 1D convolutional layers, followed by flattening and fully connected (dense) layers:

```python
class ConvModel(nn.Module):
  """Basic CNN model for binary sequence classification."""

  conv_filters: int = 64  # Number of filters for conv layers.
  kernel_size: tuple[int] = (10,)  # Kernel size for 1D conv layers.
  dense_units: int = 128  # Units in first dense fully-connected layer.

  @nn.compact
  def __call__(self, x):
    # First convolutional layer.
    x = nn.Conv(
      features=self.conv_filters, kernel_size=self.kernel_size, padding="SAME"
    )(x)
    x = nn.gelu(x)
    x = nn.max_pool(x, window_shape=(2,), strides=(2,))

    # Second convolutional layer.
```

```
x = nn.Conv(
    features=self.conv_filters, kernel_size=self.kernel_size, padding="SAME"
)(x)
x = nn.gelu(x)
x = nn.max_pool(x, window_shape=(2,), strides=(2,))

# Flatten the values before passing them to the dense layers.
x = x.reshape((x.shape[0], -1))

# First dense layer.
x = nn.Dense(self.dense_units)(x)
x = nn.gelu(x)

# Second dense layer.
x = nn.Dense(self.dense_units // 2)(x)
x = nn.gelu(x)

# Output layer (single unit for binary classification).
return nn.Dense(1)(x)
```

This model architecture is fairly "no frills" but should already be able to capture local patterns in DNA sequences. We can instantiate our model like this:

```
model = ConvModel()
```

To initialize model parameters in JAX, we need to provide a dummy input tensor that matches the expected shape of the model's input. Although we could use an actual batch from our dataset, it is common and simpler to use a tensor of ones with batch size 1 and the same shape as a single-encoded DNA sequence. Importantly, the batch size used for this dummy input does not affect the model initialization—JAX initializes parameters based on the shape of an individual input sample (excluding the batch dimension). This means the model can later be trained or used for inference with any batch size.

```
import jax
import jax.numpy as jnp

dummy_input = jnp.ones((1, *batch["sequences"][1,].shape))
print(dummy_input.shape)

rng_init = jax.random.PRNGKey(42)
variables = model.init(rng_init, dummy_input)
params = variables["params"]
```

Output:

```
(1, 200, 4)
```

This dummy input allows JAX to infer the shapes of all model parameters, and the random key `rng_init` seeds any stochastic initialization, ensuring reproducibility.

Examining Model Tensor Shapes

Understanding and keeping track of tensor shapes is a crucial part of machine learning. Always make it a habit to verify the shapes of your data as it flows through the model.

> As a practical exercise, try adding `print(x.shape)` statements at various points inside the model's `__call__` method and then rerun the `model.init` step to observe how the shapes change as data flows through the model layers.

Here are some key operations in the model that change tensor shapes:

- Convolutional layers (`nn.Conv`): These layers can modify the channel dimension. For example, our input starts with four channels (DNA bases), but the first convolution increases this to 64 channels, effectively learning 64 different sequence features or motifs. Additionally, the convolution's `padding` option affects the sequence length dimension:
  - `adding='SAME'` preserves the sequence length.
  - `padding='VALID'` reduces it depending on the kernel size. Try switching between these to see the impact on the sequence length.
- Max pooling layers (`nn.max_pool`): These reduce the sequence length by down-sampling. In our model, each max pooling layer halves the length, from 200 → 100 → 50 bases. To reduce the spatial axis more aggressively (e.g., by a factor of 5 each time), adjust the `window_shape` and `strides` arguments accordingly (typically, these are set to the same value to avoid overlapping windows and simplify downsampling).
- Flattening (`reshape`): Before passing data to dense layers, the tensor is reshaped from (`batch_size, sequence_length, channels`) to (`batch_size, flat tened_features`). This collapses the spatial and channel dimensions into one long vector per example, preparing it for fully connected layers.

Once the model parameters are initialized, you can inspect them to confirm the layer structure and shapes. Check the parameter keys (layer names) with:

```
params.keys()
```

Output:

```
dict_keys(['Conv_0', 'Conv_1', 'Dense_0', 'Dense_1', 'Dense_2'])
```

Then, inspect each layer's kernel shape to verify the expected parameter dimensions:

```
for layer_name in params.keys():
    print(f'Layer {layer_name} param shape: {params[layer_name]["kernel"].shape}')
```

Output:

```
Layer Conv_0 param shape: (10, 4, 64)
Layer Conv_1 param shape: (10, 64, 64)
Layer Dense_0 param shape: (3200, 128)
Layer Dense_1 param shape: (128, 64)
Layer Dense_2 param shape: (64, 1)
```

Some notes on these shapes:

- For convolutional layers, the parameter shape is (kernel_size, input_chan nels, output_channels). For example, Conv_0 has kernel size 10, input channels 4 (DNA bases), and outputs 64 feature maps.
- Dense layers have shapes (input_features, output_units). For instance, Dense_0 maps 3200 flattened features (50 sequence length * 64) to 128 units.

With the model initialized and the parameter shapes explored, let's start setting up our training loop.

Making predictions with the model

We can actually already obtain predictions from the model using the randomly initialized parameters, though the predictions will be random. We make predictions by calling model.apply on a batch of sequences:

```
logits = model.apply({"params": params}, batch["sequences"])

# Apply sigmoid to convert logits to probabilities.
probs = nn.sigmoid(logits)

# Print just the first few predictions.
print(probs[0:5])
```

Output:

```
[[0.48703438]
 [0.49615338]
 [0.48638064]
 [0.4973824 ]
 [0.48106888]]
```

Since the model is untrained, predicted probabilities will be around around 0.5, reflecting pure guesswork from our randomly-parameterized model.

Defining a loss function

Let's now define a binary cross-entropy loss function using the `optax` library that we can use to train our model parameters:

```
import optax

def calculate_loss(params, batch):
    """Make predictions on batch and compute binary cross entropy loss."""
    logits = model.apply({"params": params}, batch["sequences"])
    loss = optax.sigmoid_binary_cross_entropy(logits, batch["labels"]).mean()
    return loss
```

We can use it to compute the loss for the initial batch:

```
calculate_loss(params, batch)
```

Output:

```
Array(0.69774264, dtype=float32)
```

This gives a baseline loss value before training. As the model learns, this loss should decrease substantially.

> Why use binary cross-entropy loss? In this chapter, we're predicting whether a TF binds to a given DNA sequence—a classic binary classification task. Binary cross-entropy is the standard loss function for this setting: it measures how well the model's predicted probabilities align with the true binary labels.
>
> It penalizes confident but incorrect predictions more heavily, encouraging well-calibrated outputs near 0 or 1. You can also think of it as a signal reconstruction problem: the model tries to approximate a hidden binary signal, and cross-entropy imposes a sharp cost for noisy or off-target estimates.

Defining the TrainState

To start training the model, we first need to define the optimizer. We'll use Adam with a learning rate of 1e-3, which is a common default value that tends to work well across diverse problems:

```
learning_rate = 0.001

tx = optax.adam(learning_rate)
```

With this, we now have all the components to initialize the training state. For this, we'll use Flax's `TrainState` class, which is a container that bundles all the important objects for training (the model, parameters, and optimizer):

```
from flax.training.train_state import TrainState

state = TrainState.create(apply_fn=model.apply, params=params, tx=tx)
```

For convenience, let's define a function to create the train state:

```
def create_train_state(model, rng, dummy_input, tx) -> TrainState:
    variables = model.init(rng, dummy_input)
    state = TrainState.create(
        apply_fn=model.apply, params=variables["params"], tx=tx
    )
    return state
```

Defining a single training step

Finally, putting everything together, we can write a function to run one training iteration, which performs these steps:

1. Forward pass

Takes a batch of data, makes model predictions, and computes loss based on the current parameters

2. Backwards pass:

Computes gradients of the loss with respect to the parameters

3. Update

Using the computed gradients, updates the parameters to minimize the model's loss

These steps happen in the `train_step` function:

```
@jax.jit
def train_step(state, batch):
    """Run single training step to compute gradients and update model params."""
    grad_fn = jax.value_and_grad(calculate_loss, has_aux=False)
    loss, grads = grad_fn(state.params, batch)
    state = state.apply_gradients(grads=grads)
    return state, loss
```

Let's run one training step:

```
state, loss = train_step(state, batch)
```

If our setup is working well, the training loss should already be lower on this batch. Let's check:

```
calculate_loss(state.params, batch)
```

Output:

```
Array(0.63649833, dtype=float32)
```

And indeed, the loss is lower than it was before a `train_step` was run. With the model and training loop ready, let's set up a full training run.

Training the Simple Model

Let's now run the preceding `train_step` many times in order to optimize the model. Here, we train for 500 steps and periodically evaluate on the validation set:

```python
import tqdm

# Reinitialize the model state to ensure we start fresh each time cell is run.
rng_init = jax.random.PRNGKey(42)
state = create_train_state(model, rng_init, dummy_input, tx)

# Keep track of both the training and validation set losses.
train_losses, valid_losses = [], []
train_batches = train_ds.as_numpy_iterator()

# We use tqdm, which is a progress bar.
for step in tqdm.tqdm(range(500)):
  batch = next(train_batches)
  state, loss = train_step(state, batch)
  train_losses.append({"step": step, "loss": loss.item()})

  # Compute loss on the entire validation set occasionally (every 100 steps).
  if step % 100 == 0:
    valid_loss = calculate_loss(state.params, valid_ds)
    valid_losses.append({"step": step, "loss": valid_loss.item()})

losses = pd.concat(
  [
    pd.DataFrame(train_losses).assign(split="train"),
    pd.DataFrame(valid_losses).assign(split="valid"),
  ]
)
```

In Figure 3-6, we can plot the training and validation loss curves resulting from this run:

```python
import matplotlib.pyplot as plt
import seaborn as sns

from difb.utils.metric_plots import DEFAULT_SPLIT_COLORS

sns.lineplot(
  data=losses,
  x="step",
  y="loss",
  hue="split",
  style="split",
  palette=DEFAULT_SPLIT_COLORS
);
```

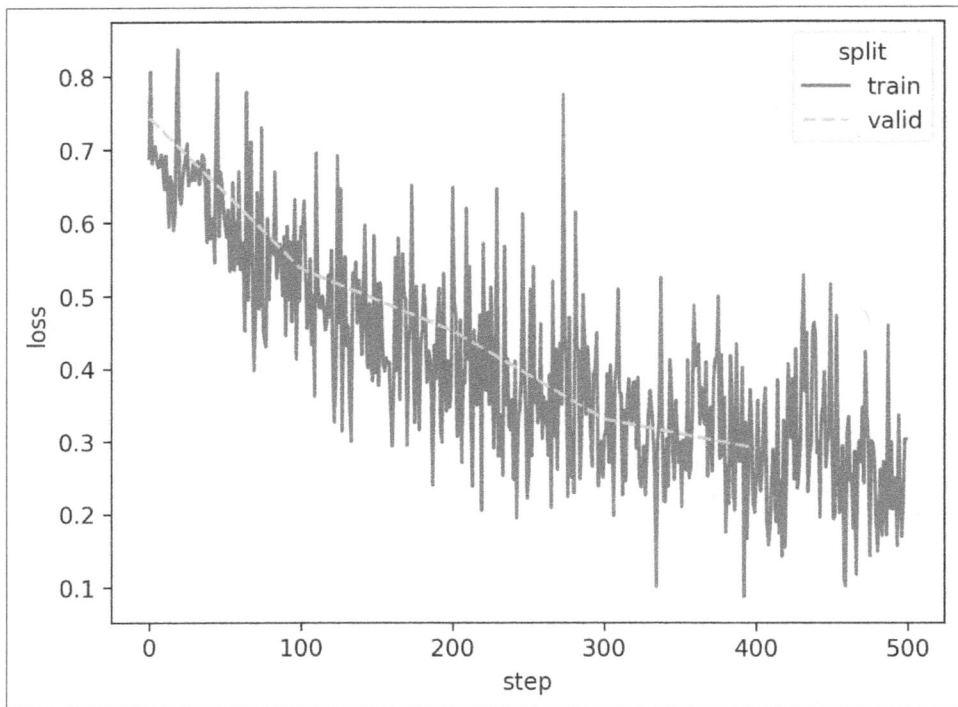

Figure 3-6. Training and validation loss over learning steps. Both decrease, indicating the model is learning signal in the data over time. Note that training loss curve is comparatively noisier due to it being computed on a small batch of the data at each step (rather than the full training set), which introduces variability.

Reaching this stage—where you have a working model, a dataset, and training with decreasing loss—is a major milestone. The rest of this chapter focuses on improving and extending this foundation.

Sanity-checking the model

Before we do anything more complicated, we should first check that the model has learned something sensible. One simple check is verifying that the trained model behaves as expected on known DNA motifs. For example, from an online search, we can see that the CTCF transcription factor is known to prefer binding DNA sequences containing motifs similar CCACCAGGGGGCGC. Let's construct the 200-base-long DNA string containing repeats of this motif and convert it to the one-hot encoded format that our model expects:

```
ctcf_motif_dna = "CCACCAGGGGGCGC" * 14 + "AAAA"
print("Length of CTCF motif-filled DNA string:", len(ctcf_motif_dna))
```

```
# We add the None here as a batch axis, since our model expects batched input.
ctcf_input = dna_to_one_hot(ctcf_motif_dna)[None, :]
ctcf_input.shape
```

Output:

```
Length of CTCF motif-filled DNA string: 200
(1, 200, 4)
```

We expect that the model will predict a very high probability of CTCF binding this sequence, since it's packed with the relevant motif. Let's check:

```
jax.nn.sigmoid(model.apply({"params": state.params}, ctcf_input))
```

Output:

```
Array([[0.9994091]], dtype=float32)
```

And this is indeed the case—the predicted probability of binding is very close to 1. This means the model has learned to identify this motif in the DNA sequence and associate it with CTCF binding.

Conversely, if we construct some pseudorandom DNA strings, the model should predict a relatively low probability of CTCF binding them:

```
random_dna_strings = [
  "A" * 200,
  "C" * 200,
  "G" * 200,
  "T" * 200,
  "ACGTACGT" * 25,
  "TCGATCGT" * 25,
  "TATACGCG" * 25,
  "CAGGCAGG" * 25,
]

probabilities = []

for random_dna_string in random_dna_strings:
  random_dna_input = dna_to_one_hot(random_dna_string)[None, :]

  probabilities.append(
    jax.nn.sigmoid(model.apply({"params": state.params}, random_dna_input))[0]
  )

probabilities
```

Output:

```
[Array([0.00025924], dtype=float32),
 Array([6.9913156e-05], dtype=float32),
 Array([6.1404e-05], dtype=float32),
 Array([2.6038255e-05], dtype=float32),
 Array([0.10472302], dtype=float32),
 Array([0.00381694], dtype=float32),
```

```
Array([0.01843039], dtype=float32),
Array([0.00171606], dtype=float32)]
```

As expected, these probabilities look relatively low, meaning that the CTCF protein is not likely to bind these sequences of random DNA. This completes a basic sanity check of our approach, but let's dig a bit deeper into what this model has already learned about DNA sequences.

Increasing Complexity

In this section, we introduce two important extensions to our modeling approach.

First, we focus on *model interpretation*. We'll apply two techniques from the earlier machine learning primer—in silico mutagenesis (ISM) and input gradients—to better understand what the model has learned. These methods produce *contribution scores* (or *saliency maps*) that assign an importance value to each base in the DNA sequence, indicating how much that base influences the model's prediction of TF binding.

Second, we'll expand the scope of our prediction task. Instead of predicting binding for just one transcription factor (CTCF), we train models for *all 10 transcription factors* in the dataset. This allows us to explore how model performance varies across TFs and how motif preferences differ between them.

Together, these steps deepen both our understanding of the model's behavior and the complexity of the biological task it's modeling. After this, we'll turn our attention to improving the model architecture itself.

Conducting In Silico Mutagenesis

Recall from the introduction that ISM is a technique in which each base in a DNA sequence is systematically mutated to all possible alternative bases one at a time, and the effect of each mutation on a given output (in this example, CTCF binding probability) is quantified. This allows us to identify which regions are important to the output—unimportant regions can be freely mutated without impacting predictions, whereas important regions significantly affect the probability of TF binding.

Before making all possible mutations, let's first check the effect of making just a single mutation. We'll start by identifying a DNA sequence in the validation set that binds the CTCF protein (i.e., has a label of 1):

```
# The first positive example of a sequence that binds the transcription factor.
first_positive_index = np.argmax(valid_ds["labels"].flatten() == 1)

original_sequence = valid_ds["sequences"][first_positive_index].copy()
print(f'This sequence has label: {valid_ds["labels"][4]}')
```

Output:

```
This sequence has label: [1]
```

Next, let's examine what the model predicts for this unmodified sequence:

```
pred = nn.sigmoid(
    model.apply({"params": state.params}, original_sequence[None, :])
)
pred
```

Output:

```
Array([[0.9581409]], dtype=float32)
```

The original sequence has a predicted binding probability of 95.8%.

Now let's create a single mutation at position 100. The original base is a G (encoded as [0, 0, 1, 0]), and we change it to a C (encoded as [0, 1, 0, 0]):

```
sequence = original_sequence.copy()
print(f"Original base at index 100: {sequence[100]}")

sequence[100] = np.array([0, 1, 0, 0])
print(f"Mutated base at index 100: {sequence[100]}")
```

Output:

```
Original base at index 100: [0 0 1 0]
Mutated base at index 100: [0 1 0 0]
```

We'll now run the model again to see if this mutation has a measurable effect on the prediction:

```
pred_with_mutation = nn.sigmoid(
    model.apply({"params": state.params}, sequence[None, :])
)
pred_with_mutation
```

Output:

```
Array([[0.93434644]], dtype=float32)
```

After the mutation, the model's predicted binding probability drops to 93.4%, a decrease of over 2.4%. This shows that even a single-base change at a sensitive position can substantially impact the model's output.

Now that we've observed how a single mutation can influence the prediction, let's extend this approach to systematically mutate every position in the sequence. This will give us a more global view of which bases matter most for the model's prediction.

Implementing in silico saturation mutagenesis

Here's the basic plan for our implementation:

Mutate
> Start with the original sequence. At each position, change the base to each of the other three possible DNA bases.

Predict
> Run the model on each mutated sequence and record the predicted binding probability.

Aggregate
> Collect all results to identify which mutations cause meaningful changes.

We begin by generating all possible single-base mutations:

```python
def generate_all_mutations(sequence: np.ndarray) -> np.ndarray:
    """Generate all possible single base mutations of a one-hot DNA sequence."""
    mutated_sequences = []
    for i in range(sequence.shape[0]):
        # At each position, one the four 'mutations' is the original base (no-op).
        for j in range(4):
            mutated_sequence = sequence.copy()
            mutated_sequence[i] = np.zeros(4)
            mutated_sequence[i][j] = 1
            mutated_sequences.append(mutated_sequence)

    sequences = np.stack(mutated_sequences)
    return sequences


mutated_sequences = generate_all_mutations(sequence=original_sequence.copy())
print(f"Shape of mutated sequences: {mutated_sequences.shape}")
```

Output:

```
Shape of mutated sequences: (800, 200, 4)
```

We now have 800 sequences: four variants for each of the 200 positions. Although only three of those per position are true mutations (since mutating, say, A to an A is a no-op), we include all four to simplify downstream logic.

We can now run predictions on all these mutated sequences in a single batch:

```python
preds = nn.sigmoid(model.apply({"params": state.params}, mutated_sequences))

# Reshape to get the shape (sequence_length, dna_bases).
preds = preds.reshape((200, 4))
```

Let's visualize the predicted binding probabilities for every mutated sequence in Figure 3-7:

```
plt.figure(figsize=(20, 3))
sns.heatmap(preds.T, cmap="RdBu_r", yticklabels=["A", "C", "G", "T"])
plt.xlabel("Position in DNA sequence")
plt.ylabel("DNA Base");
```

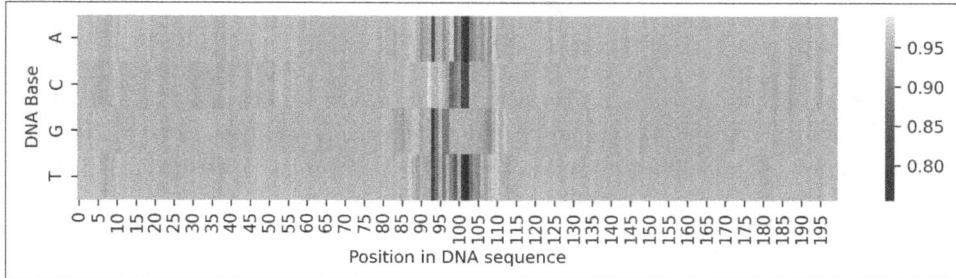

Figure 3-7. All possible variations of a 200-base pair DNA region and their corresponding probabilities of TF binding, represented as a heatmap. The x-axis indicates a position in the DNA sequence, and the y-axis represents each possible DNA base at that position. The value represents the predicted probability of CTCF binding.

This shows us that:

- Most mutations don't change the prediction much. The model predicts close to the original 95.8% probability of CTCF binding for sequences containing most mutations.

- A central region does appear to significantly affect the binding prediction.

But what we really care about is how much each mutation changes the prediction. Let's subtract the original (unmutated) predicted probability and center the color map at zero in Figure 3-8:

```
baseline_pred = nn.sigmoid(
    model.apply({"params": state.params}, original_sequence[None, :])
)
deltas = preds - baseline_pred

plt.figure(figsize=(20, 3))
sns.heatmap(deltas.T, center=0, cmap="RdBu_r", yticklabels=["A", "C", "G", "T"])
plt.xlabel("Position in DNA sequence")
plt.ylabel("DNA Base");
```

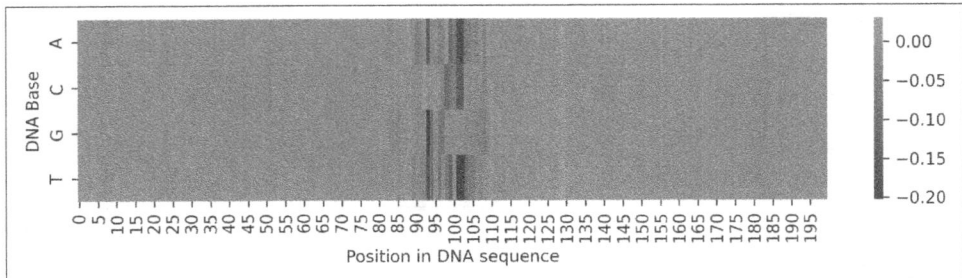

Figure 3-8. In silico mutagenesis results showing the change in predicted CTCF binding probability for each possible mutation. Light and dark shades indicate a deviation from the original prediction, with lighter colors showing mutations that increase predicted binding and darker colors showing those that decrease binding.

Now the heatmap is clearer:

- Most positions are lighter, meaning their mutations have little effect.
- A few bases in the center are dark (decreased binding) or light (increased binding), showing meaningful influence.

We can quantify mutation effects using a helper function, `describe_change`.

It allows us to have a look at the impact of mutating position 100 for all bases:

```
for i in range(4):
  print(describe_change((100, i), deltas, original_sequence))
```

Output:

```
position 100 with G→A (-4.20% decrease)
position 100 with G→C (-2.38% decrease)
position 100 with G→G (0.00% increase)
position 100 with G→T (0.24% increase)
```

Let's also summarize the overall importance of each position in the DNA sequence by summing the absolute changes across all possible base mutations at each position. The result is visualized in Figure 3-9, highlighting which regions the model considers most influential for its prediction:

```
from dlfb.dna.inspect import plot_binding_site

importance = np.sum(np.abs(deltas), axis=1)
plot_binding_site(
  panels={
    "tiles": {"label": "Deltas", "values": deltas},
    "line": {"label": "Importance", "values": importance},
  }
);
```

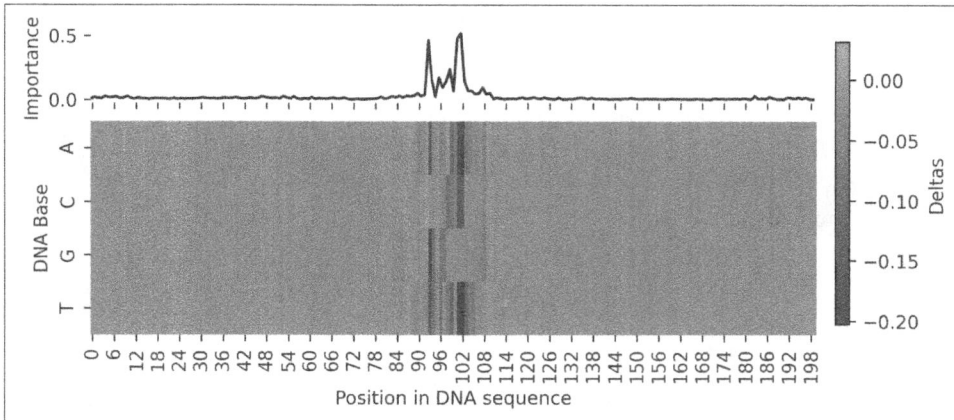

Figure 3-9. Positional importance of the TF binding motif. The bottom panel is the same as in Figure 3-8 with a linegraph superimposed.

We can identify the most impactful mutations by ranking the values. In this case, the largest *increase* comes from mutating the base at position 92 with G→C (3.11% increase), and the biggest decrease comes from mutating the base at position 102 with G→T (-20.24% decrease).

> This is a great moment to reflect on the biological meaning of these plots. The central region of the sequence likely contains the CTCF binding motif—mutations here have a strong impact on the model's prediction. In contrast, flanking regions show little effect, indicating they contribute less to binding. Interestingly, most impactful mutations in the core motif tend to reduce predicted binding, suggesting the original sequence already contains a fairly strong CTCF motif that's difficult to strengthen with single-base changes.

Verifying motif presence

To validate whether the region identified by our model corresponds to a known CTCF binding motif, we can use an external bioinformatics tool. But first, we need to convert the one-hot encoded sequence back into the standard DNA string format. This is done using the one_hot_to_dna function:

```python
def one_hot_to_dna(one_hot_encoded: np.ndarray) -> str:
    """Convert one-hot encoded array back to DNA sequence."""
    one_hot_to_base = {
        (1, 0, 0, 0): "A",
        (0, 1, 0, 0): "C",
        (0, 0, 1, 0): "G",
        (0, 0, 0, 1): "T",
```

```
    (1, 1, 1, 1): "N",  # N represents any unknown or ambiguous base.
}

dna_sequence = "".join(
    one_hot_to_base[tuple(base)] for base in one_hot_encoded
)
return dna_sequence
```

which we can use on our sequence:

```
print(one_hot_to_dna(original_sequence)[0:25], "...")
```

Output:

```
ACCCCAGGGTAGGGCCTATTGTATG ...
```

After converting the sequence, we paste it into the FIMO (Find Individual Motif Occurrences) tool, which is part of the MEME suite of motif discovery and search tools. FIMO allows for fuzzy matching of motifs—meaning the match doesn't need to be exact—which better reflects biological reality, as transcription factors often tolerate some variability in their binding sites.

For CTCF, the known binding motif we used was CCACCAGGGGGCGC. When we submitted our sequence, FIMO reported a match to the CTCF motif (specifically, the subsequence GCCTCTGGGGGCGC, spanning positions 93 to 106) with a highly significant p-value of 2.6e-05, as shown in Figure 3-10.

SECTION I: HIGH-SCORING MOTIF OCCURENCES

- There were 1 motif occurences with a p-value less than 0.0001. The full set of motif occurences can be seen in the TSV (tab-delimited values) output file fimo.tsv, the GFF3 file fimo.gff which may be suitable for uploading to the UCSC Genome Table Browser (assuming the FASTA input sequences included genomic coordinates in UCSC or Galaxy format), or the XML file fimo.xml.
- The *p*-value of a motif occurrence is defined as the probability of a random sequence of the same length as the motif matching that position of the sequence with as good or better a score.
- The score for the match of a position in a sequence to a motif is computed by summing the appropriate entries from each column of the position-dependent scoring matrix that represents the motif.
- The q-value of a motif occurrence is defined as the false discovery rate if the occurrence is accepted as significant.
- The table is sorted by increasing *p*-value.

Motif ID	Alt ID	Sequence Name	Strand	Start	End	p-value	q-value	Matched Sequence
1	CCACCAGGGGGCGC	my_sequence	+	93	106	2.58e-05	0.00661	GCCTCTGGGGGCGC

Figure 3-10. How we used FIMO to search for the known CTCF binding motif within a 200 bp DNA sequence labeled as positive. Although transcription factor motifs are usually represented more flexibly using position weight matrices (PWMs), a simple string-based search for the canonical motif CCACCAGGGGGCGC offers a quick sanity check to confirm that the expected pattern is present in the sequence.

To further verify this, we can overlay the motif region found by FIMO on our earlier saliency map of mutation impact. This helps us visually compare where mutations

---

have the strongest effect on the model's prediction versus where the detected motif actually occurs (see Figure 3-11):

```
plot_binding_site(
  panels={
    "tiles": {"label": "Deltas", "values": deltas},
    "line": {"label": "Importance", "values": importance},
  },
  highlight=(92, 106),
);
```

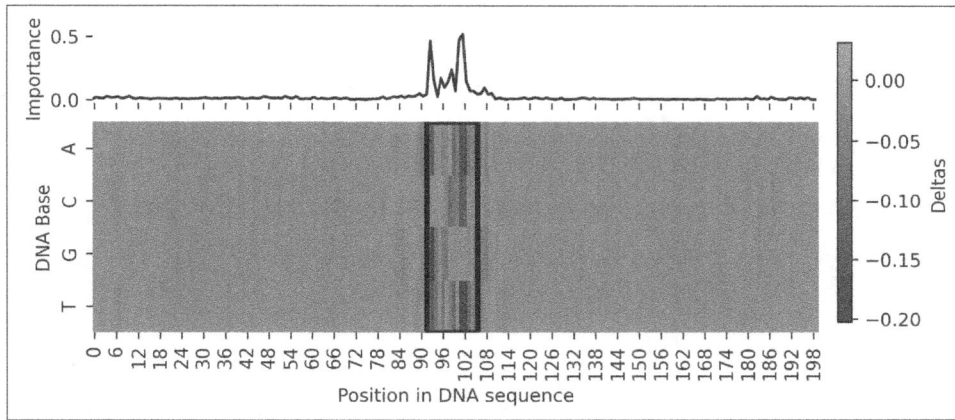

Figure 3-11. *Highlight of the TF binding site overlapping neatly with the region where mutations have the biggest influence.*

As you can see, the motif returned by FIMO lines up almost exactly with the most mutation-sensitive region identified by our model through in silico mutagenesis. This alignment gives us confidence that the model has learned to recognize meaningful biological signal—in this case, a known CTCF binding motif—directly from sequence data.

Implementing input gradients

While *in silico* mutagenesis provides highly interpretable insights, it can be computationally expensive—requiring multiple forward passes through the model for each position in the input sequence. An alternative approach that is much cheaper to compute is *input gradients*. This technique, introduced earlier in the chapter, relies on a simple idea: how much does the model output change if we nudge each input base slightly?

Mathematically, this corresponds to computing the gradient of the model's prediction with respect to its input sequence. If a small change in a particular base results in a large change in the output, that base must be important.

The implementation is straightforward. We use `jax.grad` to take the derivative of the model's predicted binding probability with respect to the one-hot input sequence:

```
@jax.jit
def compute_input_gradient(state, sequence):
    """Compute input gradient for a one-hot DNA sequence."""
    if len(sequence.shape) != 2:
        raise ValueError("Input must be a single one-hot encoded DNA sequence.")

    sequence = jnp.asarray(sequence, dtype=jnp.float32)[None, :]

    def predict(sequence):
        # We take the mean to ensure we have a single scalar to take the grad of.
        return jnp.mean(state.apply_fn({"params": state.params}, sequence))

    gradient = jax.grad(lambda x: predict(x))(sequence)
    return jnp.squeeze(gradient)
```

Let's apply it to the same sequence we used for ISM and inspect the output:

```
input_gradient = compute_input_gradient(state, original_sequence)
input_gradient.shape
```

Output:

```
(200, 4)
```

The result is a matrix of the same shape as the input sequence, where each value indicates how sensitive the model's prediction is to a small change in a particular base at a particular position.

We can visualize this as a heatmap in Figure 3-12:

```
importance = np.sum(np.abs(input_gradient), axis=1)
plot_binding_site(
    panels={
        "tiles": {"label": "Gradients", "values": input_gradient},
        "line": {"label": "Importance", "values": importance},
    },
);
```

Just like with ISM, we see that most positions have little influence on the prediction, while a central region—roughly positions 90 to 110—exhibits strong gradients. This reflects the bases that the model is most sensitive to for making its prediction. However, input gradients differ from ISM in key ways: the values are not bounded to represent discrete mutations, and even the base currently present at each position can have a nonzero gradient. This is because the gradient describes how the model's output would change with an infinitesimal increase in that base's activation—not an actual mutation. This makes gradients a bit more abstract but also more flexible and computationally efficient.

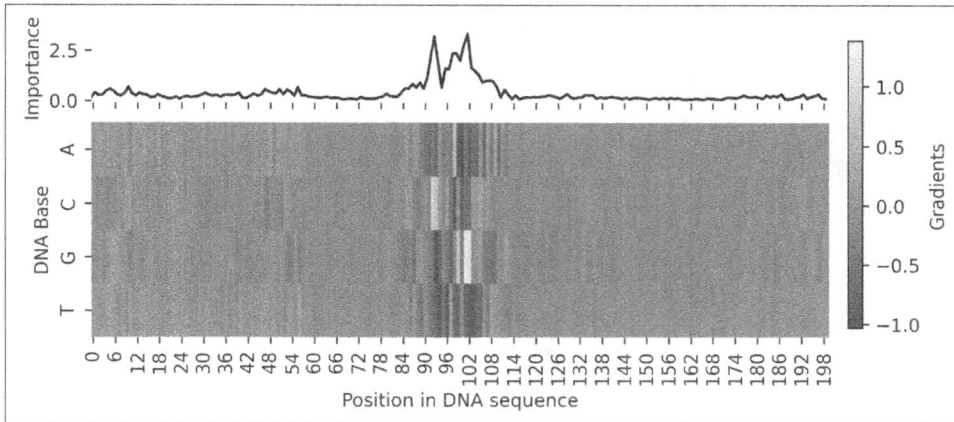

Figure 3-12. Input gradient saliency map for a CTCF-binding sequence, indicating the contribution of each DNA base to the model's prediction of CTCF binding. The bottom heatmap shows input gradients per base (A/C/G/T), while the top line shows aggregated importance across positions. A clear central region stands out as most influential for the model's decision.

To examine this region more closely, we can zoom in and label the actual bases in the 90:110 base region in Figure 3-13:

```
important_sequence = one_hot_to_dna(original_sequence)[90:110]
print("Central DNA sequence with high importance: ", important_sequence)

plt.figure(figsize=(10, 2))
sns.heatmap(
    input_gradient[90:110].T,
    cmap="RdBu_r",
    center=0,
    xticklabels=important_sequence,
    yticklabels=["A", "C", "G", "T"],
)
plt.tight_layout();
```

Output:

```
Central DNA sequence with high importance:  TGGCCTCTGGGGGCGCTCTG
```

This plot confirms what we saw using ISM and the FIMO tool: the model is placing high importance on the central region of the sequence, which corresponds to a canonical CTCF binding motif.

While input gradients are computationally much cheaper than ISM, they can be somewhat noisier or harder to interpret due to model nonlinearities or saturation effects. Still, they offer a practical way to rapidly inspect what a model is attending to—especially when analyzing many sequences at once.

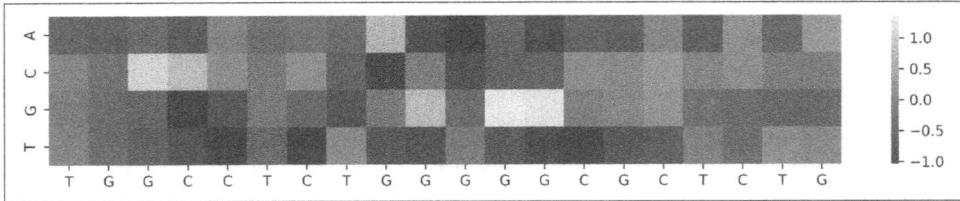

Figure 3-13. Zoomed input gradients over the CTCF motif. In this close-up view of the region identified as important by input gradients, each cell shows how the model's predicted CTCF binding probability would change in response to a small increase in a specific base at a given position. Unlike in silico mutagenesis, even the base that is already present (e.g., G→G) can have a high gradient—indicating that the model is highly sensitive to that base and relies on it for its prediction. Strongly positive or negative values reflect positions where the model has learned a motif and is using it to assess CTCF binding.

Beware How You Interpret Input Gradients

Interpreting input gradients can be unintuitive, and a common source of confusion is mistaking input gradients for mutation effects. Unlike in silico mutagenesis, input gradients do not simulate base changes. Instead, they show how the model's output would change if you slightly nudged the input toward each of the four bases—even the one that's already there.

So a high gradient at a G→G position (i.e., where the base is already G), as in Figure 3-13, doesn't mean "mutating G to G is important" (which would be meaningless). Instead, it means:

- The model is *sensitive* to the presence of that G.
- A tiny increase in the "G-ness" at that position would meaningfully affect the prediction.
- This G is contributing strongly to the model's decision.

In other words, if this G were somehow made even more G—despite already being fully G—the model's output would shift. That's not biologically plausible, but it's a side effect of how gradients work mathematically.

In short, input gradients reflect directional sensitivity, not discrete substitutions. That's why even unchanged bases can show strong signal in the heatmap.

So far, we have been analyzing a single sequence example (the first validation example). In Figure 3-14 we extend this to the first 10 sequences in the validation set that are labeled 1 (i.e., sequences that are known to bind CTCF), and visualize their input gradients:

```
from dlfb.dna.inspect import plot_10_gradients

plot_10_gradients(state, valid_ds, target_label=1);
```

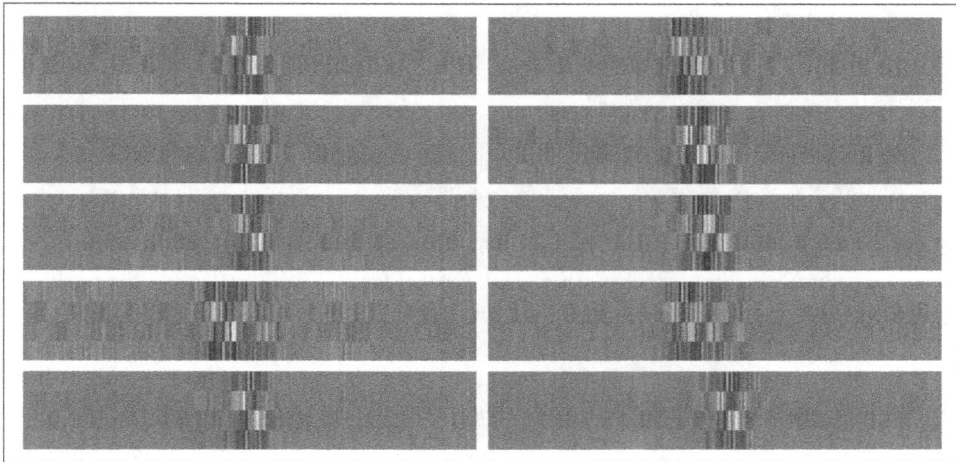

Figure 3-14. Input gradients for 10 CTCF-bound sequences show consistent central regions of high importance, suggesting a shared motif-like structure driving model predictions. Each cell shows the gradient value for mutating a specific base at a given position. Strongly positive or negative gradients indicate high sensitivity to mutations.

These gradients show a consistent central pattern across the examples, indicating that the model has likely learned a motif centered within the sequence. While the width of the important region varies slightly, the signal is strong and localized.

By contrast, Figure 3-15 shows the input gradients for 10 negative examples—sequences that the model predicts do not bind CTCF. Here, the picture is far more heterogeneous:

```
plot_10_gradients(state, valid_ds, target_label=0);
```

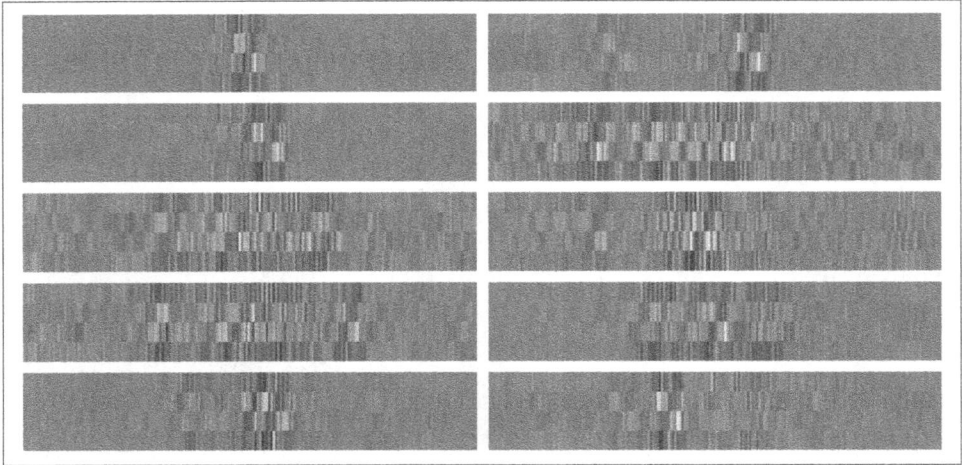

Figure 3-15. Input gradients for 10 nonbinding sequences reveal more diffuse or noisy importance patterns, but often retain a weak central focus—likely reflecting the peak-centered sampling strategy used during dataset construction.

In this plot, some sequences show a weaker but still present centered signal. Others display no clear motif or have diffuse scores spread across the entire sequence.

This variation makes sense given the way the dataset was constructed. While these sequences are labeled negative, they were carefully matched to positive examples in terms of chromatin accessibility. That means they also come from open chromatin regions and may still contain weak or partial motifs—or even motifs for other TFs. As a result, the model might still "attend" to the center of the sequence, even if it ultimately predicts no binding.

> These findings serve as a good reminder that contribution scores like input gradients don't just reflect the presence or absence of strong motifs. They also reveal where the model is looking for evidence and can surface subtle or confounding biological signals introduced by dataset design choices. In this case, the weak central patterns likely reflect a bias introduced by sampling from peak-centered open chromatin, even in negative (class 0) examples.

Now that we've explored two complementary interpretation tools—ISM and input gradients—we're in a better position to trust (and debug) what the model is learning. Let's return to the modeling task and increase both the problem complexity and model capacity.

Modeling Multiple Transcription Factors

So far, we've focused on a single transcription factor: CTCF. Let's now increase the biological scope of our modeling by predicting binding for all 10 TFs in the dataset.

Preparing a multi-TF dataset

The dataset includes binding labels for the following 10 transcription factors:

```
transcription_factors = [
  "ARID3",
  "ATF2",
  "BACH1",
  "CTCF",
  "ELK1",
  "GABPA",
  "MAX",
  "REST",
  "SRF",
  "ZNF24",
]
```

Each of these TFs has its own distinct DNA-binding preference. For instance:

- CTCF, as we've seen, binds motifs like CCACCAGGGGGCGC.
- MAX prefers the E-box motif CACGTG, important in regulating cell proliferation.
- SRF binds to CCW6GG (where W = A or T), a motif involved in muscle-specific gene expression.

These preferences are often conserved across species, and the role of each TF can be deeply rooted in specific cell type identity or developmental processes. If you're curious, it's worth looking up some of these proteins—for example, REST is a key repressor in neurons, while MAX has a known role in oncogenesis through interaction with MYC.

Our task now is to train models that can automatically discover these motif patterns and accurately associate them with TF binding.

There are two common strategies for tackling this problem:

Multitask learning
   Train a single model to output one prediction per TF, potentially learning shared representations across tasks.

Single-task learning
   Train separate binary classification models for each TF independently.

In our case, the 10 TFs are not particularly closely related, and their binding preferences are quite distinct—ranging from insulator proteins like CTCF to general transcriptional regulators like MAX and neuron-specific factors like REST. Additionally, the dataset provides separate training sets for each TF, and the original paper trained them independently.

Given these factors, we'll follow a single-task learning approach: train 10 individual binary classification models, one per TF. This also contrasts nicely with the multitask approach you saw in Chapter 2.

Defining a more complex model

Now that we're training on 10 separate TF binding tasks, it's a good opportunity to improve the training stability and generalization of our model.

We'll retain the same core convolutional architecture from earlier, but introduce three standard deep learning techniques—batch normalization, dropout, and learning rate scheduling—that are widely used in CNNs and often help even for relatively shallow models and simple datasets.

Batch normalization

We add batch normalization after each convolutional layer to improve training stability:

- Batch norm normalizes the activations across the batch, which helps smooth the optimization landscape and accelerates convergence.
- Even though our network is relatively shallow, batch norm can still improve performance and robustness.

A few implementation details:

- Batch norm behaves differently during training and inference. During training, it computes mean and variance from the current batch and updates running averages; at inference time, it uses the learned averages.
- In Flax, you control this behavior with the `is_training` flag using the `use_running_average=not is_training` pattern.

Dropout regularization

To reduce overfitting, we add dropout after the dense (fully connected) layers:

- Dropout randomly sets a portion of activations to zero during training, encouraging the model to learn redundant and robust features.

- It is typically used after dense layers, rather than convolutions, because spatially shared convolutional filters already generalize well.

- In Flax, dropout requires passing a PRNG (pseudorandom number generator) key to the forward pass.

- We use a moderate dropout rate of 0.2, which adds some regularization without significantly reducing model capacity.

Learning rate scheduling

Rather than using a fixed learning rate, we adopt a learning rate schedule to guide training:

- Dynamic learning rates usually start high (encouraging fast exploration) and decrease over time to help convergence.

- Popular options include exponential decay, step decay, and cosine annealing.

- Implemented via `optax.cosine_decay_schedule`, we use a cosine decay schedule which gradually reduces the learning rate over the course of training. The shape of this learning rate schedule is visualized in Figure 3-16:

```
num_steps = 1000

scheduler = optax.cosine_decay_schedule(
  init_value=0.001,
  decay_steps=num_steps,   # How long to decay over.
)
learning_rates = [scheduler(i) for i in range(num_steps)]

plt.scatter(range(num_steps), learning_rates)
plt.title("Learning Rate over Steps")
plt.ylabel("Learning Rate")
plt.xlabel("Step");
```

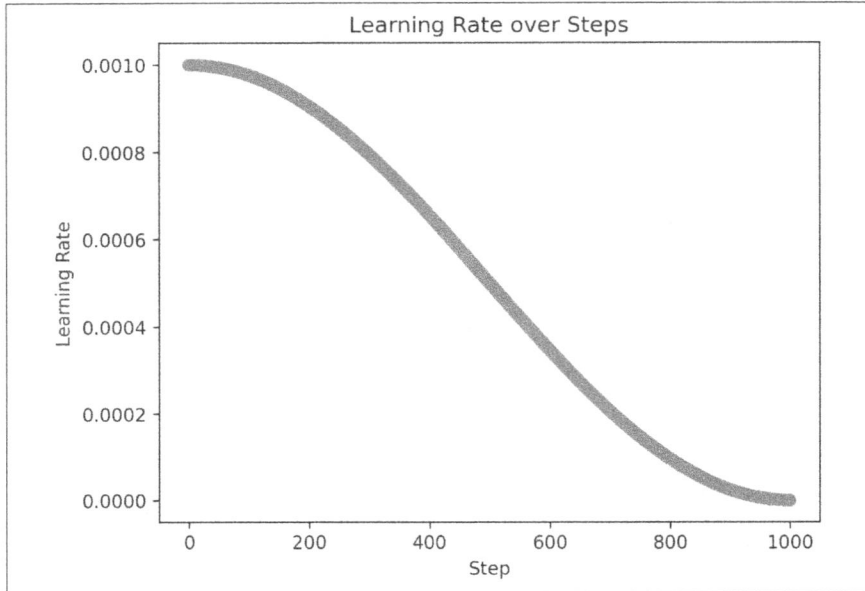

Figure 3-16. Cosine decay learning rate schedule used during training. The learning rate starts at 0.001 and gradually decreases over 1,000 steps, helping the model explore early on and converge smoothly toward the end.

Even though our dataset and architecture are fairly simple, these additions should help improve training stability and model generalization across the expanded set of transcription factor tasks.

Let's now implement the updated model definition:

```
class ConvModelV2(nn.Module):
    """CNN with batch norm and dropout for binary classification."""

    conv_filters: int = 64  # Number of filters for conv layers.
    kernel_size: tuple[int] = (10,)  # Kernel size for 1D conv layers.
    dense_units: int = 128  # Units in first dense fully-connected layer.
    dropout_rate: float = 0.2  # Proportion of dense neurons to randomly drop out.

    @nn.compact
    def __call__(self, x, is_training: bool = True):
        # First convolutional layer.
        x = nn.Conv(
            features=self.conv_filters, kernel_size=self.kernel_size, padding="SAME"
        )(x)
        x = nn.BatchNorm(use_running_average=not is_training)(x)
        x = nn.gelu(x)
        x = nn.max_pool(x, window_shape=(2,), strides=(2,))

        # Second convolutional layer.
```

```
x = nn.Conv(
  features=self.conv_filters, kernel_size=self.kernel_size, padding="SAME"
)(x)
x = nn.gelu(x)
x = nn.BatchNorm(use_running_average=not is_training)(x)
x = nn.max_pool(x, window_shape=(2,), strides=(2,))

# Flatten the values before passing them to the dense layers.
x = x.reshape((x.shape[0], -1))

# First dense layer.
x = nn.Dense(self.dense_units)(x)
x = nn.gelu(x)
x = nn.Dropout(rate=self.dropout_rate)(x, deterministic=not is_training)

# Second dense layer.
x = nn.Dense(self.dense_units // 2)(x)
x = nn.gelu(x)
x = nn.Dropout(rate=self.dropout_rate)(x, deterministic=not is_training)

# Output layer (single unit for binary classification).
return nn.Dense(1)(x)

def create_train_state(self, rng: jax.Array, dummy_input, tx):
  """Initializes model parameters and returns a train state for training."""
  rng, rng_init, rng_dropout = jax.random.split(rng, 3)
  variables = self.init(rng_init, dummy_input)
  state = TrainStateWithBatchNorm.create(
    apply_fn=self.apply,
    tx=tx,
    params=variables["params"],
    batch_stats=variables["batch_stats"],
    key=rng_dropout,
  )
  return state
```

Our ConvModelV2 implementation is starting to get a bit long and repetitive. Later in the chapter, we'll address this by refactoring out the repeated logic into Conv Block and DenseBlock components to make the model definition more concise and modular.

For now, we have all the pieces needed to create the training state:

```
rng = jax.random.PRNGKey(42)
rng, rng_init, rng_train = jax.random.split(rng, 3)
state = ConvModelV2().create_train_state(
  rng=rng_init, dummy_input=batch["sequences"], tx=optax.adam(scheduler)
)
```

The training step looks fairly similar to what we've seen before, but now it needs to handle both dropout and batch normalization.

One small additional change: the loss function is defined directly inside `train_step`. This is often done for readability and encapsulation, especially when the loss depends on extra mutable model state (like `batch_stats`) or dropout, both of which are now required in the forward pass:

```python
@jax.jit
def train_step(state, batch, rng_dropout: jax.Array):
    """Run a training step and update parameters."""

    def calculate_loss(params, batch):
        """Make predictions on batch and compute binary cross-entropy loss."""
        logits, updates = state.apply_fn(
            {"params": params, "batch_stats": state.batch_stats},
            x=batch["sequences"],
            is_training=True,
            rngs={"dropout": rng_dropout},
            mutable=["batch_stats"],
        )

        loss = optax.sigmoid_binary_cross_entropy(logits, batch["labels"]).mean()

        return loss, updates

    grad_fn = jax.value_and_grad(calculate_loss, has_aux=True)
    (loss, updates), grads = grad_fn(state.params, batch)
    state = state.apply_gradients(grads=grads)
    state = state.replace(batch_stats=updates["batch_stats"])

    metrics = {"loss": loss}

    return state, metrics
```

To confirm that everything is wired up correctly, we can overfit to a single batch— that is, run a few steps on the same batch and check that the loss decreases:

```python
# Overfit on one batch.
for i in range(5):
    rng, rng_dropout = jax.random.split(rng, 2)
    state, metrics = train_step(state, batch, rng_dropout)
    print(f"Step {i} loss: {metrics['loss']}")
```

Output:

```
Step 0 loss: 0.6932974457740784
Step 1 loss: 0.25415313243865967
Step 2 loss: 0.08513301610946655
Step 3 loss: 0.02221144177019596
Step 4 loss: 0.019464049488306046
```

Looks good—the loss decreases quickly. This indicates the model is capable of fitting the data.

However, loss alone isn't the best way to evaluate classification models. We want to monitor additional metrics like accuracy and the area under the ROC curve (auROC). We compute these in an evaluation step:

```python
def eval_step(state, batch):
    """Evaluate model on a single batch."""
    logits = state.apply_fn(
        {"params": state.params, "batch_stats": state.batch_stats},
        x=batch["sequences"],
        is_training=False,
        mutable=False,
    )
    loss = optax.sigmoid_binary_cross_entropy(logits, batch["labels"]).mean()
    metrics = {
        "loss": loss.item(),
        **compute_metrics(batch["labels"], logits),
    }
    return metrics


def compute_metrics(y_true: np.ndarray, logits: np.ndarray):
    """Compute accuracy and auROC for model predictions."""
    metrics = {
        "accuracy": accuracy_score(y_true, nn.sigmoid(logits) >= 0.5),
        "auc": roc_auc_score(y_true, logits).item(),
    }
    return metrics
```

Note that since `scikit-learn` functions are not JAX compatible, the evaluation step is not decorated with `@jax.jit`.

Let's see the output of running `eval_step`:

```python
# Evaluate the batch.
metrics = eval_step(state, batch)
print(metrics)
```

Output:

```
{'loss': 0.45877009630203247, 'accuracy': 0.9375, 'auc': 1.0}
```

Now that we've implemented the training and evaluation steps, let's define a full training loop. The train function takes an initialized model state, training and validation datasets, and a few configuration parameters:

```python
@restorable
def train(
    state: TrainStateWithBatchNorm,
    rng: jax.Array,
    dataset_splits: dict[str, tf.data.Dataset],
    num_steps: int,
    eval_every: int = 100,
) -> tuple[TrainStateWithBatchNorm, Any]:
```

```
"""Train a model and log metrics over steps."""
metrics = MetricsLogger()
train_batches = dataset_splits["train"].as_numpy_iterator()

steps = tqdm(range(num_steps))  # Steps with progress bar.
for step in steps:
  steps.set_description(f"Step {step + 1}")

  rng, rng_dropout = jax.random.split(rng, 2)
  train_batch = next(train_batches)
  state, batch_metrics = train_step(state, train_batch, rng_dropout)
  metrics.log_step(split="train", **batch_metrics)

  if step % eval_every == 0:
    for eval_batch in dataset_splits["valid"].as_numpy_iterator():
      batch_metrics = eval_step(state, eval_batch)
      metrics.log_step(split="valid", **batch_metrics)
    metrics.flush(step=step)

  steps.set_postfix_str(metrics.latest(["loss"]))

return state, metrics.export()
```

The training loop is quite similar to before, but as a quick recap, here is what it does:

- Iterates through training steps using a progress bar
- Runs `train_step` to update model parameters
- Periodically evaluates on the validation set using `eval_step`
- Logs and stores metrics at each step using a custom MetricsLogger

Since we're training one model per TF, we'll use a utility function to load datasets split by TF name:

```
def load_dataset_splits(
  path, transcription_factor, batch_size: int | None = None
):
  """Load TF dataset splits (train, valid, test) as TensorFlow datasets."""
  dataset_splits = {}
  for split in ["train", "valid", "test"]:
    dataset = load_dataset(
      sequence_db=f"{path}/{transcription_factor}_{split}_sequences.csv"
    )
    ds = convert_to_tfds(dataset, batch_size, is_training=(split == "train"))
    dataset_splits.update({split: ds})
  return dataset_splits
```

This function loads training, validation, and test splits for the given `transcription_factor` and converts them into TensorFlow datasets ready for batching.

We now have everything in place to train one model per TF:

```
prefix = assets("dna/datasets")
tf_metrics = {}

# Train one model per transcription factor.
for transcription_factor in transcription_factors:
  # Load data for this TF.
  dataset_splits = load_dataset_splits(
    assets("dna/datasets"), transcription_factor, batch_size
  )
  rng = jax.random.PRNGKey(42)
  rng, rng_init, rng_train = jax.random.split(rng, 3)
  dummy_batch = next(dataset_splits["train"].as_numpy_iterator())["sequences"]

  # Create train state.
  state = ConvModelV2().create_train_state(
    rng=rng_init,
    dummy_input=dummy_batch,
    tx=optax.adam(scheduler),
  )

  # Train the model.
  _, metrics = train(
    state=state,
    rng=rng_train,
    dataset_splits=dataset_splits,
    num_steps=num_steps,
    eval_every=100,
    store_path=assets(f"dna/models/{transcription_factor}"),
  )

  # Store metrics.
  tf_metrics.update({transcription_factor: metrics})
```

With training complete, let's take a look at just CTCF transcription factor performance in Figure 3-17:

```
from dlfb.dna.inspect import import plot_learning

tf = "CTCF"
plot_learning(tf_metrics[tf], tf);
```

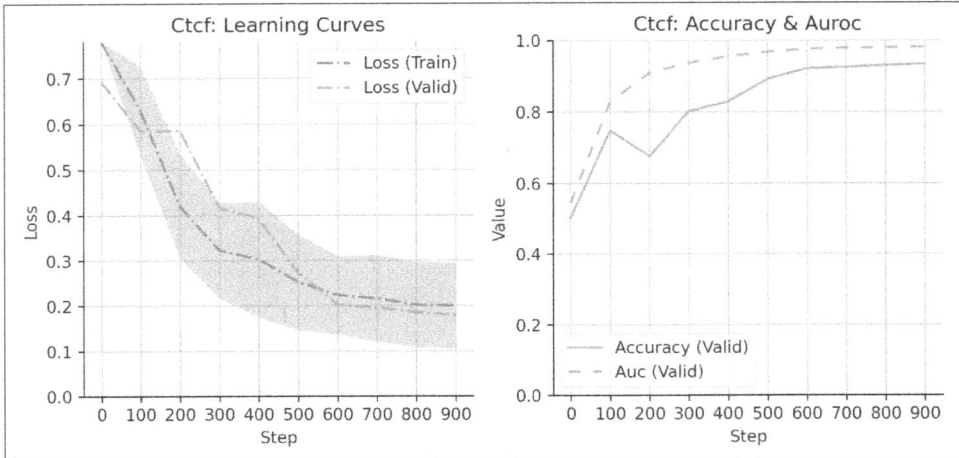

Figure 3-17. Learning curves for the CTCF transcription factor. The left panel shows the training and validation loss over time. The right panel tracks the model's classification performance using validation set accuracy and auROC. Performance improves steadily during training and converges smoothly.

Our model performs well on predicting whether CTCF binds a given DNA sequence, with training and validation losses decreasing and validation auROC approaching a perfect score of 1.0—consistent with the fact that CTCF is relatively easy to predict compared to some of the other TFs.

Let's now visualize the training curves across all 10 TFs. We first process the logged metrics:

```
from dlfb.utils.metric_plots import to_df

# Extract metrics logged per transcription factor.
tf_df = []
for tf, metrics in tf_metrics.items():
  tf_df.append(to_df(metrics).assign(TF=tf))
tf_df = pd.concat(tf_df)

# Determine order of best performance.
auc_df = tf_df[(tf_df["metric"] == "auc") & (tf_df["split"] == "valid")]
max_auc_by_tf = auc_df.groupby("TF")["mean"].max()
tf_order = max_auc_by_tf.sort_values(ascending=False).index.tolist()
tf_df["TF"] = pd.Categorical(tf_df["TF"], categories=tf_order, ordered=True)
```

We can then visualize their learning curves in Figure 3-18 sorted by auROC performance (best-performing TFs are first):

```
sns.set_context("notebook", font_scale=2, rc={"lines.linewidth": 2.5})
sns.set_style("ticks", {"axes.grid": True})
g = sns.relplot(
```

```
    data=tf_df,
    x="round",
    y="mean",
    hue="split",
    style="metric",
    kind="line",
    col="TF",
    col_order=tf_order,
    col_wrap=4,
    alpha=0.8,
    palette=DEFAULT_SPLIT_COLORS,
    dashes=True,
)
g.set_axis_labels("Step", "Value")
g.set(ylim=(0, 1));
```

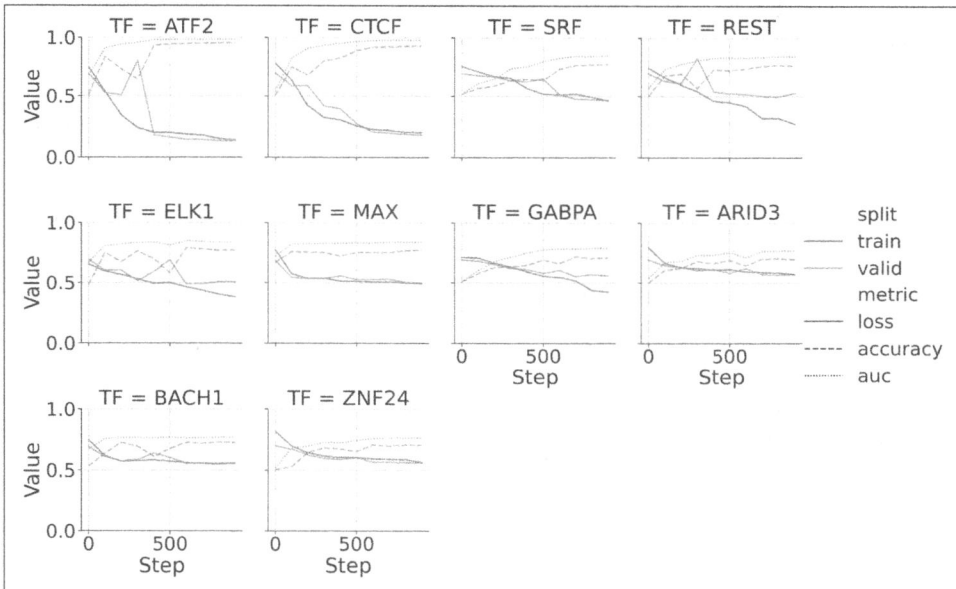

Figure 3-18. Learning curves for all 10 TFs. Each panel shows the training and valida-
tion loss (solid lines), validation accuracy (dashed lines), and validation auROC (dotted
lines) over training steps. TFs are ordered by peak auROC performance. While some
TFs, like CTCF and ATF2, reach near-perfect performance quickly, others, such as
ZNF24 and BACH1, prove more challenging to model.

Are these different TF binding models doing a good job? It can be hard to know what
constitutes a "good enough" auROC—especially in biological settings where label
noise and dataset complexity can vary widely. Fortunately, since our dataset comes
from a published benchmark, we can directly compare our model's results to those
reported in the original paper.

In Figure 3 of the source paper (*https://oreil.ly/QciKO*), we see that auROC scores vary considerably across TFs, and some are challenging to predict well—even when using information from more advanced architectures like genomic language models (gLMs) trained with large-scale pretraining. The paper's figure helps establish a performance ceiling for baseline CNNs trained on one-hot encoded DNA. Interestingly, the paper notes that pretrained gLM representations do not consistently outperform conventional approaches using one-hot inputs, suggesting that simple models can still be competitive on these TF binding tasks.

The following are key takeaways from the paper's figure:

- CTCF and ATF2 are the most predictable TFs, with both one-hot and pretrained models achieving auROC scores above 0.95. These TFs have strong, conserved binding motifs that are easy for models to learn.
- REST, MAX, and ELK1 show intermediate difficulty, with auROCs around 0.83 to 0.85.
- ZNF24, BACH1, ARID3, and GABPA tend to be more difficult, with auROCs hovering in the 0.75 to 0.80 range.

Let's now see how our own models compare. The following prints the peak validation auROC achieved by our CNNs trained independently for each TF:

```
print(max_auc_by_tf.sort_values(ascending=False))
```

Output:

```
TF
ATF2    0.985182
CTCF    0.980479
SRF     0.849944
          ...
ARID3   0.769149
BACH1   0.767122
ZNF24   0.761506
Name: mean, Length: 10, dtype: float64
```

Our results closely mirror the rankings and scores from the paper—especially in terms of which TFs are easier or harder to predict. This consistency offers reassuring external validation that our training setup is functioning correctly and our models are learning something meaningful.

Why Are Some TFs Harder to Model Than Others?

Not all TFs are equally predictable. Some—like CTCF or ATF2—bind to well-defined, highly conserved motifs that produce strong, consistent signals in the DNA. These clear patterns make it easier for neural networks to accurately predict their binding.

Others—such as ZNF24, ARID3, or BACH1—are more complex. They may bind weakly, recognize flexible or degenerate motifs, or depend on subtle contextual cues such as cofactor presence, motif spacing, or chromatin accessibility. These factors introduce biological variability and weaken the sequence-to-binding signal, making the task inherently noisier and more difficult.

In other words, variation in model performance across TFs often reflects true biological complexity—not just machine learning limitations. It's also worth noting that our benchmark is simplified: we're predicting TF binding from just 200 base pairs of sequence. In reality, binding depends on broader genomic and epigenetic context—including long-range interactions and chromatin state. Still, many of these influences are at least partially encoded in local DNA patterns, and modern deep learning models like Enformer, Borzoi, and AlphaGenome can achieve high correlation with ChIP-seq binding data using only sequence input.

Let's now explore how we might push our TF-binding prediction results further through more expressive architectures.

Advanced Techniques

Before we introduce more complex model components, let's first improve the clarity and modularity of our model architecture by refactoring it into reusable building blocks. This makes our code easier to read, extend, and maintain.

We will modularize our convolutional and MLP layers by creating two helper modules, ConvBlock and MLPBlock:

```python
class ConvBlock(nn.Module):
    """Convolutional block with batch norm, GELU and max pooling."""

    conv_filters: int
    kernel_size: tuple[int]
    pool_size: int

    @nn.compact
    def __call__(self, x, is_training: bool = True):
        x = nn.Conv(
            features=self.conv_filters, kernel_size=self.kernel_size, padding="SAME"
        )(x)
        x = nn.BatchNorm(use_running_average=not is_training)(x)
        x = nn.gelu(x)
        x = nn.max_pool(
            x, window_shape=(self.pool_size,), strides=(self.pool_size,)
        )
        return x


class MLPBlock(nn.Module):
```

```
    """Dense + GELU + dropout block."""

    dense_units: int
    dropout_rate: float = 0.0

    @nn.compact
    def __call__(self, x, is_training: bool = True):
      x = nn.Dense(self.dense_units)(x)
      x = nn.gelu(x)
      x = nn.Dropout(rate=self.dropout_rate)(x, deterministic=not is_training)
      return x
```

With these reusable blocks in place, we can now define a more compact and configurable model architecture:

```
class ConvTransformerModel(nn.Module):
    """Model combining CNN, transformer, and MLP blocks."""

    num_conv_blocks: int = 2
    conv_filters: int = 64
    kernel_size: tuple[int] = (10,)
    num_mlp_blocks: int = 2
    dense_units: int = 128
    dropout_rate: float = 0.2  # Global.
    num_transformer_blocks: int = 0
    num_transformer_heads: int = 8
    transformer_dense_units: int = 64

    @nn.compact
    def __call__(self, x, is_training: bool = True):
      for _ in range(self.num_conv_blocks):
        x = ConvBlock(
          conv_filters=self.conv_filters,
          kernel_size=self.kernel_size,
          pool_size=2,
        )(x, is_training)

      for i in range(self.num_transformer_blocks):
        x = TransformerBlock(
          num_heads=self.num_transformer_heads,
          dense_units=self.transformer_dense_units,
          dropout_rate=self.dropout_rate,
        )(x, is_training)

      x = x.reshape((x.shape[0], -1))

      for i in range(self.num_mlp_blocks):
        x = MLPBlock(
          dense_units=self.dense_units // (i + 1), dropout_rate=self.dropout_rate
        )(x, is_training)

      return nn.Dense(1)(x)
```

This architecture supports optional transformer blocks inserted between the convolutional feature extractors and the MLP classifier. You'll also notice the introduction of TransformerBlock, which we'll explore next.

Adding Self-attention and Transformer Blocks

Our final architecture combines convolutional layers with transformer blocks. While convolutional layers are excellent at extracting local motif-like patterns from DNA, transformer blocks can capture long-range dependencies—patterns that span larger regions of the sequence. These two types of layers are highly complementary.

In our case, the sequences are relatively short (just 200 base pairs), so the benefits of long-range attention may be modest. However, attention mechanisms become increasingly useful as input length increases. For longer genomic contexts—such as full gene bodies, or interactions between potentially distant regulatory elements such as promoter and enhancer sequences—transformers can integrate signals across a broader range than is possible with fixed-size convolutional kernels.

Flax includes a built-in SelfAttention module, which we can use as the foundation for our TransformerBlock:

```python
class TransformerBlock(nn.Module):
    """Transformer block with self-attention and MLP."""

    num_heads: int = 8
    dense_units: int = 64
    dropout_rate: float = 0.2

    @nn.compact
    def __call__(self, x, is_training: bool = True):
        # Self-attention with layer norm.
        residual = x
        x = nn.LayerNorm()(x)
        x = nn.SelfAttention(num_heads=self.num_heads)(x)
        x += residual

        # Feedforward block.
        residual = x
        x = nn.LayerNorm()(x)
        x = nn.Dense(self.dense_units)(x)
        x = nn.gelu(x)
        x = nn.Dropout(rate=self.dropout_rate)(x, deterministic=not is_training)
        x = nn.Dense(self.dense_units)(x)  # No GELU after this Dense.
        x += residual
        return x
```

A few notes on this design:

Self-attention
The core operation is `nn.SelfAttention`, which enables each position in the sequence to attend to every other position, capturing dependencies across the input.

Residual connections
The skip connections help stabilize training by allowing gradients to flow more easily through deep architectures.

Layer normalization
Transformers typically use layer norm rather than batch norm, as it tends to be more stable for sequence-based models and is invariant to batch size.

Position information
In our current setup, we omit explicit positional encodings—this is a simplification that may be acceptable for short sequences (e.g., 200 bp), where the relative arrangement of motifs can still be learned through the convolutional layers preceding attention. For longer inputs or if attention is applied very early, consider adding learned or sinusoidal positional encodings.

This modular `TransformerBlock` can now be optionally included between convolutional and MLP layers in our `ConvTransformerModel`.

Defining Various Model Architectures

Now that our model is modular, we can experiment with different architectural settings to better understand their impact on performance. This kind of architectural exploration is common when tuning deep learning models—there's rarely a single obvious "best" model, and trying multiple variants can reveal helpful insights.

In the following code, we define several models with different settings:

```
models = {
    # Our standard 2-layer CNN with dropout and MLP.
    "baseline": ConvTransformerModel(),
    # Ablations: Remove or reduce certain components.
    # Only a single convolutional block.
    "single_conv_only": ConvTransformerModel(
        num_conv_blocks=1, num_transformer_blocks=0, num_mlp_blocks=0
    ),
    # Reduced capacity by lowering conv filters.
    "fewer_conv_channels": ConvTransformerModel(conv_filters=8),
    # Drop the MLP layers to test if they help.
    "remove_MLP": ConvTransformerModel(num_mlp_blocks=0),
    # Potential improvements: Add more expressive capacity.
    # Add a transformer block after convolutions.
    "add_one_transformer_block": ConvTransformerModel(num_transformer_blocks=1),
```

```
  # Stack two transformer blocks.
  "add_two_transformer_block": ConvTransformerModel(num_transformer_blocks=2),
}
```

To test these variants, we'll focus on the most difficult TF from our earlier results: ZNF24, which achieved a peak validation auROC of 0.76 in our initial model. This makes it a great candidate to explore whether architectural improvements can lead to better predictive performance.

Sweeping Over the Different Models

With all our model components now modularized, we can easily run systematic experiments to compare architectural choices.

We'll train several model variants—ranging from simpler ablations to transformer-enhanced architectures—on the most challenging TF in our benchmark: ZNF24. This allows us to explore which components help or hurt performance on a relatively difficult task.

With all of this in place, we can train the different models in the `model` dict:

```
# Train and evaluate multiple model architectures on the ZNF24 dataset.
transcription_factor = "ZNF24"
dataset_splits = load_dataset_splits(
  assets("dna/datasets"), transcription_factor, batch_size
)

# Prepare a dummy input for model initialization.
dummy_input = next(dataset_splits["train"].as_numpy_iterator())["sequences"]

# Initialize PRNGs.
rng = jax.random.PRNGKey(42)
rng, rng_init, rng_train = jax.random.split(rng, 3)

# Dictionary to store metrics for each model variant.
model_metrics = {}

# Train each model variant and store its metrics.
for name, model in models.items():
  state = model.create_train_state(
    rng=rng_init,
    dummy_input=dummy_input,
    tx=optax.adamw(
      optax.cosine_decay_schedule(
        init_value=learning_rate,
        decay_steps=num_steps,
      )
    ),
  )
  _, metrics = train(
    state=state,
```

```
        rng=rng_train,
        dataset_splits=dataset_splits,
        num_steps=num_steps,
        eval_every=100,
        store_path=assets(f"dna/models/{name}"),
    )
    model_metrics.update({name: metrics})
```

After training each model, we extract logged metrics over training time and visualize
the results:

```
# Extract metrics logged per transcription factor.
model_df = []
for model, metrics in model_metrics.items():
  model_df.append(to_df(metrics).assign(model=model))
model_df = pd.concat(model_df)

# Determine order of best performance.
auc_df = model_df[
  (model_df["metric"] == "auc") & (model_df["split"] == "valid")
]
max_auc_by_model = auc_df.groupby("model")["mean"].max()
model_order = max_auc_by_model.sort_values(ascending=False).index.tolist()
model_df["model"] = pd.Categorical(
  model_df["model"], categories=model_order, ordered=True
)
```

With this code, we can plot our learning curves and metrics over time in Figure 3-19:

```
sns.set_context("notebook", font_scale=1.2, rc={"lines.linewidth": 2.5})
sns.set_style("ticks", {"axes.grid": True})
g = sns.relplot(
  data=model_df,
  x="round",
  y="mean",
  hue="split",
  style="metric",
  kind="line",
  col="model",
  col_order=model_order,
  col_wrap=2,
  alpha=0.8,
  palette=DEFAULT_SPLIT_COLORS,
  dashes=True,
)
g.set_axis_labels("Step", "Value")
g.set(ylim=(0.4, 0.9));
```

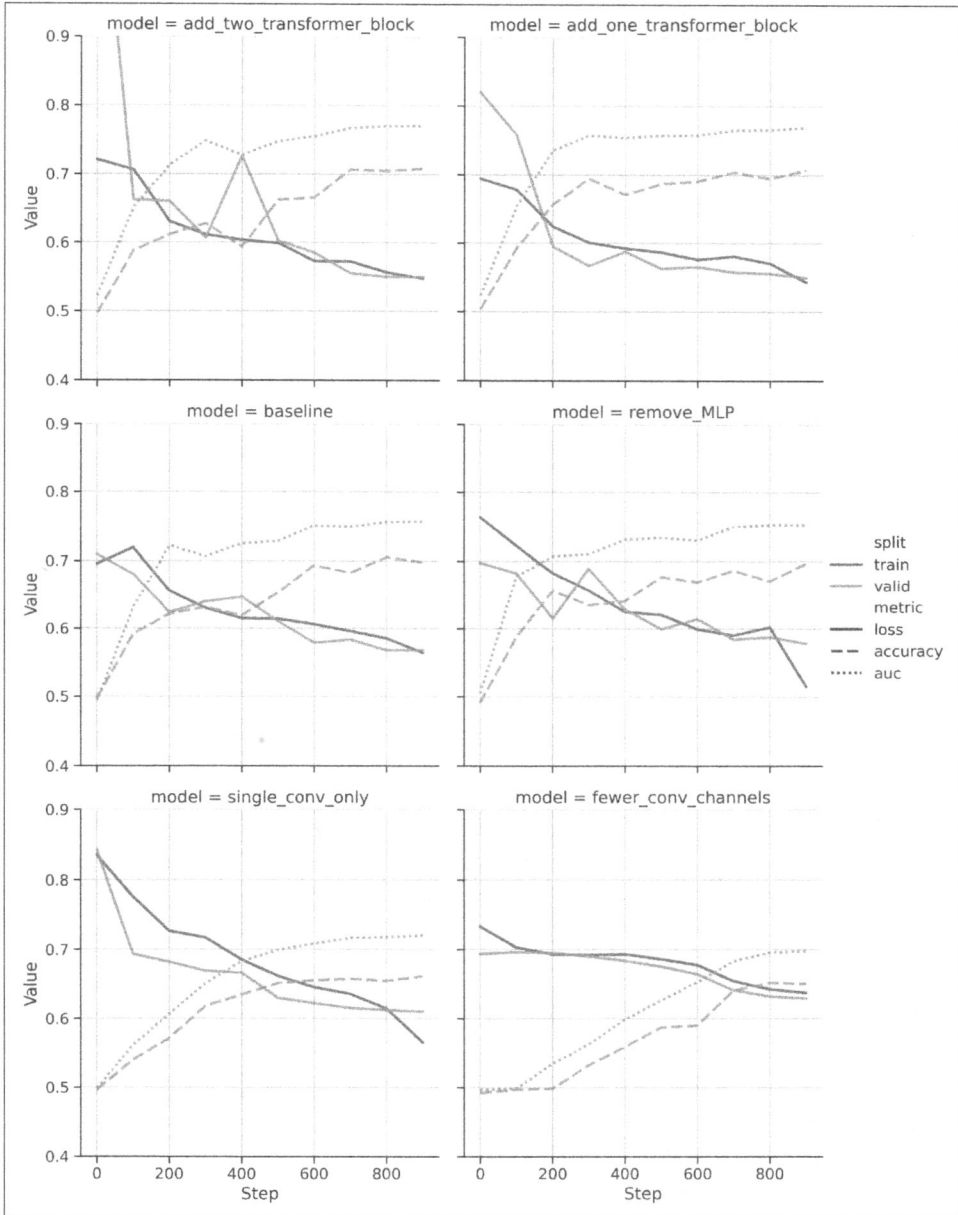

Figure 3-19. Learning curves for different model architectures on ZNF24 binding prediction. Each panel shows training and validation loss, accuracy, and auROC. We see that
adding transformer blocks slightly improves performance, while removing capacity (e.g.,
fewer conv filters) hurts it.

To better isolate model differences, we can plot just the validation auROC over time in Figure 3-20:

```
g = sns.lineplot(
    data=model_df[(model_df["metric"] == "auc")],
    x="round",
    y="mean",
    hue="model",
    style="model",
    alpha=0.8,
)
g.set_xlabel("Step")
g.set_ylabel("auROC");
```

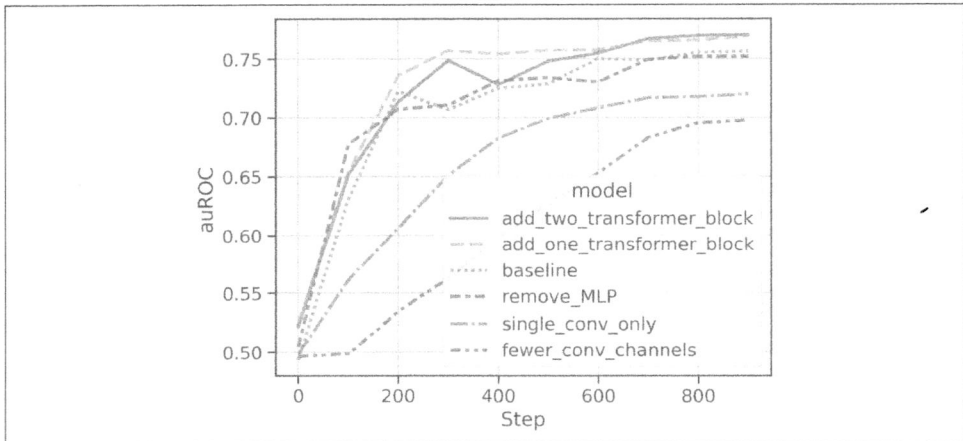

Figure 3-20. Validation auROC across training steps for each model variant. Models with transformer blocks outperform simpler baselines, while reducing convolutional capacity or removing MLPs tends to hurt performance.

And here are the max auROC values for each model (by step 1,000):

```
print(max_auc_by_model.sort_values(ascending=False))
```

Output:

```
model
add_two_transformer_block    0.770343
add_one_transformer_block    0.769196
baseline                     0.756317
remove_MLP                   0.752024
single_conv_only             0.719931
fewer_conv_channels          0.697637
Name: mean, dtype: float64
```

Some observations and hypotheses based on these results:

Convolutions are critical
> The `single_conv_only` model underperforms relative to the baseline, and reducing the number of convolutional filters degrades performance further. This suggests that having multiple convolutional layers with sufficient capacity is important for learning meaningful sequence features. It may be worth exploring deeper convolutional stacks or wider kernels.

Transformer blocks help
> Adding one or two self-attention blocks gives a modest but consistent improvement in auROC, despite the input sequences being only 200 bp long. This supports the idea that even short DNA windows can benefit from modeling long-range dependencies.

MLPs may be unnecessary
> Surprisingly, removing the MLP blocks doesn't drastically hurt performance. This suggests that most of the representational power is coming from earlier convolutional layers, and the additional MLP layers may be redundant.

Training dynamics
> Models with transformer blocks exhibit noisier validation loss during early training. This instability might be reduced with a smaller initial learning rate.

While we won't exhaustively optimize hyperparameters here, these results demonstrate the value of modular model exploration and raise intriguing questions for further study.

Evaluating on the Test Split

The final step is to evaluate our best-performing model on the held-out test set. This ensures that the model's performance generalizes to completely unseen data and was not overfit to the training or validation distributions.

We reload the checkpoint for the top model—selected based on the highest validation auROC—and evaluate it on the test split:

```
# Identify best-performing model variant based on validation auROC.
top_model = model_order[0]

# Restore the trained model state from disk.
state, _ = restore(
  assets(f"dna/models/{top_model}"),
  models[top_model].create_train_state(
    rng=rng_init,
    dummy_input=next(dataset_splits["train"].as_numpy_iterator())["sequences"],
    tx=optax.adamw(
      optax.cosine_decay_schedule(
```

```
        init_value=learning_rate,
        decay_steps=num_steps,
      )
    ),
  ),
)

# Evaluate on the held-out test set.
test_batch = next(dataset_splits["test"].as_numpy_iterator())
metrics = eval_step(state, test_batch)
print(metrics)
```

Output:

```
{'loss': 0.5123618841171265, 'accuracy': 0.75, 'auc': 0.8196078431372549}
```

Success! Our best model achieves a comparable AUC on the test set to what we saw on the validation set, demonstrating good generalization to unseen data.

Extensions and Improvements

There are many ways to extend this work—in terms of both analyzing model behavior and exploring new modeling directions.

A few analysis ideas:

Failure analysis
    Inspect misclassified sequences. Do they correspond to originally noisy or lower-magnitude ChIP-seq peaks? Are there weak motif matches within them? Trace these sequences back to the raw data to understand the source of prediction errors.

Motif discovery
    Use saliency maps to extract high-signal regions and align them to known motifs from databases like JASPAR or HOCOMOCO. You can also explore tools like TF-MoDISco for automatic motif discovery based on contribution scores.

Attribution comparisons
    Compare in silico mutagenesis and input gradients across TFs. Do easier tasks yield more consistent or interpretable saliency patterns?

Cross-TF generalization
    Evaluate a model trained on one TF against the validation data for another. Which TFs generalize well to others? This could reveal shared motif features or similarities in binding logic.

Saliency reproducibility
    Compare saliency maps across different training runs or architectures to assess how reliably motif patterns are captured.

And some potential modeling extensions:

Model optimization
> Tune architectural hyperparameters such as the number of convolutional filters, kernel width, dropout rate, or transformer head count and MLP size.

Positional encodings
> Add sinusoidal or learned positional embeddings to improve transformer modeling of sequence order—especially valuable for longer DNA sequences.

Multitask setup
> Build a multitask model that predicts binding for all TFs simultaneously. This allows shared representations across tasks and may improve performance on lower-data TFs with less data.

Quantitative binding prediction
> Instead of binary classification, predict continuous binding intensity (e.g., ChIP-seq signal). This would require adapting the model for sequence-to-sequence or dense regression output.

Pretraining
> Fine-tune pretrained DNA models such as DNABERT or Nucleotide Transformer to leverage prior genomic knowledge and improve performance with less data.

Data augmentation
> Improve generalization with reverse complements, shifted windows (jittering), or synthetic motif insertions.

These directions offer exciting possibilities for improving accuracy, interpretability, and generalizability in TF binding prediction.

Summary

In this chapter, we explored the fascinating world of gene regulation through transcription factor binding. Starting with simple convolutional models, we built a foundation for understanding how neural networks can learn to recognize sequence motifs in raw DNA sequences. From there, we incrementally increased model complexity—adding batch normalization, dropout, and even transformer blocks—to investigate how architectural changes impact performance.

This gradual, modular approach to model development not only helped clarify the strengths of each component, but also made debugging and evaluation more manageable. Along the way, we touched on model interpretability, performance benchmarking, and ideas for deeper analysis and extensions.

Up to this point, we've focused on biological data with an inherent sequential structure—first with protein sequences in Chapter 2, "Learning the Language of Proteins" and now with DNA. In the next chapter, we shift gears to a very different kind of data: *graphs*. We'll explore how graph neural networks can help us reason about relationships between entities—in particular, interactions between different drugs.

## CHAPTER 4

Understanding Drug–Drug Interactions Using Graphs

Graphs are a fundamental structure found everywhere in the world around us. A familiar example is social networks, where *nodes* represent individuals and *edges* capture the relationship between them. In train systems, nodes could represent stations and edges the routes linking them. Less obvious examples include research collaborations linked by coauthorship, web pages interconnected by hyperlinks, and supermarket baskets, where frequently copurchased items are connected.

Biology, too, is filled with data that naturally lends itself to a network framework— genes interact to control cell functions, proteins physically bind to each other, and cells send signals to each other, all forming graph-like systems. Even molecules can be represented as graphs, with atoms as nodes and chemical bonds as edges, as shown in Figure 4-1. At larger biological scales, ecological food webs capture predator–prey and other species interactions, while disease transmission networks map the spread of pathogens through populations.

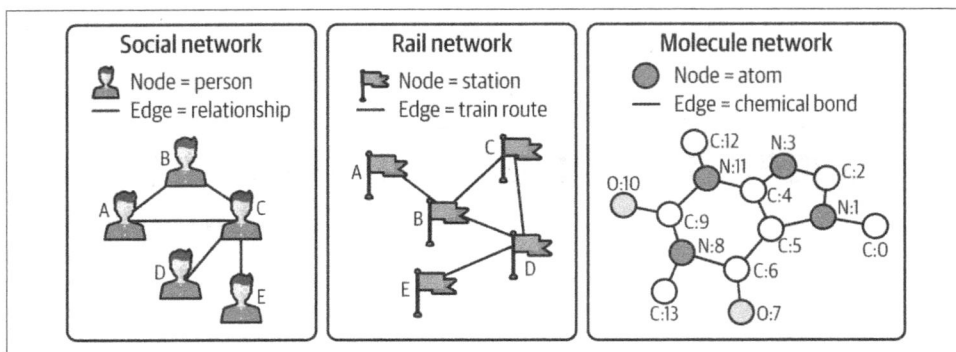

Figure 4-1. Examples of graphs from different contexts. The social network shows people as nodes connected by edges representing relationships. The rail network illustrates stations as nodes and train routes as edges. The molecule network depicts the molecular structure of caffeine, where nodes represent atoms, and edges represent chemical bonds (hydrogen atoms are not shown).

These types of network relationships can be modeled using *graph neural networks* (GNNs). Recently, deep learning on graphs has become increasingly popular and effective. In this chapter, we will explore a graph of *drug–drug interactions* (DDIs) to gain insights into its connectivity. Specifically, we aim to predict whether two nodes should connect, which is a task known as *link prediction*. Link prediction is valuable here because, while we have an existing DDI graph, it may be incomplete—some true connections between drugs might be missing due to limited research or untested combinations. By accurately predicting these links, one could improve drug safety by identifying potential negative interactions and even discover new combination therapies by predicting which drugs might interact positively.

> As in earlier chapters, we recommend keeping this chapter's companion Colab notebook open as you read. Running the code yourself helps reinforce the material and gives you a place to immediately experiment with new ideas.

Biology Primer

DDIs occur when the effects of one drug are altered by the presence of another. DDIs can amplify each drug's effects, counteract them, or change the way a drug is processed in the body, which may result in either therapeutic benefits or adverse outcomes.

Beneficial Drug–Drug Interactions

In some cases, DDIs can be harnessed for therapeutic advantage. In cancer treatment, for example, combination therapies pair drugs that target different pathways in cancer cells. One drug may inhibit tumor growth, while another restricts the tumor's blood supply, weakening it further. This multitargeted approach not only improves patient outcomes but also reduces the likelihood of drug resistance.

Similarly, certain antibiotics work better in combination. For instance, penicillin and gentamicin are often combined to treat infections like endocarditis. Penicillin weakens the bacterial cell wall, allowing gentamicin to penetrate the cell and disrupt protein synthesis, leading to a more effective antibiotic treatment.

Harmful Drug–Drug Interactions

Harmful DDIs are generally much more common than beneficial ones—most drugs are not designed with other drugs in mind, which often leads to unintended side effects in patients taking multiple medications. Additionally, many drugs influence similar biological pathways, increasing the likelihood that one drug will amplify or counteract the effects of another. For example:

Amplification example
Aspirin, commonly used as a pain reliever or blood thinner, can amplify the effects of other anticoagulants, such as warfarin. When taken together, both drugs thin the blood more than intended, raising the risk of excessive bleeding or bruising.

Counteraction example
Ibuprofen can reduce the effectiveness of antihypertensive drugs, such as ACE inhibitors and beta-blockers. Ibuprofen causes the body to retain sodium and fluid, which raises blood pressure and counteracts these medications.

Most negative DDIs are actually more *indirect*. For instance, many drugs are metabolized in the liver by the cytochrome P450 enzyme system, so drugs that inhibit this system can impact a wide range of other medications. Grapefruit, though not a "drug" in the traditional sense, contains compounds that inhibit the cytochrome P450 system. One of the most serious grapefruit interactions occurs with certain statins used to control cholesterol. Grapefruit compounds inhibit an enzyme that would normally break down these statins, causing higher-than-expected drug levels to accumulate in the bloodstream. This buildup can lead to very severe side effects, including liver damage and muscle tissue breakdown.

DrugBank

DrugBank is one of the largest databases of drug interactions, providing detailed information on drugs and their known interactions. It has been widely used in various DDI studies. For example, in Figure 4-2, an early study from 2016 clustered DrugBank DDIs (at the time, the database contained around 1,000 nodes; in this chapter, we work with a more recent version containing over 4,000 nodes) to reveal major drug clusters, including those related to cytochrome P450 interactions discussed earlier.[1]

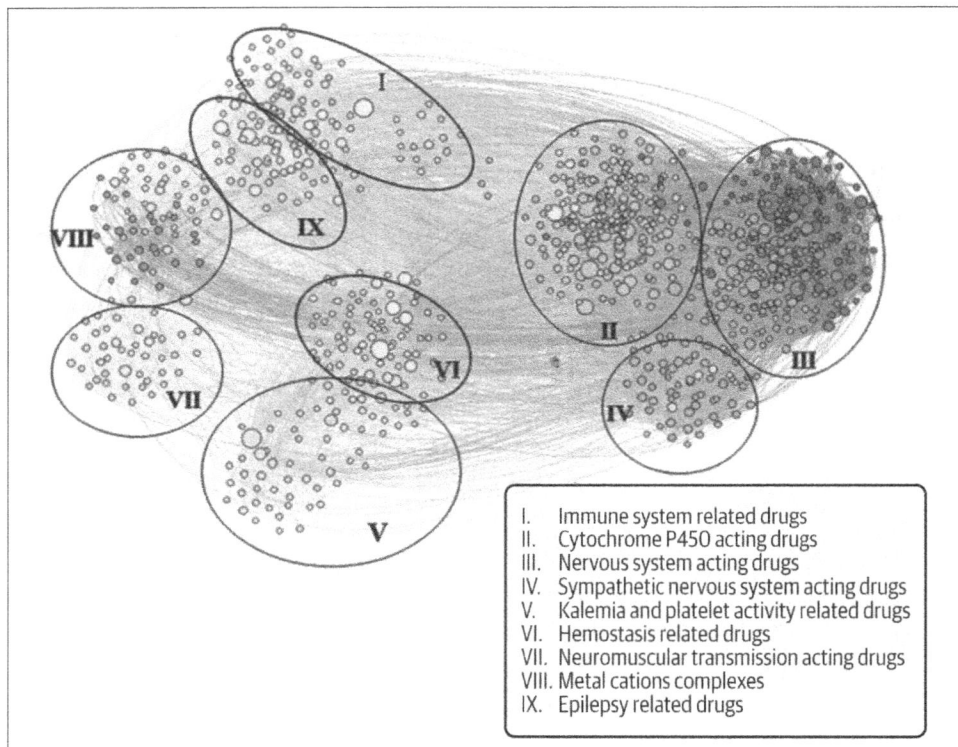

I.	Immune system related drugs
II.	Cytochrome P450 acting drugs
III.	Nervous system acting drugs
IV.	Sympathetic nervous system acting drugs
V.	Kalemia and platelet activity related drugs
VI.	Hemostasis related drugs
VII.	Neuromuscular transmission acting drugs
VIII.	Metal cations complexes
IX.	Epilepsy related drugs

Figure 4-2. Community-based drug–drug interaction network using data from Drug-Bank 4.1, containing 1,141 nodes (drugs) and 11,688 edges (drug–drug interactions). Clustering was performed using the Force Atlas 2 layout algorithm, which simulates a physical system to position nodes closer together based on their interactions, with colors assigned to highlight distinct communities of interacting drugs.

---

1 Udrescu, L. et al. "Clustering Drug–Drug Interaction Networks with Energy Model Layouts: Community Analysis and Drug Repurposing" (*https://oreil.ly/rdShs*). *Scientific Reports*, 6 (2016): 32745.

In this chapter, we will use a processed version of DrugBank's DDI data, available through a publicly accessible benchmark dataset from the Open Graph Benchmark (*https://oreil.ly/S_wR-*) resource by researchers from Stanford University.[2] Before diving into the dataset and its applications, let's begin with a brief primer on machine learning on graphs.

Machine Learning Primer

You probably already have an intuitive sense of what a graph is, but to be more precise, a graph is a structure that represents relationships between pairs of objects. It consists of two main components:

Nodes (or vertices)
> These represent individual entities, like people in a social network or proteins in an interaction network.

Edges
> These are the connections between nodes, indicating relationships or interactions. In a social network, for example, an edge might represent a friendship, while in a protein interaction network, an edge represents an observed physical interaction between two proteins.

Graphs can be *directed* (where edges have a direction, showing a one-way relationship) or *undirected* (indicating a two-way relationship). An example of a directed biological graph is predator–prey relationships between species in an ecosystem—an owl preys on mice; it's usually not the other way around. An example of an undirected biological graph is gene coexpression networks, where the nodes are genes and the edges are correlations between the expression levels of each gene pair.

Edges can have *attributes* such as *weights*, which reflect the strength of a connection. Nodes can also have attributes that capture additional information. For example, in the predator–prey example, edge weights might represent the number of times one species predates another, and each node might contain additional information about that species such as its estimated population size. Graphs vary in connection density (sparse versus dense), may include self-loops (nodes connected to themselves), and can be dynamic (changing over time, like social networks) or static.

Certain graph properties have significant computational implications. For instance, *graph size* can pose a challenge, as large graphs may need to be distributed across multiple processing units to avoid memory overload. *Graph sparsity*—which is the

---

2 Hu, Weihua, Matthias Fey, Marinka Zitnik, Yuxiao Dong, Hongyu Ren, Bowen Liu, Michele Catasta, and Jure Leskovec. 2020. "Open Graph Benchmark: Datasets for Machine Learning on Graphs" (*https://oreil.ly/eK20s*). arXiv.Org. May 2, 2020.

proportion of existing edges relative to the total possible edges in the graph—affects storage and computation efficiency, with specialized techniques designed to handle sparsely connected networks. Additionally, sparse graphs allow for more efficient convolution operations, as fewer neighbors need to be considered (explained further later in this chapter). Finally, the level of *connectivity* plays a crucial role. While graphs with many small, disconnected subgraphs can often be processed in parallel, densely connected graphs are more challenging to parallelize.

Representing Graph Structures

In Figure 4-3, we see an undirected graph containing five nodes (N0, N1, N2, N3, N4) and five edges:

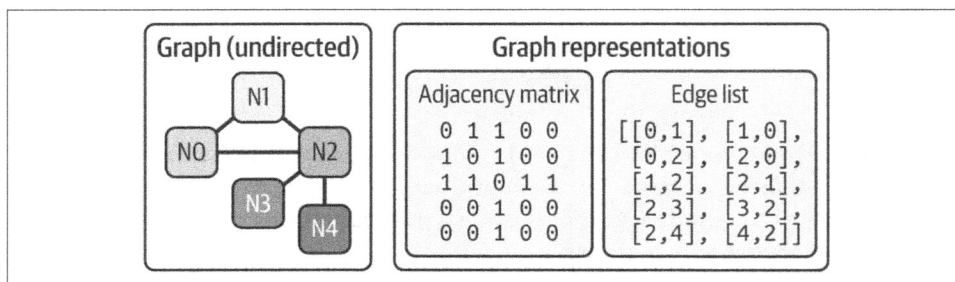

Figure 4-3. Visual representation of an undirected graph. The same graph is represented as an adjacency matrix and as a bidirectional edge list, where each undirected edge is shown in both directions. In practice, many GNN libraries require such bidirectional edge lists. Self-edges (not shown here) are also often included by default to help preserve node identity. Each node is often associated with a feature vector (not shown), but not always (as in this chapter).

We can numerically represent the graph structure in two main ways:

Adjacency matrix
> Each node is listed along the rows and columns of a matrix, with edges indicated by values in the corresponding cells.

Edge list
> Each row in this list represents an edge by specifying its start and end nodes.

The choice of representation impacts memory usage, especially depending on graph sparsity. An adjacency matrix has fixed high memory usage, as it accounts for all possible edges that could exist, while an edge list is more compact because it only stores the edges that exist. For sparse graphs, where the number of edges is much smaller than the total possible, an edge list is typically much more memory efficient.

Graph Neural Networks

With this foundational understanding of graphs, we can explore how graph neural networks learn from them. GNNs are a class of models designed to operate directly on graph structures, capturing information from both nodes and their connections. At a high level, GNNs work by iteratively aggregating information from a node's neighbors, producing rich representations (embeddings) that reflect both the node's features and its position within the broader graph structure. We'll break down this process in more detail shortly—but first, why do we need GNNs in the first place? What kinds of graph-related problems can they solve?

GNNs are commonly used for these main tasks:

Node classification
  Predicting the type or property of a node within a graph; for example, determining the category of a drug within a DDI network (e.g., antidepressant, antihistamine, or antibiotic).

Edge classification
  Predicting the type or existence of a connection between two nodes; for example, determining whether two drugs are likely to interact.

Edge regression
  Estimating a continuous value for a connection between nodes. In the context of DDI networks, this could involve predicting the severity or strength of an interaction rather than just its presence or type.

Graph classification
  Predicting a property of an entire graph; for example, identifying whether a drug molecule, represented as a graph of atoms and bonds, has a specific property, such as being water soluble or binding to a specific disease-associated protein.

These tasks all rely on the GNN's ability to extract meaningful representations from the graph structure. Whether the goal is to classify nodes, predict edges, or assess whole-graph properties, the core mechanism remains the same: learning expressive embeddings through iterative information exchange. This leads us to the central idea behind most GNN architectures: *message passing*.

Graph Embeddings and Message Passing

A primary goal in GNNs is to learn the structure around each node by generating a per-node embedding vector that captures information from its neighborhood. Unlike in images, where pixels have a fixed spatial arrangement, graph connections lack inherent order, making traditional convolutional approaches less applicable.

To address this, modern GNNs use a framework known as *message passing*, where each node iteratively exchanges messages with its neighbors and aggregates their information to update its own representation. This idea was formalized in the Message Passing Neural Network (MPNN) framework by Gilmer *et al.*, which has become the foundation for many contemporary GNN architectures.[3] Earlier forms of GNNs were introduced by Scarselli *et al.*, who proposed recursive neural models for learning on graphs, though without the modular message-passing abstraction seen today.[4]

Message passing is a flexible framework that underpins many GNN models. It often refers to the interaction between *sender* and *receiver* nodes, where the sender transmits information and the receiver aggregates it to update its own representation. *Graph convolution* is one specific implementation of message passing, where nodes aggregate information from their neighbors using functions such as summation, mean, or max. In contrast, nonconvolutional approaches such as graph attention networks (GATs) use attention mechanisms to assign different weights to neighbors based on their relative importance. The choice of aggregation function—whether sum, mean, max, or attention—affects the types of patterns the GNN can learn.

Increasing the number of message-passing layers (i.e., the number of hops a node can "see") expands each node's receptive field, enabling it to incorporate information from more distant parts of the graph. However, deeper GNNs can run into two key challenges:

Over-smoothing
> As the number of message-passing layers increases, each node incorporates information from a broader neighborhood. While this can be beneficial up to a point, stacking too many layers causes node embeddings to become increasingly similar—eventually collapsing to near-identical representations regardless of a node's local structure or features. This degrades the model's ability to distinguish between nodes, especially in classification tasks where fine-grained differences can be very important.

Over-squashing
> When long-range information must pass through a limited number of intermediate nodes or edges, it becomes overly compressed. This bottleneck prevents distant signals from being accurately preserved—especially in graphs with long, narrow paths, such as trees or hierarchies often seen in biology (e.g., gene

---

3 Gilmer et al., "Neural Message Passing for Quantum Chemistry," *Proceedings of the 34th International Conference on Machine Learning* (Sydney, NSW, Australia), vol. 70 (2017): 1263–1272, JMLR.org.

4 F. Scarselli et al., "The Graph Neural Network Model," *IEEE Transactions on Neural Networks* 20, no. 1 (2009): 61–80.

regulatory networks or phylogenetic trees). As a result, important context from far-apart nodes gets "squashed" before it can meaningfully influence predictions.

To mitigate these issues, common strategies include adding skip connections, incorporating attention mechanisms, or rewiring the graph to shorten path lengths. In practice, using just two or three message-passing layers—capturing information from a few hops away—often provides a good balance between expressivity and stability.

In GNNs, "layers" are better understood as message-passing iterations rather than traditional neural network layers. Unlike MLPs, where each layer applies a distinct learned transformation, each GNN layer aggregates information from a node's immediate neighbors. Thus, a model with three layers enables each node to incorporate information from up to three hops away in the graph.

Cold-Start Problem

A significant challenge in GNNs is predicting on *unseen nodes*, often referred to as the *cold-start problem*. Many traditional graph models operate in a *transductive* setting, where training occurs on a fixed graph, limiting predictions to relationships among nodes seen during training.

However, real-world applications often involve dynamic graphs where new nodes are introduced. For example:

- In a social network, a new user joins, and the platform needs to predict their potential connections.

- In a recommendation system, a newly released product must be matched to relevant users based on their preferences.

- In drug discovery, a newly synthesized compound must be evaluated for interactions with existing molecules.

To address the cold-start problem, GNNs can adopt an *inductive learning* approach, enabling generalization to new, unseen nodes. This capability is essential for dynamic graphs where new nodes are frequently added, as it eliminates the need to retrain the model whenever the graph changes. It is achieved by learning patterns and relationships that are transferable across the graph. For example:

- Instead of memorizing specific connections, the model identifies structural similarities (e.g., the role of a node in its local neighborhood) or shared features (e.g., common attributes across nodes).

- When a new node is added, its feature vector and immediate connections to existing nodes provide enough context for the model to embed it within the graph and make predictions.

Notable frameworks like GraphSAGE focus on inductive learning by sampling neighborhoods and aggregating local features to generate embeddings for unseen nodes. Techniques such as feature propagation and attention mechanisms further enhance this capability, making GNNs highly adaptable to evolving, real-world graphs.

> For a more detailed (but highly accessible) introduction to GNNs, we recommend the excellent lecture "Theoretical Foundations of Graph Neural Networks" (*https://oreil.ly/GXy3v*) by Petar Veličković on YouTube. It offers clear GNN explanations from one of the leading experts in the field.

GraphSAGE

In this chapter, we implement a GraphSAGE model,[5] an inductive approach that can predict the properties of nodes it has never seen before by aggregating information from their neighbors. In the original paper, GraphSAGE was evaluated on tasks like classifying academic papers into six biology-related categories using citation graphs, assigning Reddit posts to 50 communities based on user interactions, and predicting protein functions across multiple protein–protein interaction graphs. These benchmarks demonstrated GraphSAGE's ability to generalize to unseen nodes and outperform traditional methods, highlighting its versatility in dynamic, real-world graphs.

A key advantage of GraphSAGE is its scalability to massive graphs. Training on large graphs can be resource intensive because embedding updates for each node requires iterating over its neighbors. GraphSAGE addresses this challenge by using *subsampling*, where only a small, fixed number of neighbors is sampled for each node. These subgraphs are processed in mini-batches, significantly reducing memory and computation costs.

As illustrated in Figure 4-4 (from the original paper (*https://oreil.ly/wz_mG*)), GraphSAGE has two main components: sampling a subgraph and aggregating neighborhood information for each node. The resulting embeddings can be used for downstream tasks such as node classification or link prediction. While GraphSAGE can incorporate edge or node annotations, it does not depend on them, and for most of this chapter, we will focus solely on the graph structure.

---

5 Hamilton, W. L., Ying, Z., & Leskovec, J. (2017). Inductive representation learning on large graphs (*https://oreil.ly/8mNRd*). *Neural Information Processing Systems*, 30, 1024–1034.

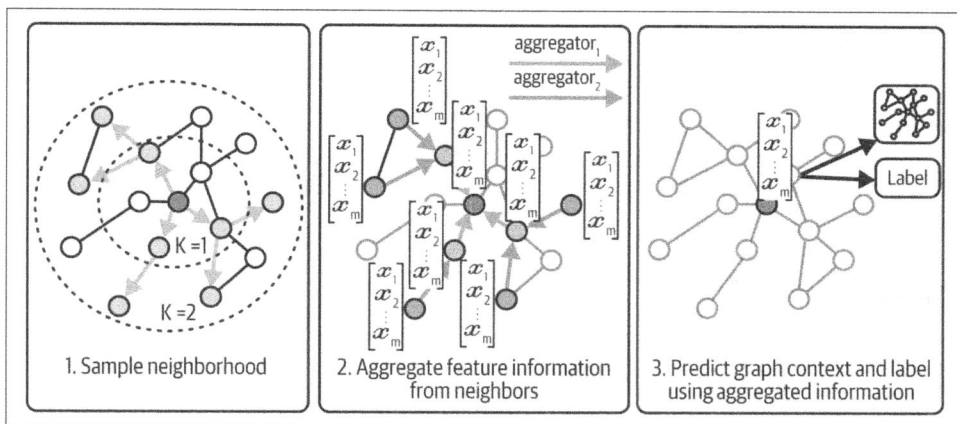

Figure 4-4. GraphSAGE stands for Graph SAmple and AggreGatE, representing its two main steps: (1) sampling a node's neighbors and (2) aggregating their features to generate an embedding. These embeddings can be used for downstream tasks, such as (3) predicting node properties or relationships within the graph.

Selecting a Dataset

In this chapter, we'll work with a unique data source: the Open Graph Benchmark (OGB) dataset of processed DrugBank DDIs called `ogbl-ddi` (*https://oreil.ly/WWr52*). This dataset is particularly convenient for two reasons:

1. It is well studied, providing a wealth of existing research to draw inspiration from.

2. It enables us to compare our model's performance with other approaches using the leaderboard (*https://oreil.ly/VPv1R*).

Additionally, OGB simplifies the workflow by offering built-in data loaders compatible with various deep learning frameworks and an `Evaluator` class for computing problem-specific metrics. This allows us to focus on building and refining our model rather than spending a long time on data preparation.

Describing the Dataset

We have already discussed DDI networks in general. The OGBL DDI dataset in particular is an unweighted, undirected graph of DDIs, where each node is an FDA-approved or experimental drug and edges represent either beneficial or harmful interactions between drugs.

To make the problem more challenging, the dataset is split in an interesting way—by the proteins that each drug targets. This "protein–target split" ensures that the test set contains drugs that primarily bind to different proteins than those in the training and validation sets, meaning they are more likely to operate through distinct biological mechanisms. This forces the model to learn more generalizable biology. If we created our own split—such as a random split of drugs—there would likely be greater overlap in biological mechanisms between the training and test sets, which would make the problem easier but would ultimately reduce the model's ability to generalize to unseen drugs in real-world scenarios.

Exploring the Dataset

As always, let's start by doing some exploratory analysis of the dataset to get a feel for what we're dealing with. We start by loading the data:

```
from ogb.linkproppred import LinkPropPredDataset

from dlfb.utils.context import assets

# Quite a large graph, may take a few minutes to load.
dataset = LinkPropPredDataset(name="ogbl-ddi", root=assets("graphs/datasets"))
```

This downloads the `ogbl-ddi` dataset and neatly packs it into an object ready for inspection. The full graph is accessible with `.graph`:

```
dataset.graph
```

Output:

```
{'edge_index': array([[4039, 2424, 4039, ...,  338,  835, 3554],
       [2424, 4039,  225, ...,  708, 3554,  835]]),
 'edge_feat': None,
 'node_feat': None,
 'num_nodes': 4267}
```

The graph is stored in *edge-list* format under the key `edge_index`. Both `edge_feat` and `node_feat` are `None`, meaning the graph includes only the structure—without additional edge features such as interaction strengths or node features such as drug properties. Next, let's examine the number of nodes and edges in the graph:

```
print(
    f'The graph contains {dataset.graph["num_nodes"]} nodes and '
    f'{dataset.graph["edge_index"].shape[1]} edges.'
)
```

Output:

```
The graph contains 4267 nodes and 2135822 edges.
```

We can plot the *degree distribution*, or the distribution of the number of connections per node, to get a sense of the high-level graph structure (depicted in Figure 4-5):

```
import matplotlib.pyplot as plt
import numpy as np
```

```
import seaborn as sns

degrees = np.bincount(dataset.graph["edge_index"].flatten())

sns.histplot(degrees, kde=True)
plt.xlabel("Degree")
plt.ylabel("Frequency")
plt.title("Degree Distribution");
```

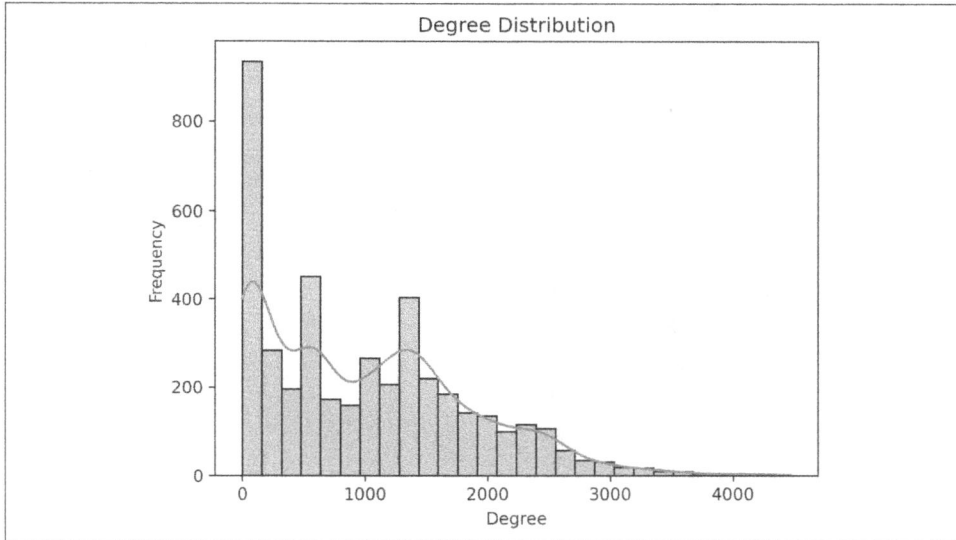

Figure 4-5. The degree distribution of nodes in the DDI network follows a power-law distribution where a few drugs interact with many others, but the majority is more isolated.

We observe that a few drugs act as *hubs*, exhibiting a high degree of interaction with many other drugs, while most drugs have a low degree, interacting with only a few other drugs. This pattern is consistent with a *power-law distribution* (*https://oreil.ly/pcVKt*), commonly seen in biological and social networks, where a small number of elements have very high connectivity (hubs) while the majority have low connectivity. However, it is important to note that this characteristic might be specific to this dataset and may not generalize to all DDI networks.

We can compute the *density* of the graph, or the ratio of edges to the number of possible edges, to quantify how densely interconnected our graph is:

```
num_nodes = dataset.graph["num_nodes"]
num_observed_edges = dataset.graph["edge_index"].shape[1]

# Since each edge in an undirected graph can be represented in two ways, we
# multiply by 2 to account for the bidirectionality.
num_observed_edges = 2 * num_observed_edges
```

```
# For any graph with n nodes, the maximum number of edges (assuming no
# self-loops) is n * (n-1).
num_possible_edges = num_nodes * (num_nodes - 1)

density = num_observed_edges / num_possible_edges

print(
    f"There are {num_observed_edges} observed edges and {num_possible_edges} "
    f"possible edges,\ngiving a graph density of {round(density, 2)}"
)
```

Output:

```
There are 4271644 observed edges and 18203022 possible edges,
giving a graph density of 0.23
```

This shows that while the dataset contains a seemingly large number of edges, it is not extremely dense, as 77% of possible connections are absent. With a density of 23%, the graph might be considered moderately interconnected, though this label is a bit subjective and depends on the specific context.

The dataset comes with its own methods to extract useful information. For example, .get_edge_split will list the graph's edges across the different data splits:

```
data_split = dataset.get_edge_split()
data_split
```

Output:

```
{'train': {'edge': array([[4039, 2424],
        [4039,  225],
        [4039, 3901],
        ...,
        [ 647,  708],
        [ 708,  338],
        [ 835, 3554]])},
  'valid': {'edge': array([[ 722,  548],
        [ 874, 3436],
        [ 838, 1587],
        ...,
        [3661, 3125],
        [3272, 3330],
        [1330,  776]]),
   'edge_neg': array([[   0,   58],
        [   0,   84],
        [   0,   90],
        ...,
        [4162, 4180],
        [4168, 4260],
        [4180, 4221]])},
  'test': {'edge': array([[2198, 1172],
        [1205,  719],
        [1818, 2866],
```

```
      ...,
      [ 326, 1109],
      [ 911, 1250],
      [4127, 2480]]]),
  'edge_neg': array([[   0,    2],
      [   0,   16],
      [   0,   42],
      ...,
      [4168, 4259],
      [4208, 4245],
      [4245, 4259]])}}
```

We can see that the valid and test splits actually contain two types of edges:

- The edge key holds the positive data, representing known drug interactions. Here, *positive* refers to the fact that these interactions are known, not whether they are beneficial or harmful.

- The edge_neg key contains negative edges, representing drug pairs with no known interactions. However, because some interactions may simply be undiscovered, this data is considered *weakly labeled* and may include inaccuracies (false negatives).

Importantly, the training dataset does not include explicit negative edges (i.e., there's no predefined edge_neg list in train). However, since most node pairs in a sparse graph are unconnected, negative samples can be drawn from this large set of nonedges during training. The method used to sample these negatives is an important hyperparameter, as it can significantly affect performance. Some negative edges are trivially easy to distinguish, which can lead to inflated metrics. In contrast, the validation and test datasets include a predefined edge_neg key that specifies which unconnected node pairs to use for evaluation.

Let's now examine the relative sizes of the train, valid, and test splits:

```
print(
    f'Number of edges in train set: {data_split["train"]["edge"].shape[0]}\n'
    f'Number of edges in valid set: {data_split["valid"]["edge"].shape[0]}\n'
    f'Number of edges in test set: {data_split["test"]["edge"].shape[0]}'
)
```

Output:

```
Number of edges in train set: 1067911
Number of edges in valid set: 133489
Number of edges in test set: 133489
```

In this dataset, the training set contains roughly 10 times more positive edges than the validation and test sets.

Another important consideration is whether all nodes in the validation and test sets also appear in the training set. This determines whether the model will encounter completely unseen nodes during evaluation—a key distinction between transductive and inductive learning. You can check this with the following code:

```
train_nodes = np.unique(data_split["train"]["edge"])
valid_nodes = np.unique(data_split["valid"]["edge"])
test_nodes = np.unique(data_split["test"]["edge"])

# Check if all nodes in valid and test sets are present in train set.
valid_in_train = np.isin(valid_nodes, train_nodes).all()
test_in_train = np.isin(test_nodes, train_nodes).all()

print(f"All validation nodes are in training nodes: {valid_in_train}")
print(f"All test nodes are in training nodes: {test_in_train}")
```

Output:

```
All validation nodes are in training nodes: True
All test nodes are in training nodes: True
```

In our case, all nodes in the validation and test sets are indeed present in the training graph. This defines a transductive setting, where the model sees all nodes during training and only needs to predict whether specific edges exist between them. This setup is simpler than the inductive case, where the model must make predictions involving entirely unseen nodes.

Starting with transductive evaluation allows us to assess model performance in a controlled setting before tackling the more complex inductive scenario. Models like GraphSAGE are well suited to inductive tasks because they generate node embeddings based on local neighborhoods. This means that even unseen nodes can be embedded meaningfully, provided they connect to known parts of the graph.

For now, we'll focus on the transductive case and ensure that the model performs well when all nodes are known.

Examining Drug Names

Although not immediately available in the graph object, there is additional annotation data that comes with the `ogbl-ddi` dataset. Let's examine this information:

```
import pandas as pd

ddi_descriptions = pd.read_csv(
    assets("graphs/datasets/ogbl_ddi/mapping/ddi_description.csv.gz")
)
print(ddi_descriptions)
```

Output:

```
           first drug id    first drug name  second drug id    second drug name  \
0                DB00001            Lepirudin        DB06605             Apixaban
1                DB00001            Lepirudin        DB06695  Dabigatran etexi...
2                DB00001            Lepirudin        DB01254            Dasatinib
...                  ...                  ...            ...                  ...
2669761          DB15657      Ala-geninthiocin       DB14055          (S)-Warfarin
2669762          DB15657      Ala-geninthiocin       DB00581            Lactulose
2669763          DB15657      Ala-geninthiocin       DB14443  Vibrio cholerae ...

                     description
0            Apixaban may inc...
1            Dabigatran etexi...
2            The risk or seve...
...                          ...
2669761      The risk or seve...
2669762      The therapeutic ...
2669763      The therapeutic ...

[2669764 rows x 5 columns]
```

We can see that each row is a DDI, with each drug having an ID (an accession in the DrugBank database (*https://oreil.ly/CfISy*)) and a description of the nature of the interaction.

> This chapter's dataset is ultimately derived from DrugBank (*https://oreil.ly/MYb6A*), which provides extensive information about drugs and their interactions. While some of this information is included in the benchmark dataset, much more could be added, such as chemical properties, target genes, and other drug-specific details. However, access to the full DrugBank resource is not free for nonacademic users.

When working with our graph, we will mostly be dealing with node indices, but we can always look up the mapping between node ID and DrugBank drug IDs:

```
node_to_dbid_lookup = pd.read_csv(
    assets("graphs/datasets/ogbl_ddi/mapping/nodeidx2drugid.csv.gz")
)
print(node_to_dbid_lookup)
```

Output:

```
      node idx   drug id
0            0   DB00001
1            1   DB00002
2            2   DB00004

...        ...       ...
4264      4264   DB15617
4265      4265   DB15623
4266      4266   DB15657
```

```
[4267 rows x 2 columns]
```

This lookup allows us to look a bit deeper at the degree distribution observation from earlier. What are the drugs that bind many other drugs? Let's examine the drugs with the highest number of edges. Since all but 14 drug interactions are represented twice in this dataframe (once as *A-B* and once as *B-A*), we can count on the first drug name column to get the most frequently binding drugs:

```
ddi_descriptions["first drug name"].value_counts().head(10)
```

Output:

```
first drug name
Quinidine          2477
Chlorpromazine     2431
Desipramine        2345
Amitriptyline      2338
Clozapine          2324
Doxepin            2273
Clomipramine       2269
Haloperidol        2269
Carbamazepine      2267
Imipramine         2260
Name: count, dtype: int64
```

Figure 4-6 visualizes the structure of these top interacting drugs. Interestingly, many of these drugs, such as desipramine, amitriptyline, and clomipramine, share a common three-ring (tricyclic) core structure, which may contribute to their similar interaction profiles.

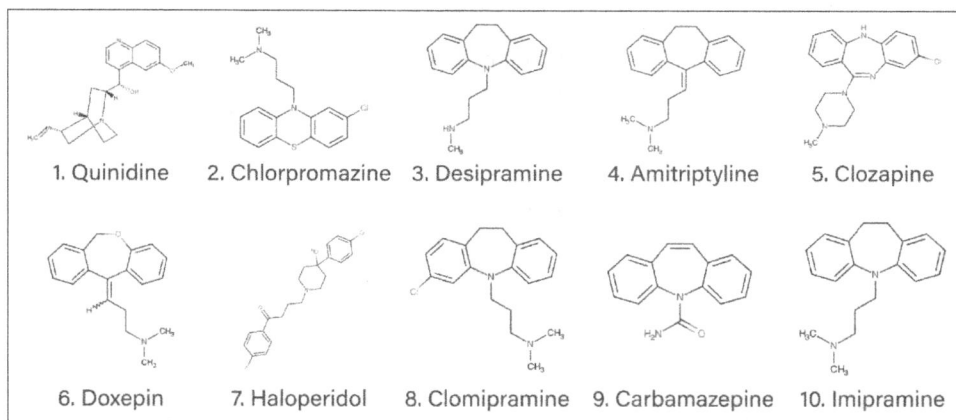

Figure 4-6. Chemical structures of the top 10 drugs with the highest number of drug–drug interactions in the dataset. Interestingly, many of these drugs, such as desipramine, amitriptyline, and clomipramine, share a common three-ring (tricyclic) core structure, which may contribute to their similar interaction profiles by promoting broad target binding and extensive metabolism via cytochrome P450 enzymes. Structures were acquired from DrugBank.

This list of drugs may seem a bit obscure if you're not accustomed to memorizing drug names, but there are a few emergent patterns here:

Affecting transporter proteins

The drug with the highest number of interactions (2,477) is quinidine, used to treat certain heart arrhythmias. Like other drugs on this list, such as clozapine and carbamazepine, quinidine interacts strongly with transporter proteins (with the most famous one being a protein called P-glycoprotein), which regulate the absorption and transport of many drugs across cells. This broad influence on drug levels largely explains its high interaction count in this dataset.

Affecting drug metabolism

Many of these drugs, like the antidepressants (desipramine, amitriptyline, clomipramine, imipramine), the antipsychotics (chlorpromazine, clozapine, haloperidol), and the mood stabilizer carbamazepine, are metabolized by the cytochrome P450 enzyme family in the liver. This system, introduced earlier, plays a major role in drug metabolism and is central to many drug interactions, because drugs that inhibit or activate cytochrome P450 enzymes can alter the metabolism of other drugs taken simultaneously.

Dosage sensitivity

Finally, these top interacting drugs also tend to have narrow *therapeutic ranges*, meaning even small changes in blood concentrations can lead to adverse effects. This makes interactions more likely to occur and be noticed.

From this additional table of drug information, we can construct a lookup table of node_id to DrugBank dbid to drug names, allowing us to bring more biological context to our project as we start modeling:

```
first_drug = ddi_descriptions[["first drug id", "first drug name"]].rename(
  columns={"first drug id": "dbid", "first drug name": "drug_name"}
)
second_drug = ddi_descriptions.loc[
  :, ["second drug id", "second drug name"]
].rename(columns={"second drug id": "dbid", "second drug name": "drug_name"})
dbid_to_name_lookup = (
  pd.concat([first_drug, second_drug]).drop_duplicates().reset_index(drop=True)
)

drugs_lookup = pd.merge(
  node_to_dbid_lookup.rename(
    columns={"drug id": "dbid", "node idx": "node_id"}
  ),
  dbid_to_name_lookup,
  on="dbid",
  how="inner",
)

drugs_lookup.iloc[935]
```

Output:

```
node_id           935
dbid          DB01043
drug_name    Memantine
Name: 935, dtype: object
```

For example, using this lookup table, we can see that node ID 935 corresponds to the drug memantine, which has the DrugBank ID DB01043.

Visualizing Graphs

Now let's take a look at what a portion of this graph data actually looks like. The entire graph is too large to meaningfully visualize all at once, but we can sample a subgraph and visualize that. The strategy here is to select a subset of nodes from the original training graph and then subset the split dataset by these nodes:

```
import numpy as np

np.random.seed(42)


def get_subgraph(edges: np.ndarray, node_limit: int) -> np.ndarray:
    """Gets a subgraph by sampling nodes and their edges."""
    nodes = np.unique(edges)
    sampled_nodes = np.random.choice(nodes, size=node_limit, replace=False)
    filtered_edges = edges[
```

```
      np.isin(edges[:, 0], sampled_nodes) & np.isin(edges[:, 1], sampled_nodes)
   ]
   print(f"Subgraph has {filtered_edges.shape[0]} edges")
   return filtered_edges


# Sample 50 nodes from the training set.
subgraph = get_subgraph(node_limit=50, edges=data_split["train"]["edge"])
```

Output:

```
Subgraph has 152 edges
```

This extracts a subgraph containing 50 nodes from the training set. We can visualize it in Figure 4-7 using the `plot_ddi_graph` function, which leverages the popular `networkx` library—a widely used Python tool for creating and visualizing graph structures:

```python
import networkx as nx
from adjustText import adjust_text


def plot_ddi_graph(graph: np.ndarray, drugs_lookup: pd.DataFrame) -> plt.Figure:
    """Plots a drug-drug interaction graph with labeled nodes."""
    fig = plt.figure(figsize=(15, 15))
    G = nx.Graph()
    G.add_edges_from(graph)
    pos = nx.spring_layout(G)
    nx.draw(
        G=G,
        pos=pos,
        with_labels=False,
        node_color="lightgray",
        edge_color="gray",
        node_size=10,
        alpha=0.3,
    )
    names = (
        drugs_lookup[drugs_lookup["node_id"].isin(G.nodes)]
        .set_index("node_id")["drug_name"]
        .to_dict()
    )
    labels = nx.draw_networkx_labels(G=G, pos=pos, labels=names, font_size=20)
    adjust_text(list(labels.values()))
    return fig


plot_ddi_graph(subgraph, drugs_lookup);
```

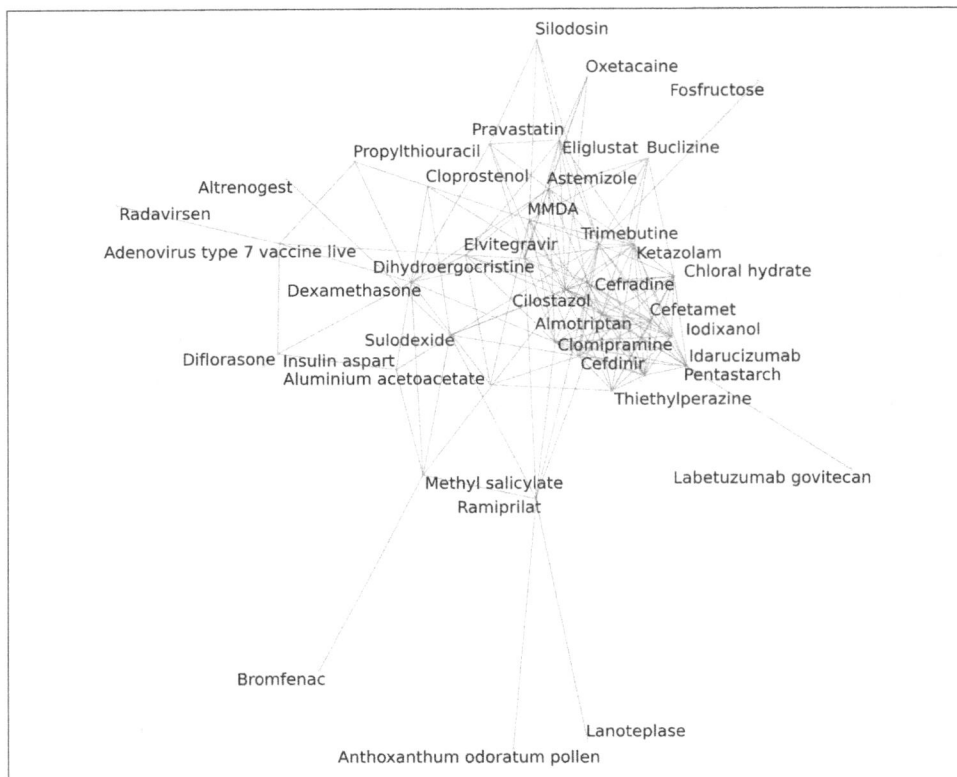

Figure 4-7. A sampled subgraph of the DDI network, with nodes labeled by drug names. While this graph was sampled to just 50 nodes for clarity, the visualization already highlights the diversity of interactions, including densely connected drugs and more isolated drugs.

Figure 4-7 highlights the diversity of interactions, including densely connected clusters (e.g., around Clomipramine, an antidepressant) and isolated or sparsely connected drugs. While this graph was sampled for clarity, it illustrates how certain drugs act as hubs, reflecting their broad interaction profiles, and others interact more selectively, potentially due to specific biological mechanisms.

With this initial data exploration complete, we're ready to move on to building the dataset.

Building a Dataset

Having explored the dataset from `LinkPropPredDataset`, we now turn our attention to the process of preparing it for use in the JAX/Flax framework. Although the dataset isn't out-of-the-box compatible, this offers a valuable opportunity to better understand the intricacies of graph processing. In this section, we'll walk through the necessary adjustments to ensure that the dataset is properly formatted for our model.

Fortunately, we don't have to start from scratch. The JAX ecosystem has `jraph`, a graph library that offers foundational, graph-aware classes and data structures, allowing us to build flexible graph processing models while benefiting from JAX's speed and efficiency. If you'd like to explore `jraph` in more detail before diving into our implementation, we recommend this excellent tutorial on graph nets with `jraph` from DeepMind.[6]

> PyTorch, particularly its extension library `pytorch-geometric`, is arguably the most comprehensive deep learning framework for working with graphs. It offers a robust toolkit that simplifies selecting graph models from a model zoo, handling efficient data loading, and working with convenient data classes. Datasets like OGBL have dedicated data loaders tailored for this framework. However, in this chapter, we are using `jraph`, as it integrates seamlessly with JAX, aligning better with our overall approach for the book.

Let's get started with building a dataset we can train models on. As mentioned, there are several ways to represent a graph, such as using an adjacency matrix or an edge list. Since we're using `jraph`, we go for the edge-list format, the default, which is much more memory efficient for sparser datasets like a DDI network.

Creating a Dataset Builder

We have packaged the dataset building into a class called `DatasetBuilder`. As we go along, you'll recognize many parts from the previous section where we explored the raw dataset. Let's go through it step-by-step, starting with the main method, `build`:

```python
def __init__(self, path):
    """Initializes the dataset builder with a path to the dataset."""
    self.path = path


def build(
    self,
    node_limit: int | None = None,
    rng: jax.Array | None = None,
    keep_original_ids: bool = False,
) -> dict[str, Dataset]:
    """Builds and returns a dictionary of dataset splits."""
    dataset_splits = {}
    n_nodes, split_pairs = self.download()
```

---

6 Google DeepMind. "Intro to Graph Neural Nets with JAX/Jraph" (*https://oreil.ly/r6MCp*).

```
annotation = self.prepare_annotation()

for name, split in split_pairs.items():
    pos_pairs, neg_pairs = split["edge"], split["edge_neg"]
    graph = self.prepare_graph(n_nodes, pos_pairs)
    pairs = self.prepare_pairs(graph, pos_pairs, neg_pairs)
    dataset_splits.update({name: Dataset(n_nodes, graph, pairs, annotation)})

if node_limit and (rng is not None):
    dataset_splits = self.subset(
        dataset_splits, rng, node_limit, keep_original_ids
    )

return dataset_splits
```

During instantiation, the builder receives a `path` to ensure that the dataset is stored in the specified location, eliminating the need to redownload it every time. The `build` method then generates a dictionary where the keys indicate data splits, each associated with a `Dataset` value. We'll examine the `Dataset` class in a little more detail shortly, but for now, think of it as a dataset bundle with convenience methods for easier handling during training. The parameters passed to `build` help with subsetting the graph, which we will also get into a little bit later.

Download the Raw Dataset

Looking at the `build` code, we can see that the raw dataset is first downloaded, leveraging the `LinkPropPredDataset` we saw before. Since the training split does not have negative pairs, we add a `neg_edges` key to simplify later handling:

```
def download(self) -> tuple[int, dict]:
    """Downloads the dataset and returns the number of nodes and edge splits."""
    raw = LinkPropPredDataset(name="ogbl-ddi", root=self.path)
    n_nodes = raw[0]["num_nodes"]
    split_pairs = raw.get_edge_split()
    split_pairs["train"]["edge_neg"] = None  # Placeholder for negative edges.
    return n_nodes, split_pairs
```

Prepare the Annotation

The dataset annotation is not directly useful when training our model in this project, but it is handy to have readily accessible to perform all sorts of sanity checks. You will recognize the implementation from before:

```
def prepare_annotation(self) -> pd.DataFrame:
    """Annotates nodes by mapping node IDs to database IDs and drug names."""
    ddi_descriptions = pd.read_csv(
        f"{self.path}/ogbl_ddi/mapping/ddi_description.csv.gz"
    )
    node_to_dbid_lookup = pd.read_csv(
        f"{self.path}/ogbl_ddi/mapping/nodeidx2drugid.csv.gz"
```

```
)
# Merge first and second drug descriptions into a single lookup.
first_drug = ddi_descriptions.loc[
  :, ["first drug id", "first drug name"]
].rename(columns={"first drug id": "dbid", "first drug name": "drug_name"})

second_drug = ddi_descriptions.loc[
  :, ["second drug id", "second drug name"]
].rename(
  columns={"second drug id": "dbid", "second drug name": "drug_name"}
)
dbid_to_name_lookup = (
  pd.concat([first_drug, second_drug])
  .drop_duplicates()
  .reset_index(drop=True)
)

# Merge with node-to-DBID lookup.
annotation = pd.merge(
  node_to_dbid_lookup.rename(
    columns={"drug id": "dbid", "node idx": "node_id"}
  ),
  dbid_to_name_lookup,
  on="dbid",
  how="inner",
)
return annotation
```

The annotation is the same for all the dataset splits; hence, we only need to prepare it once and assign it to the `Dataset`.

Prepare the Graph

Next we look at `prepare_graph`, one of the main functions of the dataset builder:

```
def prepare_graph(
  self, n_nodes: int, pos_pairs: jax.Array
) -> jraph.GraphsTuple:
  """Prepares a Jraph graph from positive edge pairs."""
  senders, receivers = self.make_undirected(pos_pairs[:, 0], pos_pairs[:, 1])
  graph = jraph.GraphsTuple(
    nodes={"gid": jnp.arange(n_nodes)},  # Optional global node ID.
    edges=None,
    senders=senders,
    receivers=receivers,
    n_node=jnp.array([n_nodes]),
    n_edge=jnp.array([len(senders)]),
    globals=None,
  )
  return graph
```

The `make_undirected` method ensures that the DDI graph is undirected, meaning the relationship between drugs *A-B* is equivalent to *B-A*. Since `jraph` does not offer a toggle between directed and undirected graphs, we need to represent all edges in both directions. This process, known as *symmetrizing* the graph, makes the adjacency relationships symmetric, effectively converting a directed graph into an undirected one. This transformation is applied across all dataset splits.

Practically speaking, implementing this transformation is straightforward. We start with the `pos_pairs` and add a corresponding set of edges where the sender and receiver nodes are swapped:

```
@staticmethod
def make_undirected(
    senders: jax.Array, receivers: jax.Array
) -> tuple[jax.Array, jax.Array]:
    """Makes an undirected graph by duplicating edges in both directions."""
    # Jraph requires undirected graphs to have both A->B and B->A edges
    # explicitly.
    senders_undir = jnp.concatenate((senders, receivers))
    receivers_undir = jnp.concatenate((receivers, senders))
    return senders_undir, receivers_undir
```

Next, we prepare the graph using the main parameters that `GraphsTuple` expects: `senders` and `receivers`, which define the edges by specifying the source and destination nodes. Each node or edge can be annotated (although they are not here), with node annotations stored in `nodes` and edge annotations in `edges`. In addition, `GraphsTuple` incorporates metadata such as `n_nodes` and `n_edges`, which indicate the number of nodes and edges, respectively, and `globals`, which can store graph-level information such as a unique graph identifier or aggregated features. While we won't use `globals` here, it remains available for scenarios requiring data applicable to the entire graph.

> You may wonder why we sometimes need to pass the number of nodes independently from the graph. Why can't this be inferred from the edges? This is because inferring node count from edges could miss isolated nodes, which have no connections to other nodes (i.e., no interactions with other drugs).

In general, `GraphsTuple` is a versatile data structure that can host data in various ways. Instead of having one `GraphsTuple` per data split, we could construct a single graph containing both training and evaluation datasets, using the `nodes` attribute to specify which set each node belongs to.

Prepare the Pairs

With the graph in place, we use the `prepare_pairs` method to obtain the drug–drug pairs—both positive and negative—that the model will classify as either connected or not:

```
def prepare_pairs(
    self, graph: int, pos_pairs: jax.Array, neg_pairs: jax.Array | None = None
) -> Pairs:
    """Prepares positive and negative edge pairs."""
    if neg_pairs is None:
        neg_pairs = self.infer_negative_pairs(graph)
    return Pairs(pos=pos_pairs, neg=neg_pairs)
```

For evaluation datasets, preparing pairs is straightforward, as we can directly use the positive and negative pairs provided by the OGBL dataset.

> You might wonder why we don't use the edges from the graph we just created to generate the positive pairs. The reason is that since we made the graph undirected, each positive pair is represented twice, which could lead to redundancy and errors during evaluation.

For the training dataset, preparing pairs is slightly more complex because negative pairs are not provided and must be inferred using the `infer_negative_pairs` method:

```
def infer_negative_pairs(self, graph: jraph.GraphsTuple) -> jax.Array:
    """Infers negative edge pairs in a graph."""
    # Initialize a matrix where all possible edges are marked as potential
    # negative edges (1).
    neg_adj_mask = np.ones((graph.n_node[0], graph.n_node[0]), dtype=np.uint8)

    # Mask out existing edges in the graph (set to 0).
    neg_adj_mask[graph.senders, graph.receivers] = 0

    # Use the upper triangular part of the matrix to avoid duplicate pairs and
    # self-loops.
    neg_adj_mask = np.triu(neg_adj_mask, k=1)
    neg_pairs = jnp.array(neg_adj_mask.nonzero()).T  # Extract indices.
    return neg_pairs
```

The method begins by constructing an adjacency matrix and initializing it with zeros using NumPy. It then marks all existing edges with ones. To identify negative edges, the method flips the matrix values so that connections become zeros and nonconnections become ones. Finally, it retains only the upper triangle (`triu`) of the matrix (excluding the diagonal) to avoid self-loops and duplicate pairs. The remaining nonzero entries are converted into an edge list of negative node pairs.

The resulting negative pairs far outnumber the positive pairs due to the graph's sparsity. This imbalance can be advantageous, as it provides more examples of negative edges to sample. However, as mentioned previously, the way negative pairs are sampled significantly affects performance, as some pairs are trivial to predict as being unconnected. We will need to carefully select a fair subset of negative pairs during training.

> Using an adjacency matrix approach assumes that it can fit into memory. If this is not feasible, alternative methods for generating negative node pairs include sampling a noncomprehensive subset or using efficient implementations that rely on sparse adjacency matrices.

The positive and negative pairs are then encapsulated in a `Pairs` data class, which we'll examine further during training. It is a simple data class that stores arrays of positive and negative pairs and includes utilities for subsampling pairs during learning and accessing pairs in batches.

Subsetting the Graph

To efficiently explore how a model learns from graphs, it is useful to be able to create a smaller subset of the dataset. This allows us to work with a more manageable graph size for experimentation. The subset method does exactly that:

```python
def subset(
    self,
    dataset_splits: dict[str, Dataset],
    rng: jax.Array,
    node_limit: int,
    keep_original_ids: bool = False,
) -> dict[str, Dataset]:
    """Creates subset of dataset splits by sampling a fixed number of nodes."""
    # Get a random subset of node_ids.
    node_ids = jax.random.choice(
        rng, dataset_splits["train"].n_nodes, (node_limit,), replace=False
    )

    # Subset every dataset split by the same node_ids.
    dataset_subset_splits = {}
    for name, dataset in dataset_splits.items():
        dataset_subset_splits[name] = dataset.subset(node_ids, keep_original_ids)

    return dataset_subset_splits
```

It selects a subset of nodes based on a specified `node_limit` and applies this subset consistently across all dataset splits (i.e., training, validation, test). By default, the subsetted graph renumbers the node IDs to create a smaller, compact graph.

However, you can retain the original node IDs from the full dataset by setting the `keep_original_ids` parameter to `True`.

The Dataset Class

Finally, we have all the pieces in place to create a `Dataset` class that will enable flexible exploration of the graph along with its annotations:

```
@dataclass
class Dataset:
    """Graph dataset with nodes, pairs, and optional annotations."""

    n_nodes: int
    graph: jraph.GraphsTuple
    pairs: Pairs
    annotation: pd.DataFrame = field(default_factory=pd.DataFrame)
```

Here, we show only the `Dataset` fields, but the class also provides several useful methods. Notably, it handles subsetting the graph and managing annotations consistently behind the scenes. While we encourage you to explore the code online to get a better sense of its functionality, understanding every detail isn't necessary to begin training.

Building a Prototype

Let's start simple. We need to build a model that predicts links between nodes. We'll do this using only the graph's connectivity—no node features or annotations. Surprisingly, this connectivity information alone can be quite powerful for learning which node pairs are likely to be connected.

While link prediction can be framed as a binary classification task (i.e., connection versus no-connection), it differs from typical classification problems in key ways. The input to the model is not a single node or feature vector, but a pair of nodes, and the prediction depends on their structural relationship in the graph.

Our prototype model will be made up of several key components, some defined directly within the model and others handled as part of the training process:

- Model components:
  - Neighborhood encoding: Generates node embeddings that reflect the local graph structure
  - Link prediction: Uses these embeddings to score the likelihood of a connection between node pairs

- Training components:
  - — Negative sampling: Selects unconnected node pairs to contrast with true edges
  - — Loss function: Computes the training signal to optimize model performance

We begin by focusing on the most crucial part: how we encode each node's local neighborhood.

Node Encoder

Arguably the most impactful choice for our model is how we encode the nodes' neighborhood. For this, we use a GraphSAGE-inspired implementation:

```python
class NodeEncoder(nn.Module):
    """Encodes nodes into embeddings using a two-layer GraphSAGE model."""

    n_nodes: int
    embedding_dim: int
    last_layer_self: bool
    degree_norm: bool
    dropout_rate: float

    def setup(self):
        """Initializes node embeddings, which cover the full graph's n_nodes."""
        self.node_embeddings = nn.Embed(
            num_embeddings=self.n_nodes,
            features=self.embedding_dim,
            embedding_init=jax.nn.initializers.glorot_uniform(),
        )

    @nn.compact
    def __call__(self, graph: jraph.GraphsTuple, is_training: bool) -> jax.Array:
        """Encodes the nodes of a graph into embeddings."""
        # Graph can be a subgraph and thus we use a subset of embeddings
        x = self.node_embeddings(graph.nodes["gid"])

        # First convolutional layer.
        x = SAGEConv(
            self.embedding_dim, with_self=True, degree_norm=self.degree_norm
        )(graph, x)
        x = nn.relu(x)
        x = nn.Dropout(rate=self.dropout_rate, deterministic=not is_training)(x)

        # Second convolutional layer.
        x = SAGEConv(
            self.embedding_dim,
            with_self=self.last_layer_self,
            degree_norm=self.degree_norm,
        )(graph, x)

        return x
```

The main input parameters to the module are:

n_nodes
: Defines the total number of nodes in the original graph.

embedding_dim
: Specifies the dimensionality of the node embeddings. This controls how richly the model can represent neighborhood information. Lower values (e.g., 16 or 32) may limit expressiveness, while higher values (e.g., 128 or 256) offer more capacity at the cost of increased computation. In practice, smaller graphs can support higher embedding dimensions, while larger graphs benefit from lower values here.

dropout_rate
: Sets the fraction of neurons randomly deactivated during training to reduce overfitting.

last_layer_self *and* degree_norm
: Configure aspects of the graph convolution behavior, which are described in more detail in the next section.

At the heart of GraphSAGE is the node_embeddings matrix. In the setup method, this is initialized as a learnable parameter using nn.Embed, with shape [n_nodes, embedding_dim]. The embeddings are initialized using the glorot_uniform method to promote stable training dynamics. During training, these embeddings are updated by aggregating information from each node's neighbors, gradually encoding higher-order structural patterns. The goal is for these embeddings to converge to representations that reflect the likelihood of connections between node pairs.

The main logic of the encoder is implemented in the __call__ method. One important (though hard-coded) design choice here is the number of SAGEConv layers, which defines how many rounds of message passing are applied:

- With one layer, each node aggregates information from its immediate neighbors.
- With two layers, each node can access information from neighbors up to two hops away.

Thus, the number of layers controls the receptive field of each node. Between layers, ReLU activation introduces nonlinearity, and dropout is applied for regularization to prevent overfitting.

Graph Convolution

We've now reached the core architectural component of our model: the SAGEConv layer. Let's dive into it:

```python
class SAGEConv(nn.Module):
    """GraphSAGE convolutional layer with optional self-loops."""

    embedding_dim: int
    with_self: bool
    degree_norm: bool

    @nn.compact
    def __call__(self, graph: jraph.GraphsTuple, x) -> jax.Array:
        n_nodes = self.get_n_nodes(graph)

        # Add self-loops if enabled.
        if self.with_self:
            senders, receivers = self._add_self_edges(graph, n_nodes)
        else:
            senders, receivers = graph.senders, graph.receivers

        # Aggregate node features from neighbors.
        if not self.degree_norm:
            x_updated = jraph.segment_mean(
                x[senders], receivers, num_segments=n_nodes
            )
        else:

            def get_degree(n):
                return jax.ops.segment_sum(jnp.ones_like(senders), n, n_nodes)

            x_updated = self.normalize_by_degree(x, get_degree(senders))
            x_updated = jraph.segment_mean(
                x_updated[senders], receivers, num_segments=n_nodes
            )
            x_updated = self.normalize_by_degree(x_updated, get_degree(receivers))

        # Combine node and neighbor embeddings by concatenation.
        combined_embeddings = jnp.concatenate([x, x_updated], axis=-1)

        return nn.Dense(self.embedding_dim)(combined_embeddings)

    @staticmethod
    def _add_self_edges(
        graph: jraph.GraphsTuple, n_nodes: int
    ) -> tuple[jax.Array, jax.Array]:
        """Adds self-loops to the graph."""
        all_nodes = jnp.arange(n_nodes)
        senders = jnp.concatenate([graph.senders, all_nodes])
        receivers = jnp.concatenate([graph.receivers, all_nodes])
        return senders, receivers

    @staticmethod
    def normalize_by_degree(x: jax.Array, degree: jax.Array) -> jax.Array:
        """Normalizes node features by the square root of the degree."""
        # We set the the degree to a minimum of 1.
```

```
    return x * jax.lax.rsqrt(jnp.maximum(degree, 1.0))[:, None]

@staticmethod
def get_n_nodes(graph):
    """Returns the number of nodes in the graph in a jittable way."""
    return tree.tree_leaves(graph.nodes)[0].shape[0]
```

When setting up a `SAGEConv` layer, we specify the embedding dimension (using `embedding_dim`), whether to add self-loops (`with_self`), and whether to apply degree normalization (`degree_norm`). The latter two options are optional, as their impact on model performance depends on the dataset's characteristics, such as size and connectivity patterns. Enabling or disabling these features can significantly influence model behavior.

Each `SAGEConv` layer performs the following key steps:

Optionally adds self-edges
Allows each node to consider its own embedding during aggregation

Aggregates neighborhood embeddings
Collects and averages the embeddings of each node's neighbors

Optionally normalizes by degree
Scales the contribution of neighbors based on their degree, reducing bias from highly connected nodes

Combines embeddings with neighbors
Merges the original node embeddings with the aggregated neighbor embeddings

> When reading the implementation of the `SAGEConv` convolutional layer, you might be wondering: *where's the convolution?*
>
> Unlike convolutional neural networks (CNNs) for images or sequences, where `nn.Conv` applies spatial filters over regular grids, graph neural networks define "convolution" more abstractly. In GNNs, convolution means aggregating information from a node's local neighborhood and combining it with its own embedding. This is implemented using operations like `segment_mean`, followed by a learnable transformation—typically via `nn.Dense`. So, while you won't see an explicit `nn.Conv` in `SAGEConv`, the dense layer at the end serves as the core trainable part of the "graph convolution."

It's worth noting that `SAGEConv` focuses solely on aggregation and linear transformation. It does not include nonlinear activations, batch normalization, or dropout—those are typically applied at the model level to allow for greater flexibility and reuse.

Adding self edges

In the `NodeEncoder`, we use two `SAGEConv` layers, and the first layer has the `with_self` parameter set to `True`. Adding self-loops ensures that a node's own embedding is included in the aggregation process during neighborhood updates (by including the node in its own list of senders). Without self-loops, a node's updated embedding would reflect only information from its neighbors, potentially biasing the representation away from its original identity. Including self-loops allows each node to contribute to its own update, balancing its existing features with neighborhood context.

This can be accomplished by adding ones to the diagonal of the adjacency matrix, effectively connecting each node to itself. In our implementation, the `with_self` and `last_layer_self` parameters control whether self-loops are included in the first and second `SAGEConv` layers, respectively.

> Even without explicit self-loops, models can preserve a node's identity by using skip connections. For example, in our implementation, we concatenate the original and updated embeddings using `combined_embeddings = jnp.concatenate([x, x_updated], axis=-1)`. This helps retain the node's original features. However, it differs from self-loops in that the original embedding is added after the neighborhood aggregation, not as part of it. Including self-loops ensures that the node's identity contributes directly to the aggregation step itself.

Aggregating the neighborhood

We now turn to the core operation of message passing: aggregating information from each node's neighborhood.

This is the first point where the graph structure is explicitly used. For each edge, the sender node's embedding (or features) is gathered and aggregated on a per-receiver basis. In other words, each receiver node updates its embedding by combining information from all its connected senders. The result is a new representation that reflects the structure and features of its local neighborhood.

In our implementation, we use `jraph.segment_mean` as the aggregation function, which computes the mean of sender embeddings for each receiver. Other common choices include sum, max, or even attention-weighted aggregation, as used in GAT-style models. The optimal aggregation method often depends on the graph's topology and the downstream task, so experimenting with different strategies can be valuable.

Normalizing by degree

Degree normalization ensures that all nodes, regardless of their connectivity, contribute more evenly during training. This can help stabilize optimization, avoiding exploding or vanishing gradients. However, excessive normalization may *over-smooth* node embeddings, potentially erasing finer details of a node's local structure.

In our implementation, we apply *symmetric degree normalization*, which works as follows:

- Messages are divided by the square root of the *sender's degree* before aggregation.
- The aggregated result is then divided by the square root of the *receiver's degree*.

This square-root scaling—common in various flavors of GNNs—balances message influence across nodes while avoiding unintended scaling effects.

Whether degree normalization improves performance depends on the dataset and its structural properties. It's especially helpful in graphs with a high degree of variability, where a few high-degree nodes might otherwise dominate message passing. It also improves stability in deeper GNNs by keeping signal magnitudes under control.

However, in graphs where raw connection strength or node centrality carries important meaning—such as physical interaction networks, citation graphs, or transportation systems—normalization may suppress informative signals. If time allows, it's often worth comparing normalized and unnormalized variants during model development.

Combining embeddings with neighborhoods

After aggregation (and optional normalization), we combine the updated node embeddings with the original embeddings—typically by concatenation. This produces a unified representation that captures both the node's initial features and the information gathered from its local neighborhood. This enrichment step ensures that embeddings reflect both individual identity and structural context.

Concatenating the original and aggregated embeddings doubles the feature dimensionality. To bring this back to the intended embedding size, the combined vector is passed through a fully connected `Dense` layer. This layer serves two key purposes:

Maintains consistent dimensionality
Projects the concatenated vector back to the original embedding size, ensuring compatibility with the next model layer

Learns better representations
Learns how to optimally fuse original and neighborhood features, enabling the model to refine what information to retain or emphasize

Because the `Dense` transformation is learnable, the model adapts over training to make the most effective use of both types of input.

> During model training, the graph structure remains fixed. Only the `node_embeddings` are updated, evolving over time as the model learns from the neighborhood aggregation and feature transformation process.

Link Prediction

We now use the learned node embeddings to predict whether a given pair of nodes is connected. If the embeddings have effectively captured the graph's structure, the model should be able to assign high scores to true edges and low scores to unrelated node pairs.

Let's look at how the embeddings are used in the `LinkPredictor` module:

```python
class LinkPredictor(nn.Module):
    """Predicts interaction scores for pairs of node embeddings."""

    embedding_dim: int
    n_layers: int
    dropout_rate: float

    @nn.compact
    def __call__(
        self,
        sender_embeddings: jax.Array,
        receiver_embeddings: jax.Array,
        is_training: bool,
    ) -> jax.Array:
        """Computes scores for node pairs."""
        x = sender_embeddings * receiver_embeddings  # Element-wise multiplication.

        # Apply MLP layers with ReLU activation and dropout.
        for _ in range(self.n_layers)[:-1]:
            x = nn.Dense(self.embedding_dim)(x)
            x = nn.relu(x)
            x = nn.Dropout(self.dropout_rate, deterministic=not is_training)(x)

        # Final output layer is a single neuron. Logit output used for binary link
        # classification.
        x = nn.Dense(1)(x)

        return jnp.squeeze(x)
```

The `LinkPredictor` takes a pair of node embeddings—one from the sender and one from the receiver—and estimates the likelihood of an edge between them. Here's how it works:

Combining embeddings

The sender and receiver embeddings are combined using element-wise multiplication. This operation captures the interaction between corresponding dimensions of each embedding, producing a fixed-size vector that reflects their pairwise compatibility.

Transforming representations

The combined vector is passed through a multilayer perceptron (MLP), consisting of several `Dense` layers with ReLU activation and dropout. These layers are defined by the `n_layers` and `embedding_dim` parameters and serve to learn increasingly abstract representations of the node pair interaction.

Output layer

The final layer is a single-neuron `Dense` layer that outputs a logit—an unnormalized score representing the likelihood of an edge. During training, this logit is passed through a sigmoid to produce a probability between 0 and 1.

The overall goal is for the model to learn to output high scores for true (positive) edges and low scores for negative edges, forming the basis for binary link prediction.

Drug–Drug Interaction Model

Let's now put everything together into a `DdiModel` that we can train:

```python
class DdiModel(nn.Module):
    """Graph-based model for predicting drug-drug interactions (DDIs)."""

    n_nodes: int
    embedding_dim: int
    dropout_rate: float
    last_layer_self: bool
    degree_norm: bool
    n_mlp_layers: int = 2

    def setup(self):
        """Initializes the node encoder and link predictor modules."""
        self.node_encoder = NodeEncoder(
            self.n_nodes,
            self.embedding_dim,
            self.last_layer_self,
            self.degree_norm,
            self.dropout_rate,
        )
        self.link_predictor = LinkPredictor(
            self.embedding_dim, self.n_mlp_layers, self.dropout_rate
        )

    def __call__(
        self,
```

```
    graph: jraph.GraphsTuple,
    pairs: dict,
    is_training: bool,
    is_pred: bool = False,
):
    """Generates interaction scores for node pairs."""
    # Compute node embeddings. The 'h' stands for hidden state or embedding.
    h = self.node_encoder(graph, is_training)

    if is_pred:
        scores = self.link_predictor(h[pairs[:, 0]], h[pairs[:, 1]], False)

    else:
        pos_senders, pos_receivers = pairs["pos"][:, 0], pairs["pos"][:, 1]
        neg_senders, neg_receivers = pairs["neg"][:, 0], pairs["neg"][:, 1]
        scores = {
            "pos": self.link_predictor(
                h[pos_senders], h[pos_receivers], is_training
            ),
            "neg": self.link_predictor(
                h[neg_senders], h[neg_receivers], is_training
            ),
        }
    return scores

def create_train_state(self, rng: jax.Array, dummy_input, tx) -> TrainState:
    """Initializes the training state with model parameters."""
    rng, rng_init, rng_dropout = jax.random.split(rng, 3)
    variables = self.init(rng_init, is_training=False, **dummy_input)
    return TrainState.create(
        apply_fn=self.apply, params=variables["params"], tx=tx, key=rng_dropout
    )

@staticmethod
def add_mean_embedding(embeddings: jax.Array) -> jax.Array:
    """Concatenates a mean embedding to the existing embeddings."""
    mean_embeddings = jnp.mean(embeddings, axis=0, keepdims=True)
    embeddings = jnp.concatenate([embeddings, mean_embeddings], axis=0)
    return embeddings
```

Let's walk through the main components of this code:

- setup method: Initializes two core submodules:

  — node_encoder: Generates node embeddings from the input graph

  — link_predictor: Scores node pairs based on their embeddings to predict the presence or absence of an edge

- __call__ method: Defines the forward pass of the model and supports both training and inference:

— The `node_encoder` computes embeddings h for all nodes in the graph. The use of h as a variable name follows a common convention, where h represents a "hidden state" or embedding.

— If `is_pred=False` (training mode):

— Positive pairs: The model passes sender and receiver embeddings through the `link_predictor` to estimate connection likelihood.

— Negative pairs: Similarly processed to estimate nonconnection scores.

— Returns a dictionary of predicted scores for both "pos" and "neg" pairs.

— And if `is_pred=True` (inference mode): Takes an arbitrary array of node pairs and returns predicted scores. This enables applying the model to new or unseen node pairs after training.

- `create_train_state` method: Sets up the training process:

— Initializes model parameters using dummy inputs and a random seed

— Constructs a `TrainState` object with the model's `apply_fn`, parameters, optimizer (`tx`), and dropout key for training

- `add_mean_embedding` static method: Appends a global mean embedding to the existing embedding matrix. This can be useful in downstream tasks where a graph-level summary representation is desired.

Together, these components define a full link prediction pipeline for drug–drug interaction graphs. Next, we'll look at how to train this model end to end.

Training the Model

From the previous sections, we have established how to prepare datasets and define a model. Now we'll proceed by creating instances of both before moving forward with training.

Create a Manageable Dataset

We will create a subset containing approximately 10% of the total graph data:

```
node_limit = 500
rng = jax.random.PRNGKey(42)
rng, rng_dataset = jax.random.split(rng, 2)

dataset_splits = DatasetBuilder(path=assets("graphs/datasets")).build(
  node_limit, rng_dataset
)
```

By reducing the dataset size, we create a graph that is easier to handle during initial experimentation. This smaller graph allows us to test the model architecture and training setup more efficiently.

We now have a graph with a set of positive and negative node pairs that we can learn from. The visualization in Figure 4-8 provides a high-level view of the training dataset's structure, where nodes represent drugs and edges represent interactions between them. The circular layout arranges all nodes around a circle, with edges connecting related nodes.

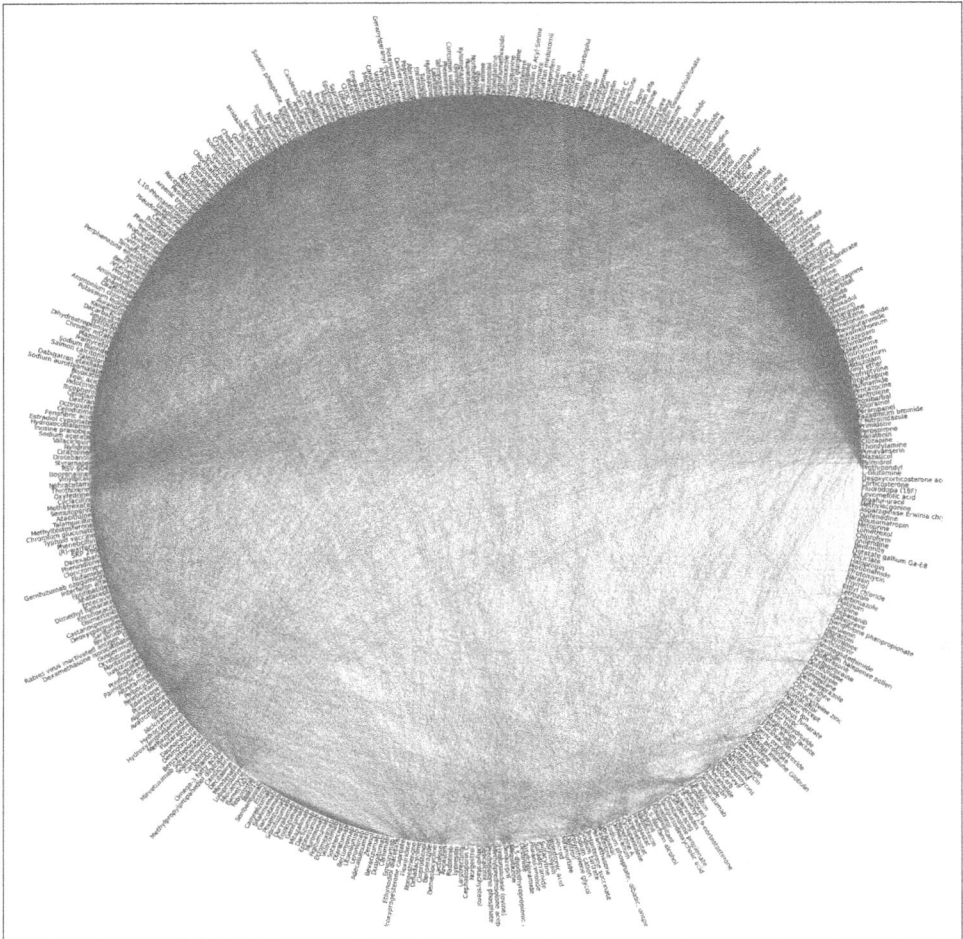

Figure 4-8. Circular layout of training dataset of 500 nodes. Each node represents a drug, and edges represent interactions between them.

While the individual node labels and details may not be legible in this plot, it offers a broad overview of key graph properties. These include the density of connections, overall sparsity, and presence of clusters or isolated nodes. This visualization illustrates the graph's complexity, despite being a small subsampled set.

```python
from dlfb.graphs.inspect import plot_graph

plot_graph(dataset_splits["train"]);
```

Create the Training Loop

Next, let's examine the training loop:

```python
@restorable
def train(
    state: TrainState,
    rng: jax.Array,
    dataset_splits: dict[str, Dataset],
    num_epochs: int,
    loss_fn: Callable,
    norm_loss: bool = False,
    eval_every: int = 10,
) -> tuple[TrainState, dict[str, dict[str, list[dict[str, float]]]]]:
    """Training loop for the drug-drug interaction model."""
    # Initialize metrics and estimate optimal batch sizes.
    metrics = MetricsLogger()
    batch_size = optimal_batch_size(dataset_splits)

    # Epochs with progress bar.
    epochs = tqdm(range(num_epochs))
    for epoch in epochs:
        epochs.set_description(f"Epoch {epoch + 1}")
        rng, rng_shuffle, rng_sample = jax.random.split(rng, 3)

        # Training loop.
        for pairs_batch in dataset_splits["train"].pairs.get_train_batches(
            batch_size, rng_shuffle, rng_sample
        ):
            rng, rng_dropout = jax.random.split(rng, 2)
            state, batch_metrics = train_step(
                state,
                dataset_splits["train"].graph,
                pairs_batch,
                rng_dropout,
                loss_fn,
                norm_loss,
            )
            metrics.log_step(split="train", **batch_metrics)

        # Evaluation loop.
        if epoch % eval_every == 0:
            for pairs_batch in dataset_splits["valid"].pairs.get_eval_batches(
```

```
      batch_size
    ):
      batch_metrics = eval_step(
        state, dataset_splits["valid"].graph, pairs_batch, loss_fn, norm_loss
      )
      metrics.log_step(split="valid", **batch_metrics)

  metrics.flush(epoch=epoch)
  epochs.set_postfix_str(metrics.latest(["hits@20"]))

return state, metrics.export()
```

The `train` function coordinates the entire training process, broken down into several key stages:

1. *Initialization*

   - Initializes a `MetricsLogger` to track training and evaluation metrics
   - Estimates an optimal batch size using the `optimal_batch_size` function
   - Displays training progress using a `tqdm` progress bar

2. *Training over epochs*

   For each epoch:

   - Sets up random number generators (`rng`) for shuffling and sampling training pairs.
   - The training set is iterated over in batches using `get_train_batches`, which samples positive and negative node pairs.
   - Each batch is passed to `train_step`, which updates model parameters based on the current loss.
   - Training metrics (e.g., loss and hits@20) are logged after each batch via `metrics.log_step`.

3. *Evaluation*

   - At intervals defined by `eval_every`, the model is evaluated on the validation set using `get_eval_batches` and `eval_step`.
   - Evaluation metrics are logged via `metrics.log_step`, and summary statistics are printed using `metrics.latest()`.

Additional features of this training loop include:

Support for custom loss functions

   A user-defined `loss_fn` can be passed in to control the optimization objective.

Optional loss normalization

Controlled by the `norm_loss` flag. When `True`, the loss is averaged over examples to ensure scale invariance across batch sizes.

Restorable training state

The function is decorated with `@restorable`, enabling checkpointing and resumption of training mid-run.

The dataset is processed in batches using the `Pairs` class, which handles consistent sampling of node pairs for both training and evaluation. We'll take a closer look at how this class works next.

Create the Pairs Class

The `Pairs` class is a utility that simplifies the handling of positive and negative node pairs during training and evaluation. You've already seen it in action within the training loop. Here, we'll break down its functionality more explicitly:

```python
@dataclass
class Pairs:
    """Represents positive and negative pairs of drug-drug interactions."""

    pos: jax.Array
    neg: jax.Array

    def get_eval_batches(
        self, batch_size: int
    ) -> Generator[dict[str, jax.Array], None, None]:
        """Generates evaluation batches of positive and negative pairs."""
        indices = jnp.arange(self._n_pairs())
        for i in range(self._n_batches(batch_size)):
            batch_indices = jnp.array(indices[i * batch_size : (i + 1) * batch_size])
            yield Pairs(
                pos=self.pos[batch_indices], neg=self.neg[batch_indices]
            ).to_dict()

    def _n_batches(self, batch_size: int) -> int:
        """Calculates number of batches in the dataset given a batch size."""
        return int(np.floor(self._n_pairs() / batch_size))

    def _n_pairs(self) -> int:
        """Returns the smaller number of positive or negative pairs."""
        return int(min(self.pos.shape[0], self.neg.shape[0]))

    def get_train_batches(
        self, batch_size: int, rng_shuffle: jax.Array, rng_sample: jax.Array
    ) -> Generator[dict[str, jax.Array], None, None]:
        """Generates shuffled training batches with sampled negative pairs."""
        # Shuffle indices for positive pairs.
        indices = jax.random.permutation(rng_shuffle, jnp.arange(self._n_pairs()))
```

```
# Get sample of negative pairs.
neg_sample = self._global_negative_sampling(rng_sample)

for i in range(self._n_batches(batch_size)):
  batch_indices = jnp.array(indices[i * batch_size : (i + 1) * batch_size])
  yield Pairs(
    pos=self.pos[batch_indices], neg=neg_sample[batch_indices]
  ).to_dict()

def _global_negative_sampling(self, rng_sample: jax.Array) -> jax.Array:
  """Samples negative pairs from the entire set to match positive set size."""
  return jax.random.choice(
    rng_sample, self.neg, (self.pos.shape[0],), replace=False
  )

def get_dummy_input(self) -> dict[str, jax.Array]:
  """Returns a small dummy subset of positive and negative pairs."""
  return Pairs(pos=self.pos[:2], neg=self.neg[:(2)]).to_dict()

def to_dict(self) -> dict:
  """Converts the Pairs object back to a dictionary."""
  return {"pos": self.pos, "neg": self.neg}
```

The class provides several key methods for batching and sampling:

get_eval_batches

Returns evaluation batches of positive and negative pairs, ensuring shape alignment and balanced sizes. It slices the pos and neg arrays using the same indices, up to the size of the smaller set.

get_train_batches

Returns shuffled training batches. Positive pairs are shuffled using rng_shuffle, and fresh negative pairs are sampled using _global_negative_sampling. This introduces variation between epochs and improves generalization.

get_dummy_input

Returns a very small batch (two positive and two negative pairs). This is useful to get the correct shape of data that we can use to initialize the model parameters.

Together, these methods enable consistent and efficient batch generation for both training and evaluation, while introducing enough variability to improve learning.

Batching by pairs

During each epoch, we must process a large number of positive and negative node pairs. As we've seen in the introduction to graph convolution layers, calculating node embeddings for large networks can become computationally expensive. To address this, we use batching—processing subsets of the data one at a time. This strategy is applied to both training and evaluation data, with some key differences:

- Training batches

  — The `get_train_batches` method provides batches of positive and negative pairs, shuffling the data at the start of every epoch to introduce diversity in the order of pairs.

  — Negative pairs are resampled once per epoch using the `_global_nega tive_sampling` method. This variation helps the model learn from a more diverse set of examples.

- Evaluation batches

  — The `get_eval_batches` method returns batches of positive and negative pairs according to the specified `batch_size`.

  — To ensure compatibility, the `_n_pairs` method limits the batch size to the smaller of the positive or negative sets, so the two arrays match in shape.

  — Evaluation batches are processed in a fixed order for reproducibility and consistent metrics across runs. This deterministic behavior simplifies debugging and ensures that order-sensitive metrics like Hits@K and MRR are stable.

To maintain uniform batch sizes, the `Pairs` class drops the final incomplete batch during both training and evaluation. This avoids irregularities in computation. Over multiple epochs, shuffling ensures that all data points are eventually seen, even if some are skipped in a single run.

To maximize efficiency, we use the `optimal_batch_size` utility function:

```python
def optimal_batch_size(
    dataset_splits: dict[str, Dataset], remainder_tolerance: float = 0.125
) -> int:
    """Calculates optimal batch size for optimizing JAX compilation."""
    # Calculate the minimum length of positive and negative pairs for each
    # dataset.
    lengths = [
        min(dataset.pairs.pos.shape[0], dataset.pairs.neg.shape[0])
        for dataset in dataset_splits.values()
    ]

    # Determine the allowable remainders per split based on the remainder
    # tolerance.
    remainder_thresholds = [
        int(length * remainder_tolerance) for length in lengths
    ]
    max_possible_batch_size = min(lengths)

    for batch_size in range(max_possible_batch_size, 0, -1):
```

```
    remainders = [length % batch_size for length in lengths]
    if all(
      remainder <= threshold
      for remainder, threshold in zip(remainders, remainder_thresholds)
    ):
      return batch_size
  return max_possible_batch_size
```

This function computes the largest batch size that minimizes dropped data and ensures consistency across training and evaluation. It balances computational efficiency and data utilization by selecting sizes that are both large and compatible with the dataset structure.

Consistent batch sizes are critical for optimizing jitted functions, which rely on static input shapes. They prevent unnecessary recompilation, improve memory and compute efficiency on accelerators like GPUs and TPUs, and reduce the complexity of handling irregular input.

Sampling negative pairs

An important aspect of training is how we sample negative pairs. Since there are many more pairs without connections than with connections, we cannot use all negative examples; doing so would create a highly imbalanced training dataset. Instead, we select a subset of negative pairs to balance the dataset. This is where `_global_negative_sampling` comes in.

The subset of negative samples can significantly impact training. In this implementation, we use the simplest approach: *global sampling*, where we uniformly sample from all possible negative pairs. This strategy is suitable when we are broadly interested in potential node connections across the entire graph:

```
def _global_negative_sampling(self, rng_sample: jax.Array) -> jax.Array:
    """Samples negative pairs from the entire set to match positive set size."""
    return jax.random.choice(
      rng_sample, self.neg, (self.pos.shape[0],), replace=False
    )
```

While global sampling is straightforward and effective, many alternative strategies exist that drive the model to learn different patterns. For example:

Local sampling
  Ensures that negative pairs share at least one sender node, focusing on pairs that are structurally similar to positive pairs. This can help the model learn more fine-grained distinctions.

Hard negative sampling
  Selects negative pairs that the model struggles to classify as negatives (i.e., pairs with a high predicted likelihood of being connected, even though they are not).

This approach forces the model to improve on challenging cases and can accelerate learning.

Adversarial negative sampling
Generates challenging negative pairs using an adversarial approach, where a secondary model selects negatives that maximize the main model's loss. While computationally expensive, it can lead to robust embeddings and improved performance.

Ratio of positive to negative pairs
Balances the number of positive and negative pairs in the dataset. While a 1:1 ratio is common, some tasks may benefit from a higher ratio of negatives (e.g., 1:5). In our DDI problem, we explored varying the ratio, but it did not significantly impact performance (not shown) and introduced unnecessary complexity.

Create the Train Step Function

The `train_step` function is where learning actually happens during training. It performs a forward pass, computes the loss, and applies gradients to update the model's parameters. This function is applied to batches of node pairs throughout each epoch.

```python
@partial(jax.jit, static_argnames=["loss_fn", "norm_loss"])
def train_step(
  state: TrainState,
  graph: jraph.GraphsTuple,
  pairs: dict[str, jax.Array],
  rng_dropout: jax.Array,
  loss_fn: Callable = binary_log_loss,
  norm_loss: bool = False,
) -> tuple[TrainState, dict[str, jax.Array]]:
  """Performs a single training step, updating model parameters."""

  def calculate_loss(params):
    """Computes loss and hits@20 metric for the given model parameters."""
    scores = state.apply_fn(
      {"params": params},
      graph,
      pairs,
      is_training=True,
      rngs={"dropout": rng_dropout},
    )
    loss = loss_fn(scores)
    metric = evaluate_hits_at_20(scores)
    return loss, metric

  # to additional variables (e.g., state, graph, pairs) without requiring them
  # to be explicitly passed, while maintaining compatibility with
  # jax.value_and_grad.
  grad_fn = jax.value_and_grad(calculate_loss, has_aux=True)
  (loss, metric), grads = grad_fn(state.params)
```

```
state = state.apply_gradients(grads=grads)

metrics = {"loss": loss, "hits@20": metric}
if norm_loss:
  metrics["loss"] = metrics["loss"] / (
    pairs["pos"].shape[0] + pairs["neg"].shape[0]
  )

return state, metrics
```

You'll notice that the function is decorated using @partial(jax.jit, static_arg names=["loss_fn", "norm_loss"]). This pattern allows these arguments—the loss function (loss_fn) and whether to normalize the loss (norm_loss)—to be treated as static during JAX's just-in-time (JIT) compilation. By marking these arguments as static, JAX avoids recompiling the function every time they change.

Inside the function, a nested calculate_loss method defines the core of the computation:

- The model is applied to the current batch using state.apply_fn(...), producing predictions (scores) for positive and negative node pairs.
- The specified loss_fn is used to compute the training loss.
- The evaluate_hits_at_20 function computes the hits@20 metric to track how well the model ranks correct node pairs near the top.

JAX's value_and_grad is then used to calculate both the loss and its gradient with respect to the model parameters. These gradients are applied to the training state using state.apply_gradients.

If norm_loss=True, the total loss is divided by the number of training pairs in the batch (positive + negative), ensuring that loss magnitudes remain comparable across different batch sizes or sampling ratios.

The default loss function used in training is the binary log loss function:

```
@jax.jit
def binary_log_loss(scores: dict[str, jax.Array]) -> jax.Array:
    """Computes the binary log loss for positive and negative drug pairs."""
    # Clip probabilities to avoid numerical instability.
    probs = jax.tree.map(
      lambda x: jnp.clip(nn.sigmoid(x), 1e-7, 1 - 1e-7), scores
    )

    # Compute positive and negative losses.
    pos_loss = -jnp.log(probs["pos"]).mean()
    neg_loss = -jnp.log(1 - probs["neg"]).mean()

    return pos_loss + neg_loss
```

This loss function performs the following steps:

Sigmoid transformation
>Converts the raw logits (unbounded scores) into probabilities in the range (0, 1). These probabilities represent the model's confidence that a given node pair is a positive class (i.e., a true drug–drug interaction).

Clipping
>Ensures numerical stability by constraining probabilities to lie slightly within (0, 1). This avoids issues such as log(0), which can cause the loss to diverge or return NaN during training.

Loss calculation
>Ensures that positive pairs are penalized when their predicted probabilities are far from 1, and negative pairs are penalized when their predicted probabilities are far from 0. The total loss is computed as the average of the positive and negative log losses.

This loss provides a simple yet effective objective for training the model to distinguish between interacting and noninteracting drug pairs.

Create the Evaluation Metric

Finally, we want to evaluate the model's performance. This follows a similar approach to `train_step`:

```
@partial(jax.jit, static_argnames=["loss_fn", "norm_loss"])
def eval_step(
  state: TrainState,
  graph: jraph.GraphsTuple,
  pairs: dict[str, jax.Array],
  loss_fn: Callable = binary_log_loss,
  norm_loss: bool = False,
) -> dict[str, jax.Array]:
  """Performs an evaluation step, computing loss and hits@20 metric."""
  scores = state.apply_fn(
    {"params": state.params}, graph, pairs, is_training=False
  )
  metrics = {"loss": loss_fn(scores), "hits@20": evaluate_hits_at_20(scores)}
  if norm_loss:
    metrics["loss"] = metrics["loss"] / (
      pairs["pos"].shape[0] + pairs["neg"].shape[0]
    )

  return metrics
```

The notable differences in `eval_step` are:

No training mode

The is_training flag is set to False, which disables behaviors like dropout and ensures that the model is evaluated deterministically.

Evaluation metric

Rather than just returning the loss, the function also computes hits@20, a ranking-based metric commonly used in link prediction tasks.

Hits@20 evaluates how well the model ranks positive node pairs compared to negative ones, giving an intuitive signal of ranking quality. Specifically, it identifies the 20th highest score among the negative pairs as a threshold and calculates the proportion of positive scores that exceed this threshold; then it calculates the proportion of positive scores that exceeds this threshold. A higher Hits@20 indicates that the model correctly ranks many true interactions above even one of the most confident false ones.

Here's the implementation:

```
@jax.jit
def evaluate_hits_at_20(scores: dict[str, jax.Array]) -> jax.Array:
    """Computes the hits@20 metric capturing positive pairs ranking."""
    # Implementation inspired by the OGB benchmark: https://oreil.ly/Oej2Y
    # Find the 20th highest score among negative edges.
    kth_score_in_negative_edges = jnp.sort(scores["neg"])[-20]

    # Compute the proportion of positive scores greater than the threshold.
    return (
        jnp.sum(scores["pos"] > kth_score_in_negative_edges)
        / scores["pos"].shape[0]
    )
```

We could have imported the ogb.linkproppred.Evaluator from the Open Graph Benchmark (OGB) library, which computes Hits@20. However, by directly implementing the metric, we make the evaluation process more transparent and tailored to our specific use case. This approach provides greater flexibility for modifications and extensions while clearly showing how the model is evaluated.

Train the Simplest Model

We are finally ready to train the model. We'll start with a relatively simple architecture and monitor its performance:

```python
import optax

rng, rng_init, rng_train = jax.random.split(rng, 3)

model = DdiModel(
    n_nodes=dataset_splits["train"].n_nodes,
    embedding_dim=128,
    last_layer_self=False,
    degree_norm=False,
    dropout_rate=0.3,
)

state, metrics = train(
    state=model.create_train_state(
        rng=rng_init,
        dummy_input={
            "graph": dataset_splits["train"].graph,
            "pairs": dataset_splits["train"].pairs.get_dummy_input(),
        },
        tx=optax.adam(0.001),
    ),
    rng=rng_train,
    dataset_splits=dataset_splits,
    num_epochs=500,
    eval_every=1,
    loss_fn=binary_log_loss,
    norm_loss=False,
    store_path=assets("graphs/models/initial_model"),
)
```

The learning curves in Figure 4-9 show the training process over 500 epochs. The training loss decreases steadily and the Hits@20 metric improves on both training and validation sets. However, the growing divergence between the training and validation curves—particularly visible in the later epochs—suggests moderate overfitting. This indicates that while the model is capturing meaningful patterns, its generalization to unseen data could be improved.

```python
from dlfb.graphs.inspect import plot_learning

plot_learning(metrics);
```

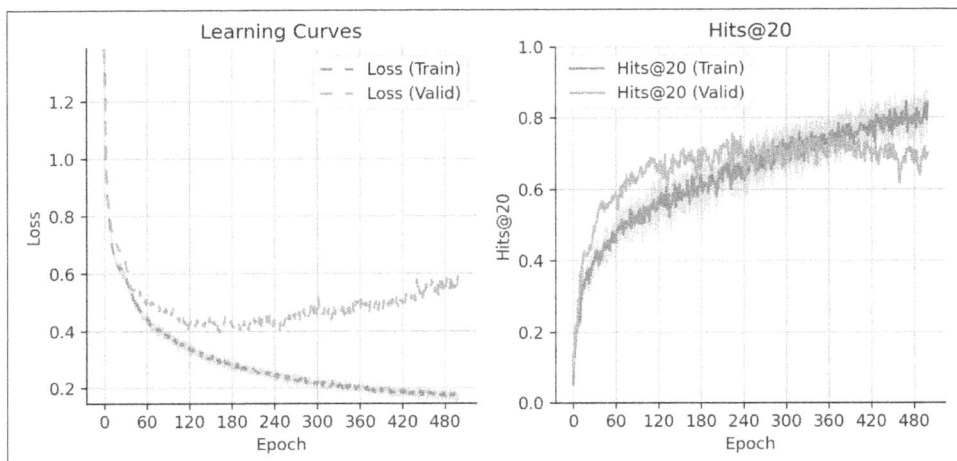

Figure 4-9. Learning curves showing the training loss (left) and Hits@20 metric (right) over 500 epochs. While both metrics improve on the training set, a growing gap between training and validation suggests moderate overfitting.

A word of warning: two sources of variance can lead to surprisingly different metrics curves in this plot—even with the same model and training setup.

Stochastic node sampling

Graph neural networks that use neighborhood sampling can be highly sensitive to the random seed. In drug–drug interaction prediction, some nodes are connected to many easily predictable links, while others are not. Which nodes get included in your training data can have a major impact on the learned representations, loss, and evaluation metrics.

Discontinuous metrics

Hits@20 is a ranking-based metric that compares each positive score to the 20th highest negative score. This thresholding introduces discontinuity: a small change in any score near the threshold—especially among the top-ranked negatives—can flip the outcome for many positives from success to failure or vice versa. This makes Hits@20 unusually sensitive to minor score shifts, even when the model or loss appears unchanged. This doesn't mean it's a bad metric—just something to be aware of.

To reduce this variability, we can increase the number of nodes we include in the dataset (as shown in "Train on a Larger Dataset" on page 232).

Now that we've trained a first working model, we'll explore strategies to reduce overfitting and improve overall performance.

Improving the Model

Everything is working end-to-end—we have prepared a dataset, built a model, and trained it. However, the model's performance is suboptimal. In this section, we'll explore some tweaks to see if we can achieve better results.

Change to AUC Loss

So far, we've used binary log loss to train our model. However, for our task, the primary goal is to ensure that positive pairs are ranked higher than negative pairs. While probabilities can also be used to prioritize, they often saturate near 1 or 0 for confident predictions, making it harder to differentiate between highly ranked pairs. In contrast, ranking-based metrics focus on the relative ordering of scores, which better aligns with the task of identifying and prioritizing the most promising drug interactions. This is particularly valuable in DDI prediction, where the goal is often to focus on the top-scoring pairs for further investigation.

Inspired by a paper on pairwise learning for neural link prediction (PLNLP),[7] which outlines key stages of a link prediction pipeline, we will swap the loss function to better align with our objective. Instead of focusing on binary classifications, we adopt a ranking-based approach that encourages the model to score connected pairs higher than unconnected ones, aligning conceptually with the area under the curve (AUC) metric.

AUC measures the probability that a randomly chosen positive instance (connected node pair) has a higher score than a randomly chosen negative instance (nonconnected pair). While directly optimizing AUC would be ideal, it is computationally challenging because its gradients are often undefined or zero. To address this, we use a *surrogate loss function* that mimics AUC's properties while remaining easy to optimize.

A simple and effective surrogate is the squared loss, which penalizes deviations from the target score difference of 1 between positive and negative pairs. This means the model is penalized both when the difference is less than 1 (underestimation) and when it is greater than 1 (overestimation). By minimizing this penalty, the model learns to consistently assign higher scores to connected pairs while maintaining an appropriate margin over unconnected ones. Here's the implementation:

```
@jax.jit
def auc_loss(scores: dict[str, jax.Array]) -> jax.Array:
    """Computes AUC-based loss for positive and negative drug pairs."""
    return jnp.square(1 - (scores["pos"] - scores["neg"])).sum()
```

---

7 Wang, Z., Zhou, Y., Hong, L., Zou, Y., Su, H., & Chen, S. (2021). Pairwise learning for neural link prediction (*https://doi.org/10.48550/arxiv.2112.02936*). arXiv (Cornell University).

This loss function encourages the model to score linked pairs higher than nonlinked pairs, improving its ranking performance. The motivation for discarding the cross-entropy loss is that the Hits@N metric, commonly used by Open Graph Benchmark (OGB) for evaluating link prediction benchmarks, does not measure the quality of predicted probabilities. Instead, it focuses solely on ensuring that true edges are ranked higher than false edges. Although the difference between these approaches is subtle, it has significant practical implications. Let's explore the effect of the loss function switch on the model:

```
rng, rng_init, rng_train = jax.random.split(rng, 3)

model = DdiModel(
  n_nodes=dataset_splits["train"].n_nodes,
  embedding_dim=128,
  last_layer_self=False,
  degree_norm=False,
  dropout_rate=0.3,
)

_, metrics = train(
  state=model.create_train_state(
    rng=rng_init,
    dummy_input={
      "graph": dataset_splits["train"].graph,
      "pairs": dataset_splits["train"].pairs.get_dummy_input(),
    },
    tx=optax.adam(0.001),
  ),
  rng=rng_train,
  dataset_splits=dataset_splits,
  num_epochs=500,
  eval_every=1,
  loss_fn=auc_loss,
  norm_loss=True,
  store_path=assets("graphs/models/initial_model_auc"),
)
```

From the learning curves in Figure 4-10, we can see that changing the loss function to an AUC-based objective has led to better results. First, training is noticeably smoother: the training and validation losses fall in near-lockstep, with virtually no gap. Second, Hits@20 climbs more quickly and saturates at a higher level on the validation set than before, reflecting stronger generalization. In short, aligning the objective with our ranking metric directly translates into better and more reliable performance.

```
plot_learning(metrics);
```

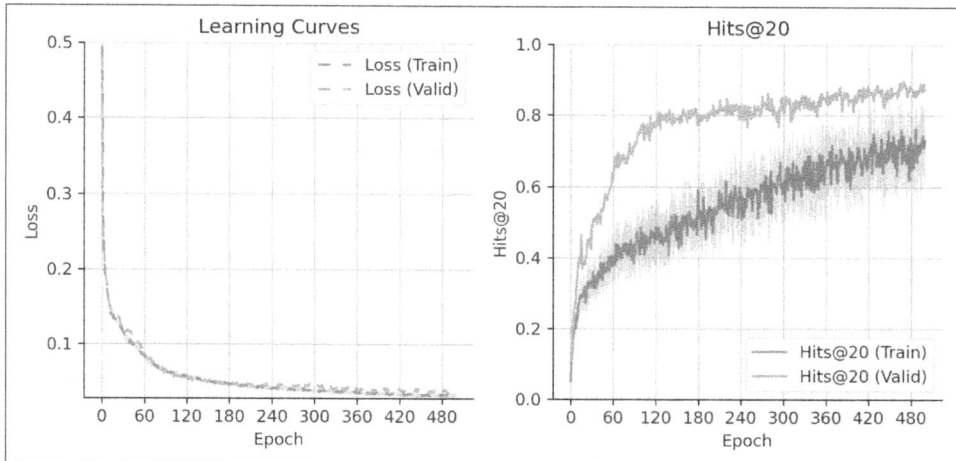

Figure 4-10. Learning curves for loss (left) and Hits@20 (right) after replacing binary-cross-entropy with an AUC-optimizing loss. The new objective keeps train and validation losses tightly coupled and pushes the validation Hits@20 to higher, stabler values.

It appears that this model is already fairly strong as measured by our choice of metric, but let's see if we can push performance even higher and reduce overfitting further by exploring hyperparameter sweeps.

Set Model Sweeping and Training Parameters

Our model and training loop include several hyperparameters, and it's not immediately clear which combinations will yield the best performance. One natural starting point is the embedding dimension, which directly controls the capacity of the model to richly represent graph structure.

Varying embedding dimensions

We will vary the `embedding_dim` parameter and train new models for each value to evaluate its impact on performance. Since previous experiments showed improved results with longer training, we will also extend the number of epochs in this sweep:

```
embedding_dims = [64, 128, 256, 512]
model_params = {
  "n_nodes": dataset_splits["train"].n_nodes,
  "last_layer_self": False,
  "degree_norm": False,
  "dropout_rate": 0.3,
}
training_params = {
  "rng": rng_train,
  "dataset_splits": dataset_splits,
  "num_epochs": 500,
```

```
    "eval_every": 25,
    "loss_fn": auc_loss,
    "norm_loss": True,
}
```

The following loop automates training across a range of `embedding_dim` values, storing the evaluation metrics for later analysis:

```
from dlfb.utils.metric_plots import to_df

all_metrics = []
for embedding_dim in embedding_dims:
    model = DdiModel(**{"embedding_dim": embedding_dim, **model_params})
    _, metrics = train(
        state=model.create_train_state(
            rng=rng_init,
            dummy_input={
                "graph": dataset_splits["train"].graph,
                "pairs": dataset_splits["train"].pairs.get_dummy_input(),
            },
            tx=optax.adam(0.001),
        ),
        **training_params,
        store_path=assets(f"graphs/models/sweep_embedding_dim:{embedding_dim}"),
    )
    df = to_df(metrics).assign(**{"embedding_dim": embedding_dim})
    all_metrics.append(df)
all_metrics_df = pd.concat(all_metrics, axis=0)
```

As shown in Figure 4-11, there's a clear relationship between embedding size and model performance: larger embedding vectors tend to yield better validation metrics. This is likely because higher-dimensional embeddings provide greater capacity to capture complex relationships and patterns in the graph:

```
import seaborn as sns
from matplotlib import pyplot as plt

from dlfb.utils.metric_plots import DEFAULT_SPLIT_COLORS

data = all_metrics_df
data = data[(data["metric"] == "hits@20")]
data = data.groupby(["metric", "split", "embedding_dim"], as_index=False)[
    "mean"
].max()
data = data.sort_values(by=["split", "mean"])

plt.figure(figsize=(7, 3.5))
sns.barplot(
    data=data,
    x="embedding_dim",
    y="mean",
    hue="split",
    palette=DEFAULT_SPLIT_COLORS,
```

```
)
plt.ylim(0.5, 1)
plt.xlabel("Embedding Dimensions")
plt.ylabel("Maximum Hits@20");
```

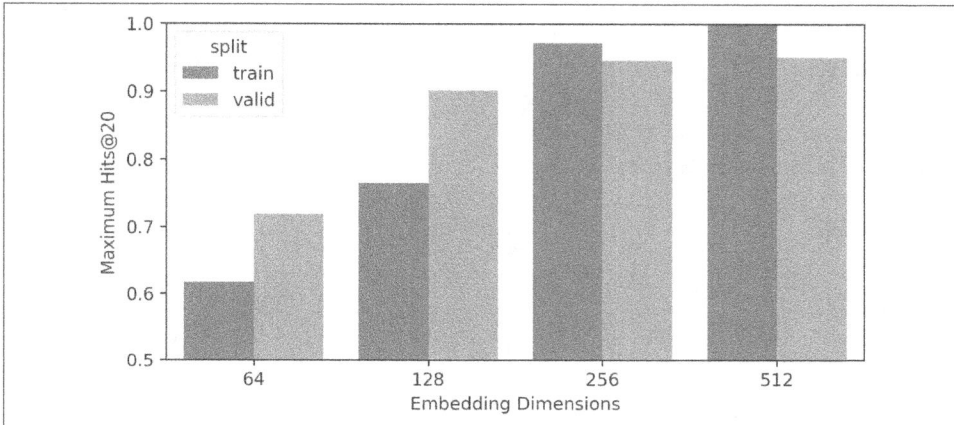

Figure 4-11. Maximum Hits@20 achieved by models with varying embedding dimensions, highlighting the impact of embedding size on performance.

The model with embedding dimensions set to 512 seem to achieve a perfect Hits@20 of ~1 on the training set, suggesting that it has sufficient capacity to fully (and even overly) fit the training data.

> Increasing the embedding size and training the model for longer, as we have done in this section, comes with higher computational costs. Whether this trade-off is worthwhile depends on the resources available and the relative importance of improving model performance in your specific use case.

Varying multiple hyperparameters

For this experiment, we broaden our search by sweeping over multiple model hyperparameters simultaneously. This builds on the previous experiment but explores a wider region of the hyperparameter space in the hopes of discovering better-performing configurations.

We will vary the following parameters:

- Dropout rate: 0, 0.3, or 0.5
- Self-edges in the last convolutional layer: Whether to include self-edges (`last_layer_self`: True or False)

- Degree normalization: Whether to normalize node embeddings by their degree (`degree_norm`: True or False)
- Number of MLP layers in the link predictor: 1, 2, or 3 (`n_mlp_layers`)

For this experiment, we fix the embedding dimension at 512, as it was among the best-performing settings in the earlier sweep. While an embedding size of 256 also performed reasonably well and would reduce computational cost, we opt for 512 in anticipation of scaling to larger graphs. A larger embedding size offers greater capacity to model complex relational patterns.

Let's set up the sweep:

```python
import itertools

model_params = {
    "n_nodes": dataset_splits["train"].n_nodes,
    "embedding_dim": 512,
}

model_params_sweep = {
    "dropout_rate": [0, 0.3, 0.5],
    "last_layer_self": [True, False],
    "degree_norm": [True, False],
    "n_mlp_layers": [1, 2, 3],
}
keys, values = zip(*model_params_sweep.items())
model_param_combn = [
    dict(zip(keys, combo)) for combo in itertools.product(*values)
]
print(pd.DataFrame(model_param_combn))
```

Output:

```
    dropout_rate  last_layer_self  degree_norm  n_mlp_layers
0   0.0           True             True         1
1   0.0           True             True         2
2   0.0           True             True         3
..  ...           ...              ...          ...
33  0.5           False            False        1
34  0.5           False            False        2
35  0.5           False            False        3

[36 rows x 4 columns]
```

To train models with each parameter combination, we use the following approach:

```python
def name_from_params(params: dict) -> str:
    """Generates a string from a parameters dictionary"""
    return "_".join([f"{k}:{v}" for k, v in params.items()])


all_metrics = []

for combn in model_param_combn:
    model = DdiModel(**{**combn, **model_params})
    _, metrics = train(
        state=model.create_train_state(
            rng=rng_init,
            dummy_input={
                "graph": dataset_splits["train"].graph,
                "pairs": dataset_splits["train"].pairs.get_dummy_input(),
            },
            tx=optax.adam(0.001),
        ),
        **training_params,
        store_path=assets(f"graphs/models/sweep_all_{name_from_params(combn)}"),
    )
    df = to_df(metrics).assign(**combn)
    all_metrics.append(df)

all_metrics_df = pd.concat(all_metrics, axis=0)
```

Similar to the earlier experiment, this loop automates the process of training models across all parameter combinations. Each model's performance is evaluated, and metrics such as Hits@20 are recorded for later analysis.

We then calculate the maximum Hits@20 metric for each parameter combination and split. Additionally, we generate a more readable representation for the convolutional layer configurations used in the encoder. This systematic exploration helps us identify the most effective combinations of hyperparameters for optimizing the model.

We then extract the maximum Hits@20 metric for each parameter combination and split:

```
def conv_layer_annot(row):
  if row["last_layer_self"] and row["degree_norm"]:
    return "with self-edges and norm"
  elif row["last_layer_self"]:
    return "with self-edges, no norm"
  elif row["degree_norm"]:
    return "with norm, no self-edges"
  else:
    return "no self-edges and no norm"


data = all_metrics_df
data = data[(data["metric"] == "hits@20")]
data = data.groupby(
  ["metric", "split", *list(model_params_sweep.keys())], as_index=False
)["mean"].max()
data = data.sort_values(by=["split", "mean"])
data["conv_layer"] = data.apply(conv_layer_annot, axis=1)
```

Next, in Figure 4-12, we visualize an overview of model performance across different hyperparameter combinations. This plot contains a lot of information, so let's first clarify how to interpret it. Each pair of points represents the training and validation Hits@20 scores for a single model configuration. Our goal is to identify configurations where validation performance is high and closely matches training performance—suggesting good generalization without overfitting. This is particularly important because the dataset is split by protein targets, meaning that the validation set contains drugs targeting different proteins than those seen during training.

> Models that show similar performance on both training and validation sets are more likely to generalize well. Prioritizing these configurations can help guide robust model selection.

```
fig = sns.relplot(
  data=data,
  x="conv_layer",
  y="mean",
  row="dropout_rate",
  col="n_mlp_layers",
  hue="split",
  palette=DEFAULT_SPLIT_COLORS,
  facet_kws=dict(margin_titles=True, despine=False),
  height=2,
)
fig.figure.subplots_adjust(wspace=0.1, hspace=0)
for ax in fig.axes.flat:
```

```
for label in ax.get_xticklabels():
    label.set_rotation(45)
    label.set_ha("right")
fig.set_axis_labels("", "maximum Hits@20");
```

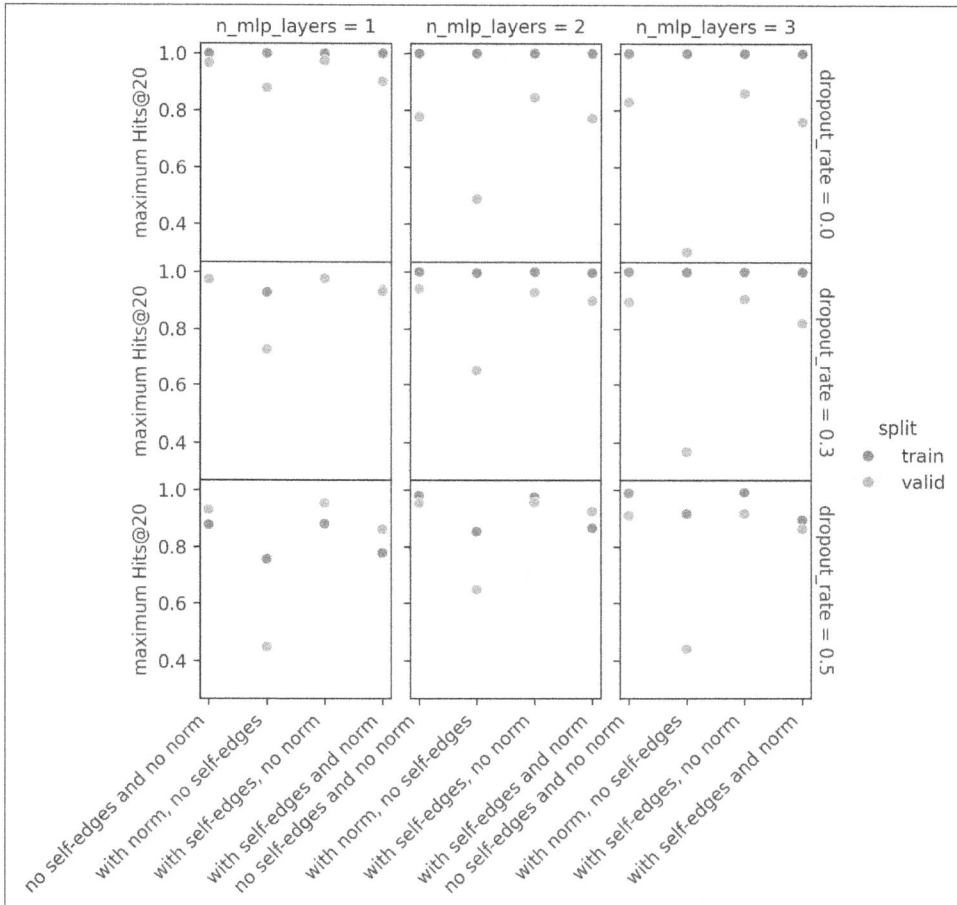

Figure 4-12. Maximum Hits@20 achieved by each configuration in our hyper-parameter grid. Rows correspond to dropout_rate (0.0, 0.3, 0.5); columns vary the number of MLP layers (1–3) in the link predictor. Each x-axis category toggles self-loops and degree normalization. The best generalization comes from the minimal model (far left, top row).

There are many models in this plot, but one stands out in the upper-left corner: the best-performing model without dropout. It uses a simple convolutional setup (no self-edges, no normalization) and a single-layer link predictor. It's notable that such a minimal configuration performs so well, suggesting that simpler models can still be highly effective.

Other observations from the plot:

- Degree normalization hurts almost everywhere. Both "with norm" columns show a noticeable drop in validation performance relative to their "no norm" counterparts.
- Deeper MLP heads overfit more. Moving from one to three MLP layers widens the train–valid gap and very rarely yields a higher peak validation score.
- Dropout does not consistently help, though some models with high dropout perform well, suggesting it's not strictly harmful either.
- Self-edges make little difference. The two "with self-edges" categories track their "no self-edges" twins closely, suggesting that this graph already conveys enough reciprocal information. Overall, the task appears quite forgiving—many configurations exceed 0.8 on validation—but the simplest, lowest-capacity model remains the most reliable choice.

Let's now inspect the learning curves for the best-performing configuration, shown in Figure 4-13.

```
from dlfb.utils.metric_plots import from_df

metrics = all_metrics_df
metrics = from_df(
  metrics[
    (metrics["dropout_rate"] == 0.0)
    & (metrics["last_layer_self"] == False)
    & (metrics["degree_norm"] == False)
    & (metrics["n_mlp_layers"] == 1)
  ]
)

plot_learning(metrics);
```

Recall that the previous metrics represented maximum values—meaning they may not have occurred at the same point in training and do not rule out eventual overfitting if training were to proceed. Plotting the full learning curves provides additional insight. In Figure 4-13, we see strong learning on the training set, but the configuration appears prone to overfitting without early stopping. This is evident from the widening gap between training and validation metrics over time.

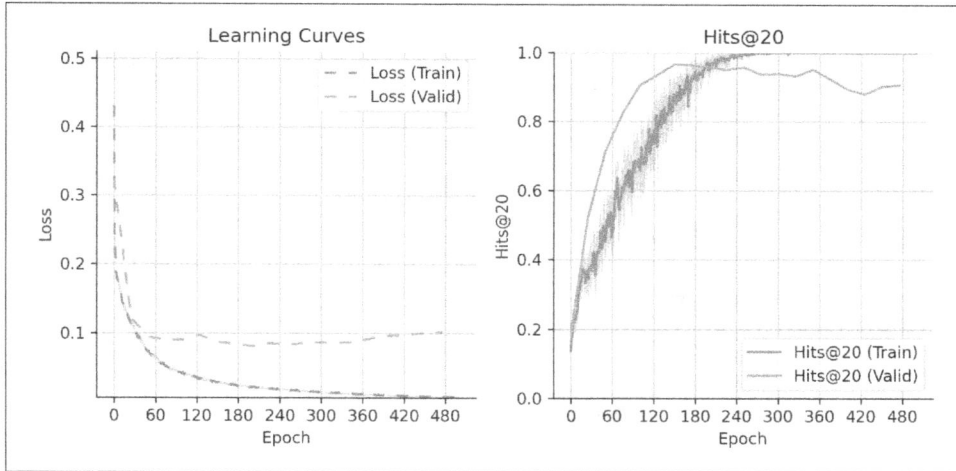

Figure 4-13. Learning curves for loss (left) and Hits@20 (right) of the best-performing configuration. The model achieves near-perfect training performance, but the validation metric plateaus and then declines, indicating overfitting.

Let's also examine the learning curves of an alternative high-performing model in Figure 4-14. This configuration comes from a different region of the hyperparameter space: it includes more layers in the link predictor and applies a high dropout rate to encourage generalization. It also uses a different convolutional setup. Notably, the training and validation curves remain closer together, suggesting that this model learns a more generalizable representation and is less susceptible to overfitting:

```
metrics = all_metrics_df
metrics = from_df(
    metrics[
        (metrics["dropout_rate"] == 0.5)
        & metrics["last_layer_self"]
        & (metrics["degree_norm"] == False)
        & (metrics["n_mlp_layers"] == 2)
    ]
)

plot_learning(metrics);
```

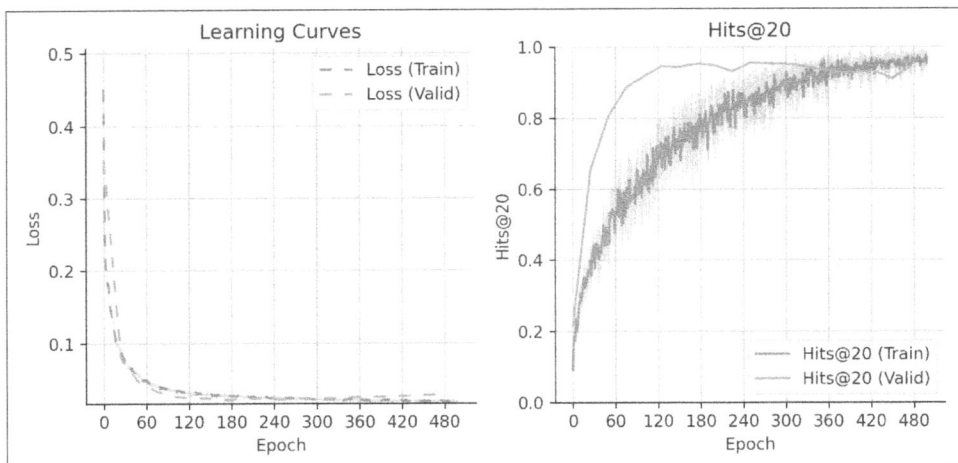

Figure 4-14. Learning curves for loss (left) and Hits@20 (right) of an alternative high-performing model. This configuration demonstrates improved generalization, with train and validation metrics remaining stable and closely aligned.

While it's tempting to focus only on the top headline metrics to identify the best-performing configuration, examining training dynamics can offer deeper insight into model performance. Based on both maximum metrics and learning curve behavior, we would recommend this latest model for its strong performance and stability.

> There are many additional hyperparameters that could be explored —including those already present in our model and training loop, as well as others not yet considered. For example, we could experiment with alternative negative sampling strategies, vary the ratio of negative to positive pairs, or increase the depth of the graph convolutional layers. However, for this smaller dataset, the current performance is strong enough that we consider further tuning unnecessary.

Train on a Larger Dataset

Finally, we train a model on a much larger DDI dataset—scaling up from using roughly one-tenth to approximately one-half of the available data.

As noted earlier, the optimal model configuration often depends on dataset size and graph connectivity. With this larger dataset, additional hyperparameter exploration revealed that *degree normalization* becomes critical for strong performance. Moreover, a moderate dropout rate of 0.3 struck the best balance between regularization and learning capacity for this setting.

Here is the full `DdiModel` setup:

```
node_limit = 2134
rng = jax.random.PRNGKey(42)
rng, rng_dataset, rng_init, rng_train = jax.random.split(rng, 4)
dataset_splits = DatasetBuilder(path=assets("graphs/datasets")).build(
  node_limit, rng_dataset
)

model = DdiModel(
  n_nodes=dataset_splits["train"].n_nodes,
  embedding_dim=512,
  dropout_rate=0.3,
  last_layer_self=True,
  degree_norm=True,
  n_mlp_layers=2,
)

_, metrics = train(
  state=model.create_train_state(
    rng=rng_init,
    dummy_input={
      "graph": dataset_splits["train"].graph,
      "pairs": dataset_splits["train"].pairs.get_dummy_input(),
    },
    tx=optax.adam(0.001),
  ),
  rng=rng_train,
  dataset_splits=dataset_splits,
  num_epochs=1000,
  eval_every=25,
  loss_fn=auc_loss,
  norm_loss=True,
  store_path=assets("graphs/models/larger_model"),
)
```

The resulting learning curves in Figure 4-15 look quite different from the smaller-set runs. Validation Hits@20 climbs to ~0.9—on par with our best earlier models—but training Hits@20 remains unexpectedly low, hovering near 0.1. This wide gap likely reflects the increased difficulty of distinguishing positives from a much larger and more diverse set of negatives in the denser graph. It may also indicate that even with an AUC-like loss, nailing top-k ranking metrics like Hits@20 remains challenging at scale. Either way, further work—such as improved negative sampling or top-k–specific objectives—may be needed to fully leverage the larger dataset, and we encourage you to explore further.

```
plot_learning(metrics);
```

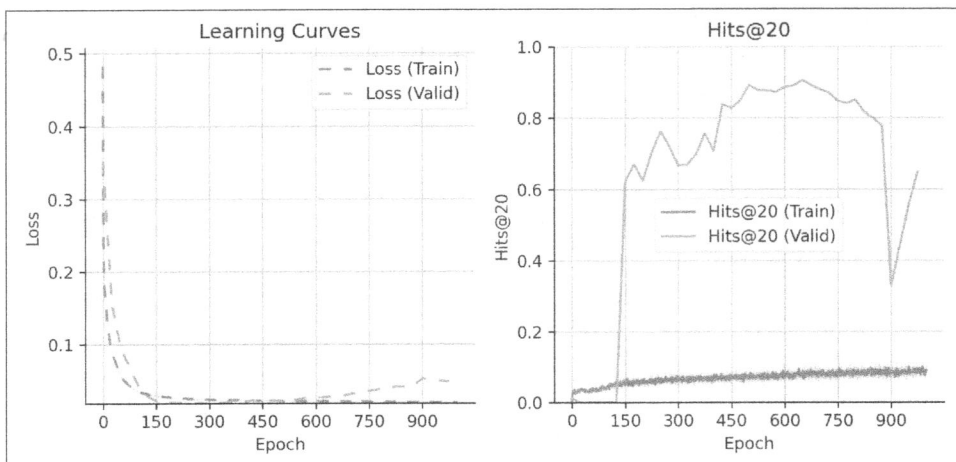

Figure 4-15. Learning curves for loss (left) and Hits@20 (right) of the best-performing model trained on the larger dataset. Maximum validation performance is high but does not show significant improvement over the performance of models trained on smaller datasets, suggesting further hyperparameter tuning or model adjustments may be needed.

The strong validation performance suggests this model could scale to even larger datasets. However, scaling beyond 50% of the dataset becomes difficult due to memory constraints. With an embedding size of 512 for all nodes, the current implementation can run into out-of-memory (OOM) issues during XLA compilation.

A better solution would be to adopt a sampling strategy, as used in GraphSAGE, where the model processes subgraphs in batches. This makes it feasible to scale to larger datasets without reducing embedding size or compromising performance. However, implementing this is beyond the scope of the current chapter.

Extensions

There are many ways to extend the model and training setup explored in this chapter:

Graph sampling for scalability
    Replace full-batch training with neighborhood sampling to scale to larger graphs without exceeding memory constraints.

Incorporate node features
    Integrate drug-specific information, such as chemical structure or pharmacological annotations, to enrich the learned embeddings and improve prediction accuracy.

Improve negative sampling

Use more informative strategies like hard negative mining, where the model is shown negative pairs it finds confusing (e.g., those with high predicted interaction scores). This encourages the model to focus on challenging distinctions and improves generalization.

Try alternative GNN architectures

Explore other models like Graph Attention Networks, which weigh the importance of neighboring nodes, which are theoretically more expressive.

Transfer to new biological problems

Apply the same modeling framework to other interaction networks, such as gene regulatory, protein-protein, or drug-target interaction graphs.

Additionally, here are some analysis ideas you could explore:

Error analysis by drug class

Break down model performance by therapeutic class or chemical category to identify where the model struggles most.

Prediction certainty

Examine the distribution of predicted probabilities. Which drug pairs is the model highly confident about (close to 0 or 1)? Which fall near 0.5? Investigate whether uncertain predictions share common characteristics (e.g., unusual structures, sparse connectivity).

Embedding visualization

Use dimensionality reduction methods (e.g., t-SNE, UMAP) to project node embeddings into 2D and inspect whether drugs with similar functions cluster together.

Temporal validation

If drug interaction timestamps are available, evaluate the model on future data after training on past interactions. This mimics a real-world deployment scenario.

Counterfactual analysis

Perturb the graph structure—for example, remove a known interaction or introduce a plausible but incorrect one—and observe how predictions change. This helps probe model sensitivity and identify influential edges.

Summary

In this chapter, we developed graph neural network models to predict links between nodes, applying them to the biologically meaningful task of drug–drug interaction (DDI) prediction. Starting with a simple architecture, we systematically explored model components and training strategies, eventually achieving strong validation performance through careful tuning of the loss function and key hyperparameters. Along the way, we introduced practical tools for evaluating performance, diagnosing overfitting, and scaling to larger datasets. These techniques are broadly applicable beyond DDIs, extending to a wide range of biological graph problems.

Our results show that even relatively simple graph models, when thoughtfully designed and optimized, can learn meaningful biological structure. As graph-based data becomes increasingly central in biology, the approaches introduced here provide a solid foundation for tackling more complex tasks—in drug discovery, genomics, systems biology, and beyond.

## CHAPTER 5

Detecting Skin Cancer in Medical Images

In previous chapters, we focused on small-scale biological phenomena, such as the molecular properties of proteins, DNA sequences, and drug molecules. In this chapter, we will zoom out to a larger biological scale, applying deep learning to analyze tissue-level and disease-related processes. Specifically, we will train a skin cancer detection model to classify images of skin into various cancerous or benign categories.

This is an exciting application because deep learning models have made significant strides in skin analysis, with studies achieving dermatologist-level accuracy in distinguishing malignant from benign lesions since at least 2018.[1] While challenges remain in integrating these models into clinical workflows—such as regulatory approval, data standardization, and prediction explainability—their potential to assist medical professionals by enhancing early detection and reducing unnecessary biopsies is highly promising.

We will be using skin cancer image data from the International Skin Imaging Collaboration (ISIC) (*https://oreil.ly/h2DiY*), a project dedicated to advancing skin cancer imaging research and providing standardized datasets. Over the years, ISIC has released a range of challenges focused on skin lesion classification and pathology, with an increasing number of images available. To learn more, you can read this review paper of ISIC datasets and benchmarks.[2]

The dataset we will use is available as the "Skin Cancer ISIC" challenge on Kaggle (*https://oreil.ly/_2jqU*), making it well prepared and relatively easy to get started with.

---

1 You can read about one such study online (*https://oreil.ly/YmXDg*).

2 Cassidy, Bill, et al., "Analysis of the ISIC Image Datasets: Usage, Benchmarks and Recommendations" (*https://doi.org/10.1016/j.media.2021.102305*), *Medical Image Analysis* 75 (January 2022).

However, to ensure that we cover important lessons on handling real-world data challenges, we have intentionally chosen a dataset that is relatively small and has significant class imbalance, allowing us to explore techniques for mitigating these issues.

One advantage of working with image data is that humans are naturally skilled at interpreting visual information, enabling us to sanity-check both the dataset and model predictions. Throughout this chapter, we will examine many images to guide our modeling decisions. This will also highlight why skin cancer classification is a challenging problem—not only for humans, but for deep learning models as well.

In terms of models, this chapter focuses on convolutional neural networks (CNNs) —specifically, ResNet CNNs, which have demonstrated strong performance across a wide range of image classification tasks. If you'd like to explore alternative approaches, consider checking out discussions and notebooks shared by other users on the Kaggle discussion board (*https://oreil.ly/bUHNt*).

As always, to get the most out of this chapter, keep the companion Colab notebook from our repo open as you read. Experimenting with the code as you go will deepen your understanding and make the concepts stick.

Biology Primer

First, let's introduce the biological phenomenon our models will address: skin cancer, its different types, and the challenges of classifying them.

Skin Cancer

Skin cancer is the most common type of cancer worldwide, with an estimated 1.5 million new cases in 2022.[3] It encompasses a wide range of conditions caused by the abnormal growth of skin cells, often driven by a combination of genetic factors and environmental carcinogens such as ultraviolet (UV) radiation.

In this chapter, we will examine both malignant (cancerous) and benign lesions. The term *lesion* broadly refers to any mark or abnormality on the skin, ranging from harmless growths to those requiring medical intervention. The following are some of the most common types:

---

3 Statistics by the World Health Organization (WHO) (*https://oreil.ly/79RmS*).

Malignant skin cancers

Basal cell carcinoma
   The most common type of skin cancer, usually slow-growing and rarely spreading to other organs.

Squamous cell carcinoma
   Another common type, typically localized but capable of becoming invasive if left untreated.

Melanoma
   The deadliest form of skin cancer, known for its ability to metastasize (spread to other parts of the body) quickly.

Actinic keratosis
   A precancerous lesion caused by sun damage. While not malignant, it can progress to squamous cell carcinoma if left untreated.

Benign skin lesions

Dermatofibroma
   A firm, benign growth often found on the legs.

Nevus (mole)
   A common benign growth that varies in size, shape, and color.

Pigmented benign keratosis
   A noncancerous pigmented lesion, often resembling seborrheic keratosis.

Seborrheic keratosis
   A benign, wartlike growth that can appear brown, black, or tan. Sometimes called "age spots" or "wisdom warts," these are often found in older adults.

Vascular lesions
   Benign vascular growths such as hemangiomas and cherry angiomas, formed by abnormal blood vessel proliferation.

Many of these conditions can look quite similar. Figure 5-1 shows example images from this chapter's training dataset, illustrating the different lesion types.

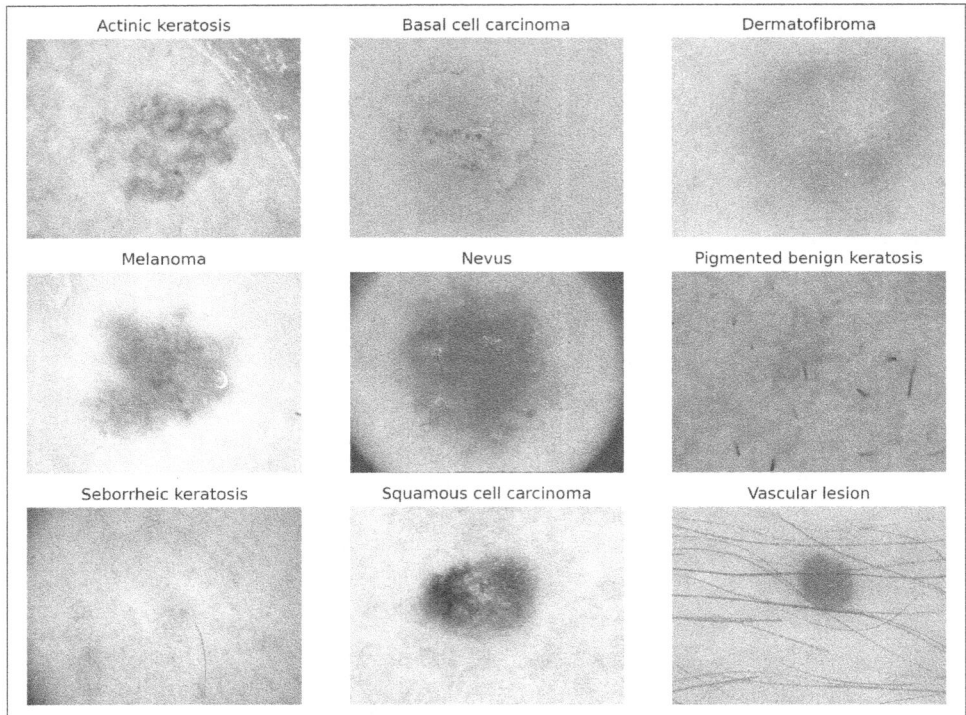

Figure 5-1. Grid displaying various types of skin lesions, both benign and malignant, from the dataset used for classification in this chapter.

As you can probably imagine from looking at these example images, misclassifications are a key challenge in skin cancer detection. These errors arise because different lesion types can share similar visual characteristics, such as pigmentation, texture, or irregular borders. The most critical errors occur when melanoma is misclassified as a benign lesion (a false negative), potentially delaying life-saving treatment. Conversely, false positives, such as benign growths mistaken for melanoma, are less serious but can lead to unnecessary biopsies and patient anxiety.

At the same time, there can be significant visual differences within a single class. For example, Figure 5-2 shows how melanomas can vary widely in appearance, making classification even more challenging.

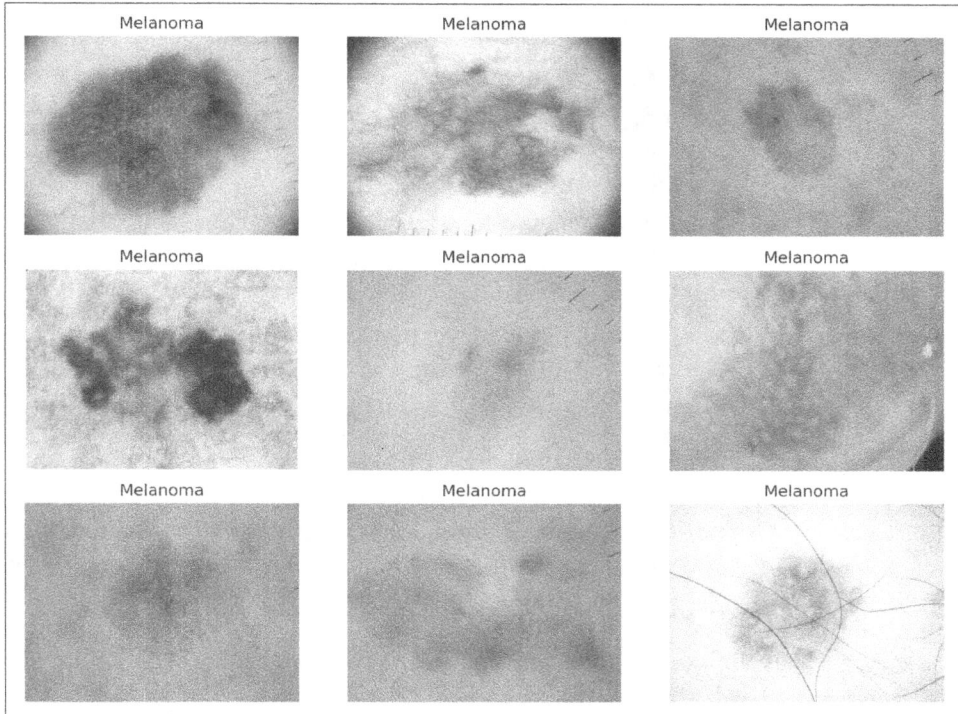

Figure 5-2. Melanomas can exhibit a wide range of visual characteristics, making consistent classification difficult. Some may appear dark brown or black with irregular borders, while others are lighter, reddish, or even patchy in color. Certain cases show a crusty texture, whereas others present as smooth, flat lesions.

This visual variability stems from underlying biological differences—including the type of cells involved, how deep or aggressively the lesion grows, and how much pigment is produced. For melanoma specifically, changes in melanin production and growth pattern can result in widely varying appearances, even within the same class.

Causes and Risk Factors

Skin cancer develops when genetic mutations disrupt the normal regulation of skin cell growth and division. The most common trigger for these mutations is UV radiation—primarily from sunlight or artificial sources like tanning beds—which damages cellular DNA over time. If this damage isn't properly repaired—for example, when the cell's DNA repair systems make errors or become overwhelmed—it can lead to uncontrolled cell growth and, ultimately, tumor formation.

Several factors increase the risk of developing skin cancer:

- *Fair skin* contains less melanin, the pigment that provides some natural protection against UV damage.

- *Frequent sunburns*, especially in childhood, suggest repeated UV-induced damage, which can accumulate over a lifetime.

- *A high number of moles* (especially atypical or dysplastic moles) can reflect underlying instability in melanocyte behavior, increasing the chance that one could turn cancerous.

- *Family history* may point to inherited genetic susceptibilities, such as mutations in tumor suppressor genes.

- *Environmental carcinogens*, like arsenic or industrial chemicals, can also contribute to mutational burden.

In addition to external factors, *specific genetic mutations*—such as in the *BRAF* gene —are commonly found in melanoma. These mutations may arise spontaneously or in response to environmental triggers like UV radiation, and they play a key role in driving tumor growth. Importantly, understanding these mutations has enabled the development of targeted therapies that are tailored to a patient's individual tumor profile, marking a shift toward more personalized cancer treatment.

How Skin Cancer Is Diagnosed

In clinical settings, dermatologists diagnose skin cancer through visual examination, dermoscopy, and biopsy. *Dermoscopy* is a noninvasive imaging technique that magnifies subsurface skin structures to help distinguish benign from malignant lesions. The images in commonly used skin cancer classification datasets primarily consist of dermoscopic images, captured using specialized *dermatoscopes* equipped with a light source and measurement markers, rather than standard cameras.

If a lesion appears suspicious, a *biopsy* is performed—a small tissue sample is taken and examined under a microscope to detect cellular abnormalities, such as irregular nuclei, atypical cell shapes, disorganized tissue architecture, or uncontrolled mitotic activity. These features help confirm whether the lesion is malignant and determine its type and stage.

The ABCDE rule (Asymmetry, Border irregularity, Color variation, Diameter >6 mm, Evolving changes) helps assess lesions for signs of melanoma. AI models can assist by analyzing dermoscopic and clinical images to flag high-risk lesions for further evaluation.

Image-Based Skin Cancer Detection

Deep learning has revolutionized image-based skin cancer detection, with AI models now achieving diagnostic accuracy comparable to expert dermatologists. A 2024 meta-analysis[4] of 53 studies found that AI consistently outperformed general practitioners and less experienced dermatologists in distinguishing melanoma from benign lesions, while performing on par with specialists. This suggests that AI can serve as a valuable diagnostic aid, enhancing early detection and decision making.

However, integrating AI into real-world clinical practice remains challenging. Most skin cancer models are trained on a handful of public datasets (e.g., ISIC, HAM10000), which lack diversity in skin types and imaging conditions, limiting generalizability. Additionally, AI performance in controlled, retrospective studies often does not translate to real-world settings, where factors like lighting, lesion presentation, and physician workflows introduce variability.

Another key barrier is explainability. Clinicians need to understand *why* a model makes a specific prediction, not just receive an isolated probability score. For example, if an AI predicts a skin lesion to be melanoma, is it due to the asymmetry, irregular borders, or color variation? And how are these different factors combined and weighted by the model? Machine learning methods like saliency maps and attention mechanisms help visualize what the model is focusing on, but they remain imperfect. Without clear reasoning, AI recommendations are difficult to trust or integrate into medical decision making.

Regulatory approval and clinical validation are major hurdles for deploying AI in healthcare. Diagnostic models must meet rigorous safety, accuracy, and transparency standards before receiving approval from regulatory bodies like the FDA or CE. This typically involves extensive clinical trials, reproducibility testing, and post-deployment monitoring. Moreover, AI tools must be integrated into existing workflows without disrupting clinician judgment or introducing new biases—all while maintaining patient privacy and data security.

While challenges remain, AI is steadily moving toward real-world deployment, with some dermatology AI systems already *CE-marked*—a certification indicating compliance with EU safety and efficacy standards—allowing for clinical use in the European Union. Additionally, smartphone apps such as SkinVision and Miiskin offer AI-based skin lesion analysis. While these apps are usually not approved for clinical decision

---

4 M. P. Salinas, et al., "A systematic review and meta-analysis of artificial intelligence versus clinicians for skin cancer diagnosis" (*https://doi.org/10.1038/s41746-024-01103-x*). *NPJ Digital Medicine* 7, no. 1 (2024): 125.

making, they can still provide risk assessments and encourage users to seek medical evaluation.

Machine Learning Primer

In Chapter 3, we applied a CNN to model 1D sequence data—specifically, DNA sequences. However, CNNs are more commonly used for 2D image processing, powering tasks such as:

Image classification
Assigning an image to a specific category, such as identifying whether it contains a dog or a cat

Object detection
Detecting and localizing objects within an image, such as drawing a bounding box around a cat

Segmentation
Partitioning an image into meaningful regions, such as labeling all pixels that belong to a cat

This section provides a brief primer on how CNNs work for images. As additional learning material, we recommend the 3Blue1Brown introductory video titled "But what is a convolution?" (*https://oreil.ly/k8zoM*), which offers a visual and intuitive explanation of CNNs. For a more in-depth exploration, the renowned Stanford CS 231n: Convolutional Neural Networks for Visual Recognition course (*https://oreil.ly/C_wPx*) provides a comprehensive introduction to the field.

Convolutional Neural Networks

CNNs are a specialized type of neural network designed for grid-like data, such as images (or sequences, as seen in Chapter 3). They automatically learn *hierarchical* patterns, meaning they extract features at different levels of abstraction:

- Early layers in the neural network detect simple patterns like edges, textures, and color contrasts.
- Middle layers recognize shapes and structures by combining these basic features.
- Deeper layers build on these to identify complex objects or meaningful categories.

For example, in skin cancer detection, early CNN layers may detect edges and color variations in skin lesions, mid-level layers might recognize irregular borders or asymmetry, and the latest layers would combine these features into high-level learned patterns to distinguish between benign and malignant lesions.

Although we describe CNNs as learning hierarchical representations, it's important to avoid *anthropomorphizing* them. CNNs don't "see" objects the way humans do. Instead, they learn statistical patterns in pixel values that maximize predictive accuracy. Techniques like *activation mapping* (highlighting which parts of an image influence a classification) and *probing* (examining what kinds of features different layers encode) help us understand and visualize correlations within the model. However, these methods provide post hoc insights for human interpretation—they don't imply that the model itself has a structured or explainable reasoning process.

Understanding a Convolution

The core building block of a CNN is the *convolutional layer*, which applies *filters* (also called kernels) to extract features from the input image. A filter is a small grid of numbers that slides across the image, detecting local patterns such as edges, textures, or color transitions.

An image is represented as a grid of pixels—in grayscale images, each pixel holds a single intensity value (ranging from 0 for black to 255 for white), while color images typically have three channels (usually red, green, and blue or *RGB*), each with its own intensity map.

As a filter moves across the image, it performs a *dot product* operation at each position: the filter values are multiplied element-wise with the corresponding pixel values in the image patch underneath, and the results are summed. This produces a single output value per position, building a new representation called a *feature map*, that highlights where the pattern encoded by the filter appears in the image.

The specific values in the filter determine what it detects. For example, the following 3×3 filter emphasizes vertical *edges* by responding strongly to changes in pixel intensity along the horizontal (x) axis:

$$\begin{bmatrix} -1 & 0 & 1 \\ -1 & 0 & 1 \\ -1 & 0 & 1 \end{bmatrix}$$

When applied to an image by sliding across it and computing the dot product with pixel values, this filter enhances areas where pixel intensity changes vertically, such as object boundaries, making it useful for edge detection.

Similarly, a filter designed to detect horizontal edges responds to changes in pixel intensity along the vertical (y) axis. It typically looks like this:

$$\begin{bmatrix} -1 & -1 & -1 \\ 0 & 0 & 0 \\ 1 & 1 & 1 \end{bmatrix}$$

This filter activates in regions where there is a strong transition from dark to light (or vice versa) from top to bottom, highlighting horizontal structures in the image.

In a way, this concept is related to the idea of "filters" in social media, which often apply simple mathematical transformations (such as increasing contrast or sharpening details) to modify an image's appearance.

> Before CNNs, in a task called *feature engineering*, researchers manually designed and optimized these filter matrices to detect edges, textures, and other features—a labor-intensive process. Now, neural networks learn these filters automatically, optimizing them for the task at hand.

A single convolutional layer doesn't just apply one filter; it typically learns multiple filters (e.g., 64) in parallel. Each filter captures different features, producing multiple *feature maps*, which are stacked together as separate channels in the layer's output.

Understanding Dimensions

If a grayscale image of size 256 × 256 (with a single channel) is passed through a convolutional layer with 64 filters, the output will have dimensions (256, 256, 64)—assuming padding is used to preserve spatial dimensions. The height and width remain unchanged, while the channel dimension expands, as each of the 64 filters extracts different feature representations from the image. This is true regardless of the filter size (e.g., 3 × 3, 5 × 5), as long as the stride and padding settings maintain spatial dimensions.

> Remember that neural network filters are usually randomly initialized, meaning they start off extracting no meaningful patterns. Through backpropagation, the filters gradually learn to detect useful visual features such as edges, textures, or shapes.

After convolution, a nonlinearity (such as the ReLU activation function) is typically applied to the feature maps. This is crucial because convolution alone is just a linear operation, meaning that without a nonlinearity, stacking multiple layers would be equivalent to just one big matrix multiplication. Adding the activation function allows the network to learn more complex, nonlinear patterns.

Now let's consider a color image with dimensions (256, 256, 3), where the three channels correspond to red, green, and blue (RGB). How does convolution work when an image has multiple input channels?

Unlike grayscale images, where each filter operates on a single channel, a convolutional filter in a color image must process all three channels simultaneously. Instead of being a simple 3 × 3 matrix, each filter is actually a 3 × 3 × 3 tensor, meaning:

- Each 3 × 3 slice of the filter is applied to the corresponding color channel (red, green, or blue) of the image.
- The results from all three channels are summed to produce a single output value per pixel. Typically, this is not a normal sum but actually a *weighted sum* where each channel's contribution is multiplied by a separate learned weight.
- This process is repeated for all filters, creating multiple feature maps in the next layer.

For example, applying a convolutional layer with 64 filters to an input of shape (256, 256, 3) results in an output of (256, 256, 64), where each filter has combined the three input channels in different ways to extract meaningful patterns. We would then apply an activation function such as a ReLU as before.

You may have noticed we called a 3 × 3 filter a *matrix*, but a 3 × 3 × 3 filter a *tensor*. To briefly clarify the terminology:

- A *scalar* is a single number and can be seen as a 0D tensor (e.g., 5).
- A *vector* is a 1D tensor (e.g., [1, 2, 3]).
- A *matrix* is a 2D tensor (e.g., a 3 × 3 filter).
- A *tensor* is a general term for arrays of any number of dimensions, including 3D+ structures like 3 × 3 × 3 filters.

In short, tensors are the fundamental data structure in deep learning, generalizing scalars, vectors, and matrices to arbitrary dimensions.

Pooling

Now that we've covered what convolutions do, let's move on to *pooling layers*. Pooling reduces the spatial dimensions of feature maps, making computations more efficient and helping to prevent overfitting by keeping only the most prominent activations.

The most common type is max pooling, which selects the highest value in a given region of the feature map. This ensures that strong activations are preserved while reducing spatial resolution (downsampling).

Consider a 4 × 4 feature map (the image-like output of a convolution) before applying 2 × 2 max pooling (with stride 2, meaning the filter moves two pixels at a time):

$$\begin{bmatrix} 1 & 3 & 2 & 1 \\ 4 & 5 & 7 & 2 \\ 3 & 2 & 9 & 6 \\ 1 & 8 & 6 & 3 \end{bmatrix}$$

After applying 2 × 2 max pooling, the highest value in each 2 × 2 block is retained, reducing the matrix to a size of 2 × 2:

$$\begin{bmatrix} 5 & 7 \\ 8 & 9 \end{bmatrix}$$

This greatly reduces the number of values while preserving the most important activations. Another common pooling type is *average pooling*, which takes the mean of each region instead of the max. This would result in this matrix:

$$\begin{bmatrix} 3.25 & 3.0 \\ 3.5 & 6.0 \end{bmatrix}$$

To summarize, pooling:

- Improves computational efficiency by shrinking the representation
- Retains key information by preserving prominent activations (like maxima or averages)
- Adds a mild form of regularization by making the model less sensitive to small shifts in the input

However, pooling is not a substitute for more robust regularization methods like dropout or weight decay.

Other Components of a CNN

Beyond convolution and pooling, several other key components help CNNs function effectively:

Activation functions
> As mentioned earlier, a nonlinearity (typically ReLU) is applied after each convolutional layer. This enhances the model's expressivity, enabling it to learn complex, nonlinear relationships.

Batch normalization
> This normalizes feature maps across a batch to keep activations in a stable range. It helps CNNs train more efficiently, reduces sensitivity to weight initialization, and enables the use of higher learning rates—especially useful for deep architectures (and many CNNs are quite deep).

·*Dropout*
> This is a regularization technique where random activations are set to zero during training to prevent overfitting. This forces the network to rely on multiple pathways for making predictions, improving generalization.

Fully connected layers
> After feature extraction, the final layers of a CNN are typically fully connected. These layers combine the learned feature representations and produce the final output—such as class probabilities in an image classification task.

With these components combined, we've covered the building blocks needed to train a fully functioning CNN. Next, we'll explore a widely used architecture that brings them all together: ResNets.

ResNets

A ResNet (residual network) is a CNN architecture introduced in 2015[5] that enables the training of very deep networks without suffering from the *vanishing gradient problem*, a common issue in which deeper networks struggle to propagate gradients effectively, slowing down learning. ResNets address this by introducing *residual connections* (also called *skip connections*), which allow information to bypass some layers. This stabilizes training and improves performance, making ResNets a go-to architecture for computer vision tasks like image classification and object detection.

The key idea behind a residual connection is simple: instead of learning a full transformation f(x), the network learns f(x) + x, where x is the original input.

---

5 He, K., Zhang, X., Ren, S., & Sun, J. (2015, December 10). Deep residual learning for image recognition (*https://oreil.ly/ZqFSo*). arXiv.org.

This means that if f(x) is small or difficult to learn, the network can still default to simply passing through the input x unchanged (an identity function). This prevents layers from degrading the performance of deeper networks, effectively "skipping" operations that don't contribute meaningful improvements.

Here's a simplified implementation of a residual block in pseudocode:

```
def ResidualBlock(x):
    """Basic building block of a ResNet."""
    identity = x  # Preserve the original input for the skip connection.
    residual = Convolution(x)  # Apply convolutional transformation.
    return identity + residual  # Add the identity (skip connection).
```

This simple skip connection allows gradients to flow more easily through the network, making training deep architectures much more feasible.

We can rewrite this pseudocode in Flax using the Linen API:

```
class ResidualBlock(nn.Module):
    """A minimal Flax CNN residual block."""
    features: int  # Number of output channels.
    kernel_size: Tuple[int, ...] = (3, 3)  # Convolution kernel size.
    strides: Tuple[int, ...] = (1, 1)  # Convolution stride.

    @nn.compact
    def __call__(self, x):
        identity = x  # Preserve the original input for the skip connection.
        residual = nn.Conv(
          self.features, self.kernel_size, self.strides, padding="SAME"
        )(x)
        return identity + residual  # Add the identity (skip connection).
```

The following are some notes on the key arguments to nn.Conv:

features
  Defines the number of learned filters. If features=100, an input (224, 224, 3) becomes (224, 224, 100), extracting 100 different feature maps.

kernel_size
  Defines the receptive field of the convolution. A small kernel like (3, 3) captures fine details, while a larger kernel like (16, 16) detects broader patterns.

strides
  Controls the step size of the filter. strides=(1, 1) preserves resolution, while larger strides downsample the feature maps. Downsampling is more commonly done via pooling layers than by increasing the stride.

padding
  Defines the padding for consistent shapes. padding="SAME" ensures that output dimensions match the input by adding necessary padding.

Kernel Size and Computational Cost

While using large convolutional kernels might seem intuitive for capturing broader context, they significantly increase computational cost—especially in early layers, where feature maps are large and convolutions must be applied many times. Although the number of parameters doesn't depend on input size, computation time scales with how often each kernel is applied.

A more efficient strategy is often to stack multiple small kernels—for example, three 3 × 3 convolutions instead of one 9 × 9 convolution. This increases the receptive field while introducing nonlinearities between layers, enabling the network to learn more expressive, hierarchical features with fewer parameters. Stacking small kernels grows the receptive field because each layer builds on the dependencies of the previous one, even if padding='SAME' keeps the output size unchanged.

Another technique is dilated convolution, which expands the receptive field without increasing parameters or downsampling the feature maps—useful when broader spatial context is important.

With this in mind, let's implement a more complete residual block with the following additions:

- Two stacked convolutional layers to extract deeper features
- A ReLU activation function to introduce nonlinearity
- Batch normalization to stabilize training
- A check for channel mismatches, applying a 1 × 1 convolution when necessary to ensure compatibility for addition

```
class ResidualBlock(nn.Module):
  """A basic Flax CNN residual block with two convolutional layers."""
  features: int  # Number of output channels.
  kernel_size: Tuple[int, ...] = (3, 3)  # Convolution kernel size.
  strides: Tuple[int, ...] = (1, 1)  # Convolution stride.

  @nn.compact
  def __call__(self, x):
    identity = x  # Preserve the original input for the skip connection.

    # First convolution + batch normalization + ReLU activation.
    x = nn.Conv(self.features, self.kernel_size, self.strides, padding="SAME")(x)
    x = nn.BatchNorm()(x)
    x = nn.relu(x)

    # Second convolution + batch normalization (no activation here).
    x = nn.Conv(self.features, self.kernel_size, self.strides, padding="SAME")(x)
    x = nn.BatchNorm()(x)
```

```
# If the input and output dimensions do not match, apply a 1 x 1 convolution.
if identity.shape[-1] != x.shape[-1]:
  identity = nn.Conv(
    self.features, kernel_size=(1, 1),
    strides=self.strides,
    padding="SAME"
  )(identity)

# Add the skip connection.
x += identity

# Final ReLU activation.
return nn.relu(x)
```

This implementation is now much closer to what you'll encounter in real-world architectures.

---

1 x 1 Convolutions

A 1 × 1 convolution (also called a pointwise convolution) operates on each spatial location independently across channels. Unlike standard convolutions that span spatial neighborhoods (e.g., 3 × 3 or 5 × 5), 1 × 1 convolutions capture cross-channel interactions while preserving the spatial resolution.

In residual blocks, a 1 × 1 convolution is often used to adjust the number of channels in the skip connection, ensuring that the input (identity) and output (x) have the same shape before addition. Without this adjustment, the addition would fail due to mismatched shapes, since tensor addition requires matching shapes across all dimensions.

Mathematically, a 1 × 1 convolution is actually equivalent to applying a fully connected (linear) transformation at each spatial location independently. However, it is more efficient than using dense layers in modern deep learning frameworks like JAX or TensorFlow, as it better utilizes GPU acceleration and tensor-level parallelism.

With this foundation in CNNs and ResNets, we can now apply these techniques to our goal: building a deep learning model for skin cancer prediction.

Exploring the Data

As always, before building a model, we first need to understand our dataset. In this chapter, we will use data from ISIC (*https://oreil.ly/g2EyR*), an initiative that fosters collaboration between medical professionals and AI researchers. ISIC regularly hosts machine learning challenges, encouraging researchers to develop and submit models for classifying skin lesion images.[6]

The ISIC archive is a freely accessible repository containing tens of thousands of skin lesion images, making it a valuable resource for developing and benchmarking AI-based diagnostic tools. You can explore the dataset through their online portal, as shown in Figure 5-3.

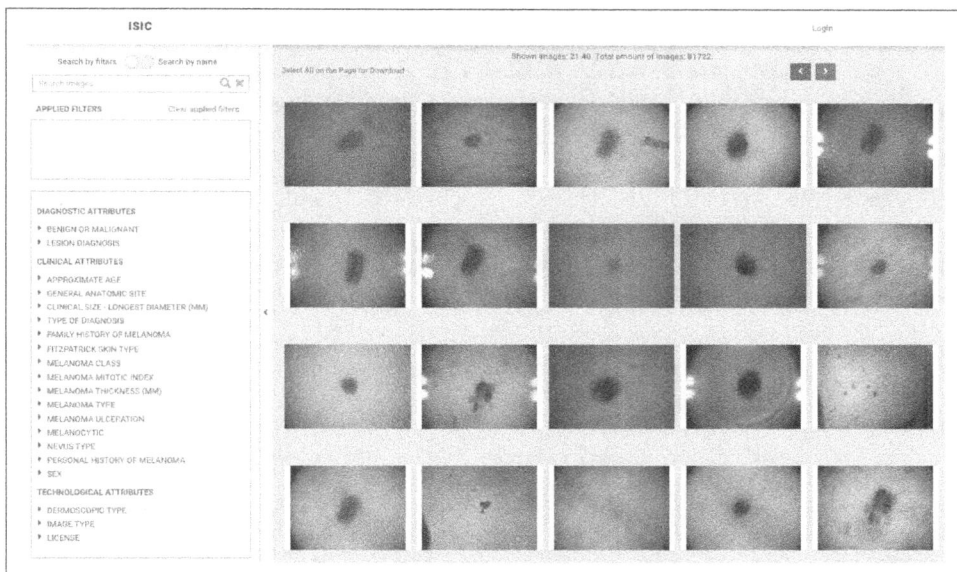

Figure 5-3. A screenshot (https://oreil.ly/F6BwK) displaying a vast collection of 81,722 public skin lesion images.

---

6 Codella, N. C. F., et al. (2018). Skin lesion analysis toward melanoma detection: A challenge at the 2017 International symposium on biomedical imaging (ISBI), hosted by the international skin imaging collaboration (ISIC). 2022 IEEE 19th International Symposium on Biomedical Imaging (ISBI), 168–172. https://doi.org/10.1109/isbi.2018.8363547

A First Glimpse

Rather than using the ISIC dataset directly, we are working with a version hosted on Kaggle (*https://oreil.ly/cez_1*). This allows us to explore the initial steps of working with a new dataset, including essential sanity checks to ensure its integrity.

> A key advantage of using a Kaggle dataset is that we can compare our approach to existing work and establish reasonable performance expectations. However, caution is needed—while Kaggle notebooks can be a great source of inspiration, they are not peer reviewed and may contain serious errors.
>
> For example, some models reported impressive performance on this dataset, but closer inspection revealed *data leakage*—where images from the training set also appeared in validation or test sets, leading to artificially inflated accuracy. In one extreme case, we even found a model that evaluated only on training images, making its results completely meaningless.

Let's ensure that we build a robust data pipeline by carefully inspecting the dataset before proceeding. We start by exploring the raw dataset directory to understand its structure. By listing all *.jpg* files, we can get an initial impression of the available labels, which correspond to different types of skin lesions. The `rglob` method is particularly useful here, as it scans directories recursively:

```python
import re
from pathlib import Path

from dlfb.utils.context import assets

image_file = next(Path(assets("cancer/datasets/raw")).rglob("*.jpg"))

print(rf"One of the images: {re.sub('^.*?datasets/', '', str(image_file))}")
```

Output:

```
One of the images: raw/Test/melanoma/ISIC_0000031.jpg
```

Examining the filepath, we can see that the dataset has already been split into `Train` and `Test` sets, with subdirectories for each skin lesion type.

Next, we will count the number of images per class across both splits. To do this efficiently, we define a helper function, `load_metadata`, that:

- Recursively collects all image filepaths
- Extracts the dataset split (`Train`/`Test`) and class label from each path
- Stores the results in a pandas `DataFrame` for easy inspection and visualization

---

We will also track `frame_ids`, which serve as a reference for quickly retrieving specific images during our later sanity checks.

```python
import pandas as pd


def load_metadata(data_dir: str) -> pd.DataFrame:
    metadata = []
    for path in Path(data_dir).rglob("*.jpg"):
        split, class_name, _ = path.parts[-3:]
        metadata.append(
            {
                "split_orig": split,
                "class_orig": class_name,
                "full_path": str(path),
            }
        )
    return pd.DataFrame(metadata).rename_axis("frame_id").reset_index()


metadata = load_metadata(assets("cancer/datasets/raw"))
print(metadata)
```

Output:

```
      frame_id split_orig       class_orig            full_path
0            0       Test         melanoma  /content/drive/M...
1            1       Test         melanoma  /content/drive/M...
2            2       Test         melanoma  /content/drive/M...
...        ...        ...              ...                  ...
2354      2354      Train  dermatofibroma  /content/drive/M...
2355      2355      Train  dermatofibroma  /content/drive/M...
2356      2356      Train  dermatofibroma  /content/drive/M...

[2357 rows x 4 columns]
```

Next, we count the number of images per class in both splits using the pandas `crosstab` function:

```python
counts = pd.crosstab(
    metadata["class_orig"], metadata["split_orig"], margins=True
)
print(counts)
```

Output:

```
split_orig            Test  Train   All
class_orig
actinic keratosis       16    114   130
basal cell carcinoma    16    376   392
dermatofibroma          16     95   111
melanoma                16    438   454
nevus                   16    357   373
pigmented benign ...    16    462   478
```

```
seborrheic keratosis        3     77    80
squamous cell car...       16    181   197
vascular lesion             3    139   142
All                       118   2239  2357
```

This dataset is relatively small, containing a total of 2,357 images, split into 2,239 for training and 118 for testing. Additionally, the class distribution is highly imbalanced, with some classes having very few examples to train on. Such an imbalance poses challenges for model generalization and requires careful handling to prevent biased predictions. It also makes evaluation less reliable, as performance metrics based on very few examples can be noisy and unrepresentative.

We visualize this distribution in Figure 5-4:

```
fig = counts.drop(["All"], axis=1).drop(["All"], axis=0).plot.barh(stacked=True)
fig.set_xlabel("Number of images")
fig.set_ylabel("Skin lesion type")
fig.set_title("Class distribution");
```

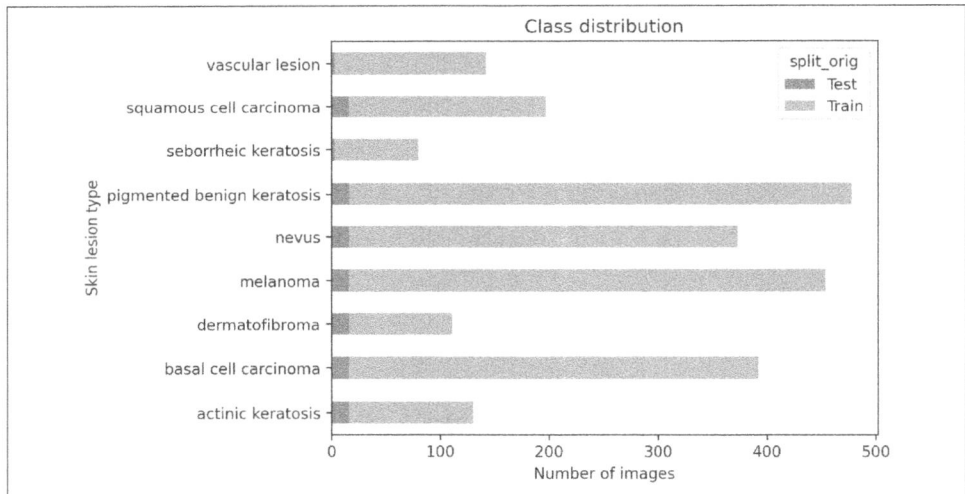

Figure 5-4. Bar plot of the distribution of classes across original training and test sets.

We can draw several key insights from the bar plot of class counts:

Class imbalance
Some categories, like pigmented benign keratosis, are overrepresented (454 total images), while others, like seborrheic keratosis, are severely underrepresented (80 total images). This imbalance could bias the model toward predicting the more frequent classes.

Small test set

Some categories have as few as three images in the test set, making it difficult to evaluate model performance across different lesion types. Such a small test set can easily lead to misleading performance metrics. We can also see that the ratio of train to test data is not even across lesion classes.

No validation set

The dataset only provides Train and Test splits, but no Valid subset. A validation set is crucial for tuning hyperparameters and assessing model improvements without touching the final test set.

To effectively train a model on this dataset, we'll need to address these points using techniques like data augmentation and resampling. Before diving into these, let's visually inspect some images to better understand the dataset. This will help us verify that the images match the lesion types introduced earlier in this chapter.

Another important consideration is that this dataset consists solely of images and labels, without additional metadata such as lesion location, patient demographics, or clinical notes. This lack of context makes the classification task more challenging, as real-world diagnosis often relies on more than just the visual appearance of a lesion. For example, a dark lesion on the scalp of a 70-year-old patient may increase the probability of melanoma, since both age and sun-exposed areas are known risk factors.

Previewing the Images

This is the first time in this book that we're working with image data, so let's take a moment to explore how to handle it in Python. A common library for loading and processing images is Pillow, the modern version of the original PIL (Python Imaging Library). It retains the PIL module name for compatibility and is widely used for image handling.

Now let's take a look at one of the images. We'll start with a filepath we retrieved earlier, as shown in Figure 5-5:

```
from PIL import Image

Image.open(metadata["full_path"].iloc[0])
```

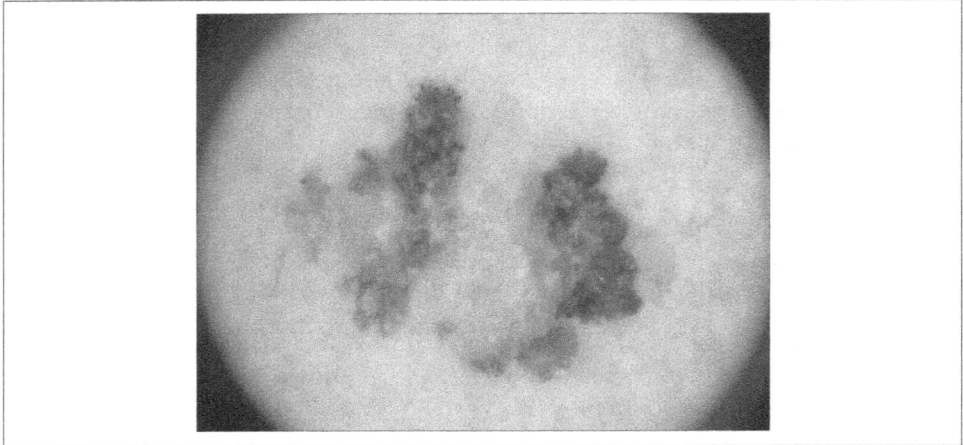

Figure 5-5. An example skin lesion image loaded using Pillow in Python.

We'd like to inspect images from specific lesion classes. To make this process easier, we'll create a function that:

- Randomly selects an image from a specified class in the dataset
- Loads the image using PIL (Pillow)
- Displays it with Matplotlib, including the class name as a title for clarity

```python
import matplotlib.pyplot as plt


def show_random_image(metadata: pd.DataFrame, class_name: str) -> plt.Figure:
    record = (
        metadata[metadata["class_orig"] == class_name]
        .sample(1)
        .to_dict(orient="records")[0]
    )
    fig = plt.figure(figsize=(4, 4))
    plt.imshow(Image.open(record["full_path"]))
    plt.title(record["class_orig"].capitalize())
    return fig
```

Let's now use this code to examine an example melanoma image, as shown in Figure 5-6:

```
show_random_image(metadata, "melanoma");
```

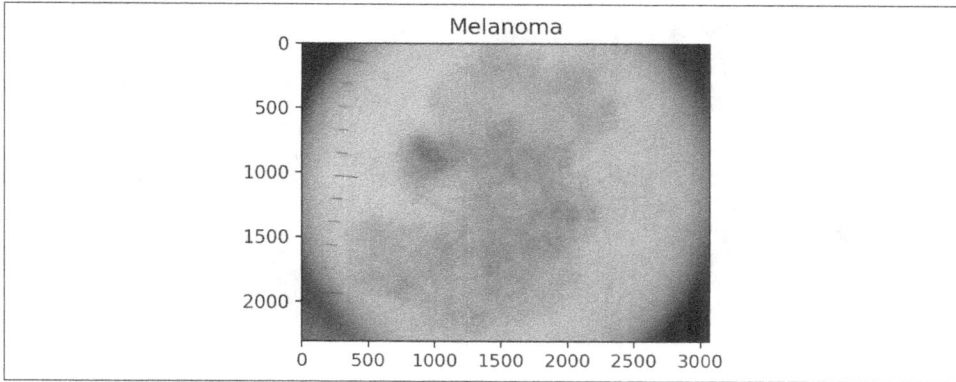

Figure 5-6. Example image from the "melanoma" class.

Next, let's write a `plot_random_image_grid` to help us visualize a sample image from each skin lesion class. This will help us quickly get a sense of the dataset's diversity and variation in lesion appearance:

```python
def plot_random_image_grid(
    metadata: pd.DataFrame, ncols: int = 3
) -> plt.Figure:
    """Display a random example image from each class in a grid."""
    records = metadata.groupby("class_orig").sample(1).to_dict(orient="records")
    nrows = (len(records) + ncols - 1) // ncols
    fig, axes = plt.subplots(nrows, ncols, figsize=(10, 2.5 * nrows))
    axes = axes.flatten()

    for record, ax in zip(records, axes):
        ax.imshow(Image.open(record["full_path"]))
        ax.set_title(record["class_orig"].capitalize())
        ax.axis("off")

    plt.tight_layout()
    return fig
```

Let's run this function (see Figure 5-7):

```
plot_random_image_grid(metadata);
```

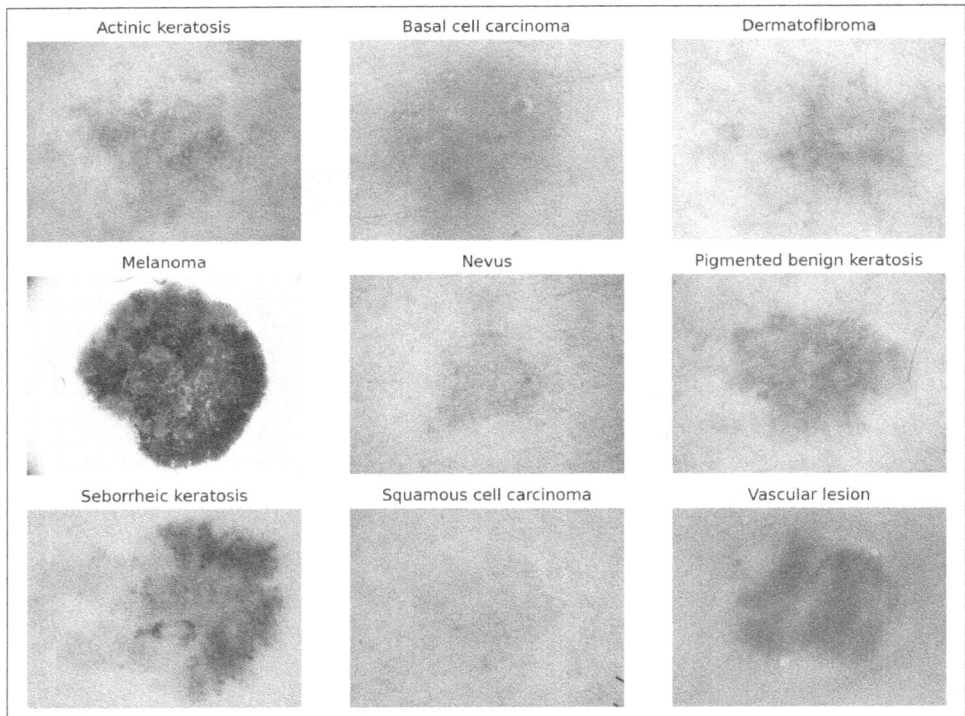

Figure 5-7. Grid of skin images with their corresponding labels.

With a better visual understanding of our dataset, we're now ready to build a flexible input pipeline that will support the rest of our experiments.

Addressing Dataset Issues

Now that we've explored the dataset, let's address some of its limitations. Since improving model performance will be an iterative process, it's helpful to create a `DatasetBuilder` class that makes it easy to experiment with different dataset configurations. We'll design it to be flexible enough to support both multiclass classification (all lesion types) and binary classification (e.g., melanoma versus non-melanoma), allowing us to explore a variety of setups as we develop our models.

First, to ensure proper evaluation, we define a 70/20/10 split for `train`, `valid`, and `test` sets:

```
import numpy as np

np.random.seed(seed=42)
splits = {"train": 0.7, "valid": 0.20, "test": 0.10}
```

```
metadata["split"] = np.random.choice(
    list(splits.keys()), p=list(splits.values()), size=metadata.shape[0]
)

counts = pd.crosstab(metadata["class_orig"], metadata["split"], margins=True)
print(counts)
```

Output:

```
split                 test  train  valid   All
class_orig
actinic keratosis       13     85     32   130
basal cell carcinoma    39    284     69   392
dermatofibroma          11     84     16   111
melanoma                55    307     92   454
nevus                   41    250     82   373
pigmented benign ...    39    358     81   478
seborrheic keratosis     9     50     21    80
squamous cell car...    14    138     45   197
vascular lesion         10     99     33   142
All                    231   1655    471  2357
```

We have visualized the class distribution across the new splits in Figure 5-8:

```
counts.drop(["All"], axis=1).drop(["All"], axis=0).plot.barh(stacked=True);
```

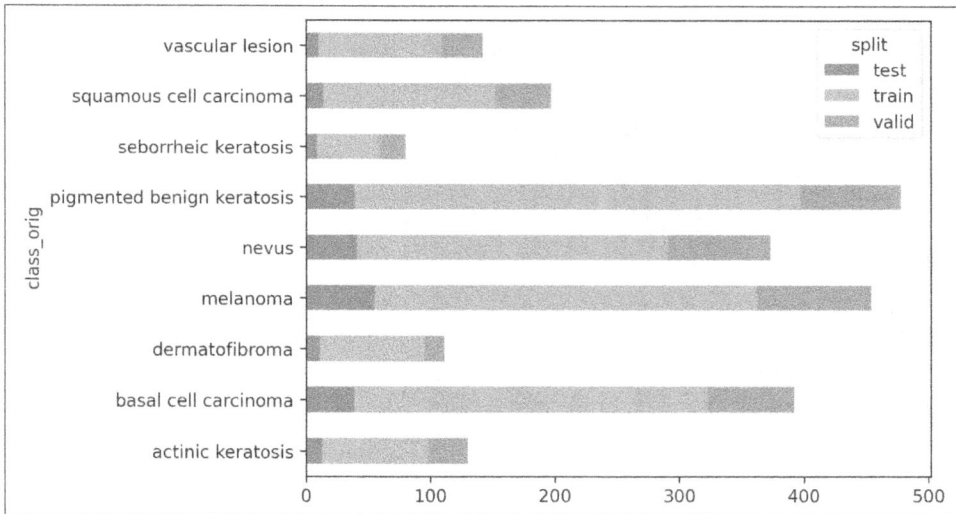

Figure 5-8. Bar plot of the distribution of classes across new training, validation, and test sets.

With this, we now have a better-structured dataset split—featuring a dedicated validation set, a larger test set, and consistent train/valid/test proportions across lesion classes. This sets the stage for building our DatasetBuilder, which will allow us to efficiently experiment with different training configurations.

Balancing the batches

Some skin lesion classes are much more common than others in our dataset. For example, pigmented benign keratosis and nevus are heavily represented, while classes like seborrheic keratosis and dermatofibroma are much rarer. If we train directly on this imbalanced data, the model may focus on predicting the dominant classes correctly while ignoring the rare ones—a bias that can hurt generalization.

To mitigate this, we implemented a *balanced sampler* during training. This ensures that each batch contains an equal number of examples from each class, giving the model more exposure to underrepresented categories. The validation and test sets, however, maintain their original class distributions to reflect real-world class frequencies during evaluation.

Although the balanced sampler didn't lead to measurable gains in our experiments for this chapter, it remains a useful tool when working with severely imbalanced datasets, which we encourage you to try in your future projects.

Augmenting the dataset

Our dataset is relatively small, and in deep learning, few things improve model performance more reliably than having access to more high-quality data. Since collecting new labeled images is expensive and time-consuming, we turn to *data augmentation*—a technique that synthetically increases dataset size by applying label-preserving transformations to existing images.

Augmentation encourages the model to generalize better by exposing it to natural variability in image appearance. For example, random rotations, flips, crops, color shifts, and brightness changes teach the model that the class of a lesion doesn't change just because its lighting or orientation does. This helps reduce overfitting and improves robustness to real-world image variation.

However, it's important to remember that augmentation does *not* introduce truly new information. Augmented examples are variations of the same underlying data—so any existing dataset biases (e.g., skin tone imbalance or limited anatomical diversity) will still be present. What augmentation does offer is a way to nudge the model toward learning more general, abstract features rather than memorizing surface-level details.

Rather than manually applying transformations, we use the DeepMind PIX library (*https://oreil.ly/3FxiK*)—a JAX-compatible image augmentation toolkit. It integrates cleanly into our pipeline and allows us to dynamically apply transformations on the fly during training, keeping memory usage low and training efficient.

In our setup, each image is randomly transformed with a fixed probability. As with other JAX operations, the random number generator (rng) is explicitly managed, ensuring that augmentations are reproducible and traceable:

```
@jax.jit
def rich_augmentor(image: jax.Array, rng: jax.Array) -> jax.Array:
  """Applies random flips, brightness, contrast, hue changes, and rotation."""
  image = pix.random_flip_left_right(rng, image)
  image = pix.random_flip_up_down(rng, image)
  image = pix.random_brightness(rng, image, max_delta=0.1)
  image = pix.random_contrast(rng, image, lower=0.9, upper=1)
  image = pix.random_hue(rng, image, max_delta=0.05)
  # Angles are provided in radians, i.e. +/- 10 degrees.
  image = pix.rotate(
    image,
    angle=jax.random.uniform(rng, shape=(), minval=-0.174533, maxval=0.174533),
  )
  return image
```

To illustrate augmentation in practice, we select a melanoma image and apply our transformation pipeline multiple times. Each variation is slightly different, simulating realistic image variability. See the result in Figure 5-9:

```
import jax
import jax.numpy as jnp

rng = jax.random.PRNGKey(seed=42)
image = Image.open(
  metadata[metadata["class_orig"] == "melanoma"]
  .sample(1)["full_path"]
  .values[0]
)

_, axes = plt.subplots(3, 3, figsize=(9, 9))
axes = axes.flatten()
axes[0].imshow(image)
axes[0].set_title("Original")
axes[0].axis("off")
image = jnp.asarray(image) / 255.0
for ax in axes[1:]:
  rng, rng_augment = jax.random.split(rng, num=2)
  ax.imshow(rich_augmentor(image, rng_augment))
  ax.set_title("Augmented")
  ax.axis("off")
plt.tight_layout()
```

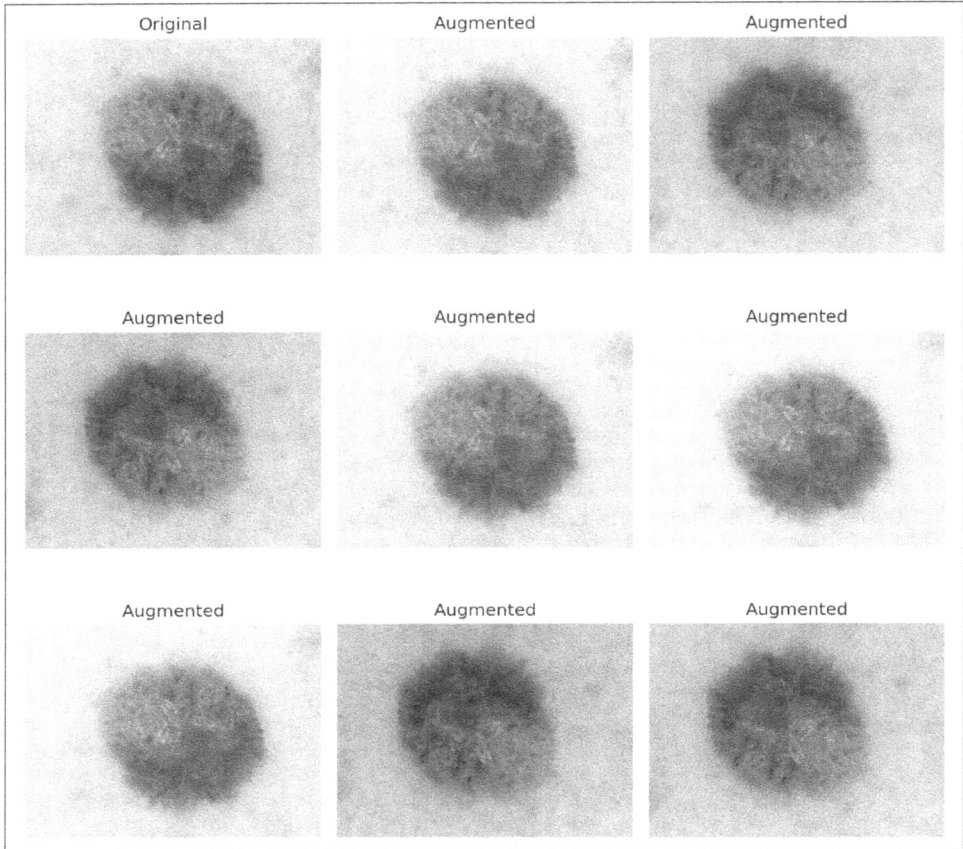

Figure 5-9. Original skin image (top left) followed by different augmented versions, demonstrating transformations like flipping, rotation, and color shifts.

This setup is a good starting point, but keep in mind that each augmentation method has tunable parameters—and the PIX documentation (*https://oreil.ly/74UjB*) lists many more options. Selecting and tuning augmentations is effectively a hyperparameter search. What works best depends heavily on the dataset and task. For a broader overview of effective augmentation strategies, see this survey.[7]

Preprocessing the images

Before feeding images into a neural network, we need to standardize them. This involves ensuring that all inputs have the same shape and pixel value range. Without

---

7 Shorten, C., & Khoshgoftaar, T. M. (2019). A survey on Image Data Augmentation for Deep Learning (*https://doi.org/10.1186/s40537-019-0197-0*). *Journal of Big Data*, 6(1).

this step, differences in resolution, aspect ratio, or intensity could prevent the model from learning consistent patterns.

Many deep learning computer vision models—including ImageNet-trained ResNets we will work with in this chapter—expect fixed-size square input images. However, skin lesion photos vary widely in shape and size. Simply resizing these directly to a square can distort important clinical features. To avoid this, we first resize the image while preserving its aspect ratio, then center-crop a square from the result. This produces consistent input shapes without stretching or compressing lesions.

```python
def resize_preserve_aspect(
    image: jax.Array, short_side: int = 256
) -> jax.Array:
    """Resize image with shorter side is `short_side`, keeping aspect ratio."""
    h, w, c = image.shape
    scale = short_side / jnp.minimum(h, w)
    new_h = jnp.round(h * scale).astype(jnp.int32)
    new_w = jnp.round(w * scale).astype(jnp.int32)
    resized = jax.image.resize(image, (new_h, new_w, c), method="bilinear")
    return resized


def center_crop(image: jax.Array, size: int = 224) -> jax.Array:
    """Crop the center square of given size from an image."""
    h, w, _ = image.shape
    top = (h - size) // 2
    left = (w - size) // 2
    return image[top : top + size, left : left + size]
```

We also need to normalize pixel values. Raw image intensities usually range from 0 to 255, but neural networks tend to train more effectively when inputs are scaled to a smaller, consistent range. A common method is *min-max scaling*, where each pixel is divided by 255 to bring values into the [0, 1] range. Another option is standardization, where each pixel is transformed to have zero mean and unit variance.

> The term *standardization* is overloaded in this context and refers to two different concepts:
>
> - *General image preprocessing*: Ensuring consistency in image size, format, and value ranges
> - *A specific normalization strategy*: Transforming pixel values to have a mean of 0 and a standard deviation of 1
>
> Throughout this book, we'll use *standardization* in the first sense—as a synonym for general input preprocessing.

For our model, we'll use min-max scaling to normalize pixel values, dividing by 255 to bring intensities into the [0, 1] range. This approach is widely used in CNN training and works well for image data:

```
def rescale_image(image: jax.Array) -> jax.Array:
    """Normalizes pixel values to the [0, 1] range by dividing by 255."""
    return image / 255.0
```

Now let's apply all preprocessing steps—reading JPEG images, resizing while preserving aspect ratio, center-cropping, and normalizing—to ensure that our dataset is standardized before training. While we could choose any image size, we'll use 224 × 224 pixels to match the expected input size of the original ResNet architecture, which we'll explore later in the chapter. Since these transformations need to be applied only once, doing them in advance improves efficiency and avoids redundant computations during training.

In Figure 5-10, you can see an example of an image before and after preprocessing:

```
original_image = Image.open(metadata.iloc[0]["full_path"])
image = np.array(original_image)
image = resize_preserve_aspect(image)
image = center_crop(image)
image = rescale_image(image)

_, axes = plt.subplots(1, 2, figsize=(6, 3))
axes = axes.flatten()
axes[0].imshow(original_image)
axes[0].set_title("Original")
axes[1].imshow(image)
axes[1].set_title("Preprocessed");
```

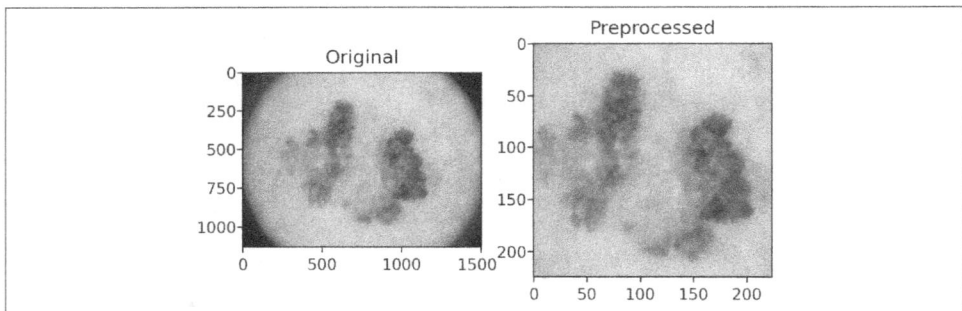

Figure 5-10. Images are standardized to ensure consistent dimensions and pixel value ranges. Note that although the preprocessed image has been rescaled from 0-255 to 0-1, imshow automatically adjusts the display scale, so the visual appearance remains unchanged.

With data preprocessing topics complete, we can now focus on efficiently storing and accessing the data—a crucial step for scaling training without overloading system memory.

Data storage with memory-mapped arrays

We've already emphasized the importance of efficiency in model training: faster data processing means faster iterations and ultimately better models. A major inefficiency we want to avoid is repeatedly preprocessing the same data every time a batch is passed to the model. At the same time, storing all processed images in memory isn't feasible. So, how can we balance efficiency and memory constraints?

A practical solution is *memory-mapped NumPy arrays*. These allow us to store pre-processed images on disk while accessing them as if they were in memory, making retrieval fast and efficient. This way, we preprocess all images once before training and then load them dynamically during training without redundant computation.

The implementation is straightforward: we first define the storage file, specify the data type (`float32`), and set the shape of the dataset: 224 × 224 pixels with three color channels. This creates an empty file on disk, preallocating the required storage space. We then preprocess and store each image in the memory-mapped array:

```python
from tempfile import TemporaryFile

images_on_disk = np.memmap(
    TemporaryFile(), dtype="float32", mode="w+", shape=(10, 224, 224, 3)
)
for i, full_path in enumerate(metadata[:10]["full_path"]):
    image = Image.open(full_path)
    image = np.array(image)
    image = resize_preserve_aspect(image)
    image = center_crop(image)
    image = rescale_image(image)
    images_on_disk[i, :, :, :] = image
images_on_disk.flush()
```

This approach gives us the best of both worlds: efficient preprocessing without excessive memory consumption.

Building a DatasetBuilder

At this point, we have tackled the major challenges of working with this dataset: splitting it into training, validation, and test sets; handling class imbalances; preprocessing images; and augmenting the data. Now we will organize these components into a structured `DatasetBuilder` class that returns `Dataset` instances that have the data iterators that will actually fetch us batches of data that we can train or evaluate a model on.

Designing a flexible dataset builder pays off when iterating on your model. It lets you quickly try different label mappings, sampling strategies, and preprocessing steps—all without rewriting core code. This modularity speeds up experimentation and helps you pinpoint which data choices actually improve performance.

Let's take a closer look at the `DatasetBuilder`, which coordinates the creation of dataset splits and preprocessing:

```python
class DatasetBuilder:
    """Builds a dataset with metadata, loaded images, and class mappings."""

    def __init__(self, data_dir: str, out_dir: str | None = None) -> None:
        self.data_dir = data_dir
        self.out_dir = out_dir or data_dir

    def build(
        self,
        rng: jax.Array,
        splits: dict[str, float],
        preprocessors: list[Callable] = [crop],
        image_size: tuple[int, int, int] = (224, 224, 3),
        class_map: dict[str, Any] | None = None,
    ) -> dict[str, Dataset]:
        """Builds the dataset splits from loaded metadata and loaded images."""
        metadata = MetadataLoader(self.data_dir, self.out_dir).load(class_map)
        images = ImageLoader(metadata, self.out_dir).load(preprocessors, image_size)

        # Shuffle the dataset and assign each example to one of the dataset splits.
        num_samples = len(metadata)
        rng, rng_perm = jax.random.split(rng, 2)
        shuffled_indices = jax.random.permutation(rng_perm, num_samples)

        # Create each dataset split using the shuffled indices and store the
        # corresponding metadata and image data in a Dataset object.
        dataset_splits, start = {}, 0
        for name, size in self._get_split_sizes(splits, num_samples):
            indices = np.array(shuffled_indices[start : (start + size)])
            dataset_splits[name] = Dataset(
                metadata=metadata.iloc[indices],
                images=images,
                num_classes=metadata["class"].nunique(),
            )
            start += size
        return dataset_splits

    def _get_split_sizes(self, splits: dict[str, float], num_samples: int):
        """Convert fractional split sizes to integer counts that sum to total."""
        names = list(splits)
        sizes = [int(num_samples * splits[name]) for name in names[:-1]]
```

```
sizes.append(num_samples - sum(sizes))  # last split gets the remainder
yield from zip(names, sizes)
```

We've already implemented each of these components, but let's highlight a few key details:

Metadata loading

MetadataLoader ensures that all image filepaths and their associated class/split information are consistently saved to a CSV file. This avoids inconsistencies from relying on rglob, which does not guarantee a consistent file order. Because the image data is later stored in a memory-mapped array in the same order as the metadata, maintaining this consistency is crucial.

Flexible class mapping

The MetadataLoader also supports binary or grouped classification setups via a customizable class_map. If no map is provided, it defaults to a standard multiclass configuration based on the directory structure.

Image preprocessing

ImageLoader handles reading, processing, and storing images in memory-mapped arrays. To avoid recomputation, images are stored in a compact raw format once, and then multiple preprocessing strategies (e.g., resizing or cropping) can be applied and stored separately as memmaps.

Dataset creation

The build method brings everything together: it loads metadata and images, shuffles the dataset, splits it into train/validation/test partitions, and wraps each into a Dataset object. Each dataset contains consistent metadata and preprocessed image views, ready for training.

Loading the Metadata and Images

You can find the code for the supporting classes for the MetadataLoader and Image Loader in the dlfb.cancer.dataset.builder library. Here we will briefly describe how they work and fulfill their responsibility to create the Dataset.

The MetadataLoader constructs a metadata CSV file if one does not already exist. It extracts the original split and class information from the folder structure, applies any desired label mappings, and provides consistent row-level access to the dataset. Image loading is a bit more involved. While MetadataLoader handles the filepaths and class labels, ImageLoader is responsible for reading, preprocessing, and storing the actual image data. Here are some notes on how ImageLoader works:

Image preprocessing and storage

To avoid redundant computations, ImageLoader first stores the raw images as a single memory-mapped array. This array is compact, storing flattened versions of each image along with their original shapes and offsets so they can later be reconstructed. Once this raw storage is complete, various preprocessing functions (such as cropping or resizing) can be applied and the results saved into additional memory-mapped files. This design allows us to support multiple preprocessing strategies without reloading the original JPEGs or repeating expensive operations.

Preprocessing flexibility

Preprocessing functions are applied in a modular way. Any number of transforms can be passed into the `ImageLoader`, and each will produce its own memmap. These preprocessed views are stored under different keys (e.g., `"crop"`, `"resize"`) inside the `Images` object, which acts as a unified interface to access them.

Dataset assembly

Finally, the `DatasetBuilder.build()` method pulls all of this together. It loads the metadata, processes the images, shuffles the dataset, splits it into training, validation, and test sets, and returns `Dataset` objects for each. These objects provide clean access to both image arrays and class labels, ready for model training.

This is all stored in the `Dataset` class itself:

```python
@dataclass
class Dataset:
    """Dataset class storing images and corresponding metadata."""

    metadata: pd.DataFrame
    images: Images
    num_classes: int

    def get_dummy_input(self) -> jnp.ndarray:
        """Returns dummy input with the correct shape for the model."""
        return jnp.empty((1,) + self.images.size)

    def num_samples(self) -> int:
        """Returns the number of samples in the dataset."""
        return self.metadata.shape[0]

    def get_images(self, preprocessor: Callable, indices: jax.Array) -> jax.Array:
        """Returns preprocessed images for the given indices."""
        return self.images.loaded[preprocessor.__name__][indices]

    def get_labels(self, indices: list[int]) -> jax.Array:
        """Returns integer class labels for the given indices."""
        return jnp.int16(self.metadata.loc[indices]["label"].values)
```

Where the `Images` are:

```
@dataclass
class Images:
    """Stores image data and size information."""

    loaded: dict[str, np.memmap]
    size: tuple[int, int, int]
```

The `Dataset` class is the primary interface for accessing data during training. It manages both the `metadata` (which tracks image paths and labels) and the memory-mapped `images`. Since we split the dataset into training, validation, and test sets, we can control their proportions using the `splits` argument.

Batching the data

To streamline dataset handling, during training, we introduce the `BatchHandler` class. This class encapsulates the components required to generate batches from a `Dataset` object:

- The `preprocessor` selects which preprocessed image view (e.g., cropped, resized) to use.

- The `sampler` determines how to draw samples for each batch (e.g., balanced versus random sampling).

- The optional `augmentor` applies real-time image transformations such as flipping, color jitter, or rotation to improve generalization.

```
class BatchHandler:
    """Helper to provide appropriately prepared batches form a dataset."""

    def __init__(
        self,
        preprocessor: Callable = crop,
        sampler: Callable = epoch_sampler,
        augmentor: Callable | None = None,
    ):
        self.preprocessor = preprocessor
        self.sampler = sampler
        self.augmentor = augmentor

    def get_batches(self, dataset: Dataset, batch_size: int, rng: jax.Array):
        """Prepare dataset batches with the requested image manipulations."""
        self._validate_batch_size(dataset, batch_size)
        rng, rng_sampler = jax.random.split(rng, num=2)

        for batch_indices in self.sampler(
            dataset.metadata, batch_size, rng_sampler
        ):
            images = dataset.get_images(self.preprocessor, batch_indices)
```

```python
    if self.augmentor is not None:
      rng, rng_augment = jax.random.split(rng, 2)
      images = jax.vmap(lambda image, r: self.augmentor(image, rng=r))(
        images, jax.random.split(rng_augment, images.shape[0])
      )

    batch = {
      "frame_ids": batch_indices,
      "images": images,
      "labels": dataset.get_labels(batch_indices),
    }
    yield batch

  @staticmethod
  def _validate_batch_size(dataset: Dataset, batch_size: int) -> None:
    """Ensures that batch_size is within feasible bounds."""
    if batch_size > dataset.num_samples():
      raise ValueError(
        f"Batch size ({batch_size}) cannot be larger than dataset size "
        "({len(frame_ids)})."
      )
    if batch_size > dataset.num_classes:
      raise ValueError(
        f"batch_size ({batch_size}) has to be larger than "
        f"number of unique labels."
      )
```

The `get_batches` method is the core of `BatchHandler`. It's a generator function, meaning it yields one batch at a time, which is memory-efficient and well-suited for training loops. Here's what happens inside:

1. It validates the requested `batch_size`.

2. It samples batch indices using the `sampler` function.

3. It fetches preprocessed images and labels from the `Dataset`.

4. If an `augmentor` is provided, it applies augmentation per image using JAX's `vmap` for efficiency.

5. It yields batches as dictionaries containing `frame_ids`, `images`, and `labels`.

This design cleanly separates dataset structure from augmentation and sampling logic, making it easy to experiment with different batching strategies during training.

Readying the Dataset

We're now ready to construct our dataset and sanity-check the outputs of the data pipeline:

```
from dlfb.cancer.train.handlers.augmentors import rich_augmentor
from dlfb.cancer.train.handlers.samplers import balanced_sampler

rng = jax.random.PRNGKey(seed=42)
rng, rng_dataset = jax.random.split(rng, 2)

builder = DatasetBuilder(data_dir=assets("cancer/datasets/raw"))

dataset_splits = builder.build(
  rng=rng_dataset,
  splits={"train": 0.7, "valid": 0.20, "test": 0.10},
  image_size=(224, 224, 3),
)

print(dataset_splits["train"].metadata)
```

Output:

```
     frame_id split_orig          class_orig          full_path      label  \
     177          177    Train    actinic keratosis   /content/drive/M...     0
     408          408    Train    basal cell carci... /content/drive/M...     1
     445          445    Train    basal cell carci... /content/drive/M...     1
     ...          ...      ...                ...                 ...       ...
     1579        1579    Train    pigmented benign... /content/drive/M...     5
     1966        1966    Train    seborrheic kerat... /content/drive/M...     6
     305          305    Train    basal cell carci... /content/drive/M...     1

                        class
     177     actinic keratosis
     408     basal cell carci...
     445     basal cell carci...
     ...                ...
     1579    pigmented benign...
     1966    seborrheic kerat...
     305     basal cell carci...

     [1649 rows x 6 columns]
```

This call initializes the full data pipeline: it loads metadata; processes images; splits the dataset into training, validation, and test partitions; and returns them as Dataset objects (as keys in the dataset_splits dict).

We can now use the BatchHandler to generate a batch from the training set:

```
rng, rng_train = jax.random.split(rng, 2)

train_batcher = BatchHandler(sampler=balanced_sampler, augmentor=rich_augmentor)
train_batches = train_batcher.get_batches(
  dataset_splits["train"], batch_size=32, rng=rng_train
)
batch = next(train_batches)
```

Each batch contains three keys—`images`, `labels`, and `frame_ids`:

- `images`: A float32 tensor with normalized pixel values.
- `labels`: Integer class labels corresponding to each image.
- `frame_ids`: Indices into the original metadata, useful for inspection and debugging.

Let's confirm the shape of the `images` tensor:

Output:

```
batch["images"].shape

(32, 224, 224, 3)
```

Each batch consists of 32 images, and we can see that each image has dimensions (224, 224, 3).

Before moving forward, let's visualize the first image in the batch to ensure our preprocessing pipeline has worked as intended. This simple sanity check helps catch any unexpected artifacts early on (see Figure 5-11):

```
plt.imshow(batch["images"][0]);
```

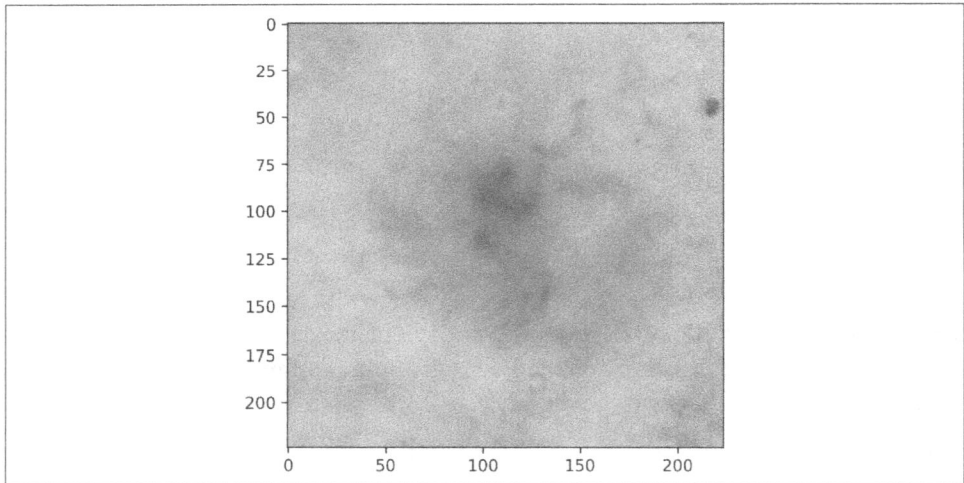

Figure 5-11. A processed skin lesion image from our dataset, confirming that our pipeline is working correctly.

Yup, it still looks like some sort of skin lesion! With a fully functional dataset pipeline prepared and validated, we're ready for the next step: training our model.

Building Skin Cancer Classification Models

With our dataset ready, we now turn to the central modeling exploration of this chapter. We'll compare a range of architectures, gradually increasing in complexity. Here's an overview of the models we'll build and evaluate:

SimpleCNN

A lightweight two-layer convolutional neural network that serves as our baseline. It shares the same classification head (SkinLesionClassifierHead) as the other models but is otherwise kept minimal. We won't tune it heavily—its purpose is to establish a performance baseline without any architectural sophistication.

ResNetFromScratch

A full ResNet50 architecture trained from randomly initialized weights. This model assesses the impact of the architecture alone, without any benefits from pretraining on ImageNet. It helps isolate the contribution of the ResNet design itself.

FinetunedResNet

A ResNet50 initialized with pretrained weights, with all layers fine-tuned on our skin lesion dataset. This approach leverages transfer learning, incorporating knowledge learned from large-scale datasets like ImageNet to improve convergence speed and generalization.

FinetunedHeadResNet

Similar to the previous item, but with the pretrained ResNet50 backbone frozen. Only the final classification head is trained. This variant tests how far we can get by leveraging pretrained features without modifying the backbone.

This progression—from scratch to full fine-tuning—lets us directly compare training efficiency and classification performance across strategies. Training from scratch gives full control but demands more data and compute. Transfer learning, especially when freezing the backbone, is typically faster and more robust in low-data regimes like ours.

ResNet50 is a widely used model convolutional architecture with 50 layers. You can easily experiment with deeper variants such as ResNet101 (*https://oreil.ly/P8Cvr*) or ResNet200. These offer greater representational power but come at the cost of increased compute and a higher risk of overfitting—especially when training data is limited. All variants are compatible with the same ImageNet preprocessor.

To implement these models, we'll follow these steps:

1. Load the ResNet50 model for Flax from transformers.

2. Extract the model backbone.

3. Attach a custom classification head (`SkinLesionClassifierHead`).

4. Train and evaluate each model variant.

Let's start by loading the pretrained ResNet50 model.

Loading the Flax ResNet50 Model

Flax offers a prebuilt ResNet implementation (*https://oreil.ly/Yaf_p*), which provides a flexible and well-optimized architecture for feature extraction. However, prebuilt models aren't always available—or pretrained—depending on your needs. To demonstrate how to work with pretrained models in Flax, we'll load a ResNet50 model from Hugging Face instead of using the Flax example directly.

The Hugging Face `FlaxResNetModel` (*https://oreil.ly/xWVhz*) provides a convenient way to import and use pretrained ResNet models, making it easy to fine-tune or extract features from standard models.

To see how this works in practice, we'll follow Hugging Face's example code and apply a pretrained ResNet50 to a sample image. We've been looking at skin lesions for quite a while, so let's take a break with something more cheerful: a photo of cats lounging on a couch (see Figure 5-12):

```
import requests

# Load the example image from the documentation.
url = "http://images.cocodataset.org/val2017/000000039769.jpg"
image = Image.open(requests.get(url, stream=True).raw)
plt.imshow(image);
```

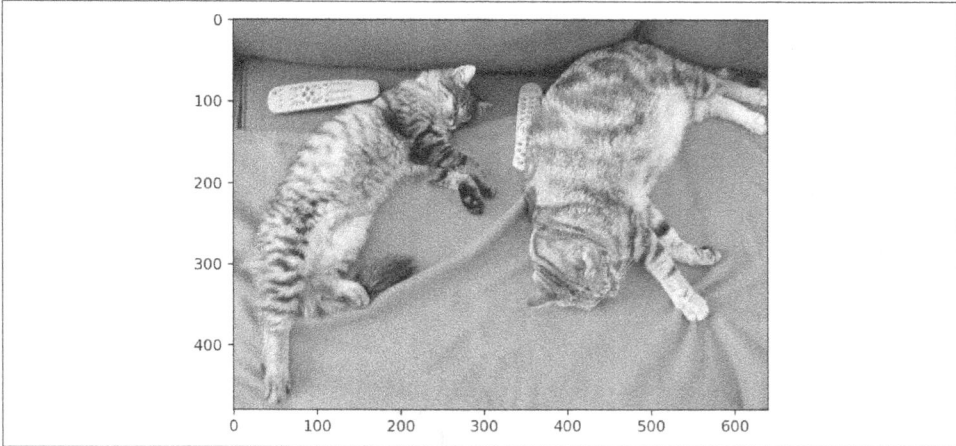

Figure 5-12. Cats on a couch.

Loading a pretrained ResNet model from Hugging Face is straightforward. We'll use a ResNet50 model trained on ImageNet and pair it with an image preprocessor to ensure that our input image is correctly formatted:

```
from transformers import AutoImageProcessor, FlaxResNetForImageClassification
resnet_model = FlaxResNetForImageClassification.from_pretrained(
    "microsoft/resnet-50"
)
image_processor = AutoImageProcessor.from_pretrained("microsoft/resnet-50")
```

Now let's preprocess the cat image to obtain the `pixel_values` input format expected by the model:

```
inputs = image_processor(images=image, return_tensors="jax")
inputs.keys()
```

Output:

```
dict_keys(['pixel_values'])
```

Let's inspect what the preprocessor does to input images in Figure 5-13:

```
plt.imshow(jnp.transpose(inputs["pixel_values"], (0, 2, 3, 1))[0]);
```

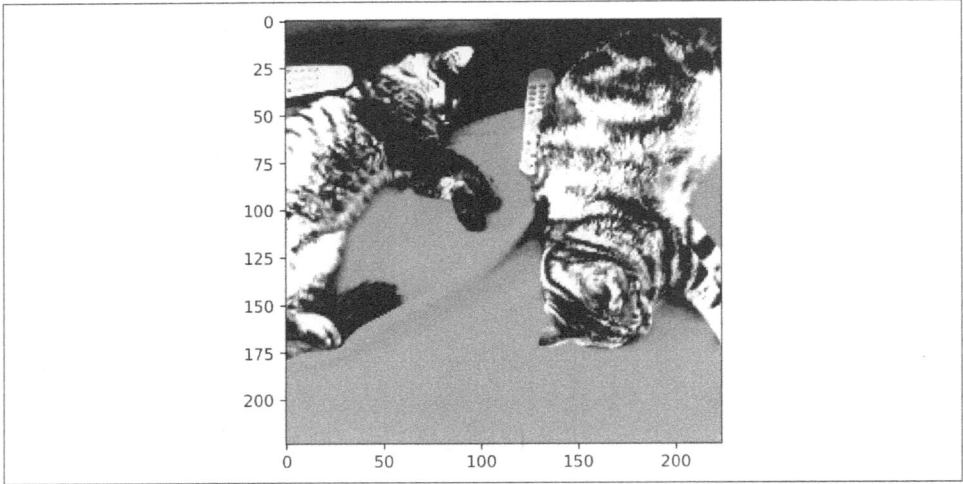

Figure 5-13. Cats on a couch after image processor standardization.

The processed image looks noticeably different from the original; the colors have shifted, and the image has been cropped to 224 × 224 pixels. This transformation is essential when using pretrained weights: models expect inputs normalized to the mean and standard deviations of the dataset they were trained on. This standardization ensures consistent performance, even if your input data (like medical images) has different color distributions.

Here are two important guidelines:

- When fine-tuning a pretrained model, normalize inputs using the original training dataset's statistics (e.g., ImageNet mean and std).
- When training from scratch, you can normalize using your own dataset's range or statistics (mean and standard deviations).

Now let's make a prediction:

```
outputs = resnet_model(**inputs)
logits = outputs.logits
predicted_class_idx = jnp.argmax(logits, axis=-1)
print(
    "Predicted class:", resnet_model.config.id2label[predicted_class_idx.item()]
)
```

Output:

```
Predicted class: tiger cat
```

Success! Well, sort of—the model classifies the image as the class "tiger cat." While not a perfect match, it's a reasonable guess given the image. This highlights both

---

the strengths and limitations of pretrained models: they excel at recognizing familiar patterns but may struggle with patterns outside of original training data distribution.

Let's now explore how we can reuse the core architecture of this model while adapting it to a new task: classifying skin lesions.

Extracting the ResNet Backbone

As mentioned, the pretrained ResNet50 model we're using has been trained on ImageNet to classify images into one of 1,000 classes such as "tiger cat," "airplane," and "fire truck." You might wonder: is melanoma one of the classes? Unfortunately, it is not. So how can we still use this model?

The key lies in the layered architecture of deep learning models. The pretrained ResNet50 consists of:

- A *backbone* (ResNet module) that extracts general features from images
- A *classifier head* that maps these features to one of ImageNet's 1,000 categories

Since we don't need the ImageNet classification head, we'll replace it with our own custom classifier for melanoma detection.

> While Hugging Face offers `FlaxResNetModel`—a ResNet variant without the classification head—for exactly this use case, such headless versions aren't available for every model. Learning how to manually extract and repurpose parts of a pretrained model is an essential skill. It gives you full control over what components are used and prepares you for working with a wider range of architectures where clean abstractions may not exist. That's why we demonstrate the manual approach here.

To separate the backbone from the classifier head:

```
module, variables = resnet_model.module, resnet_model.params
backbone_module, backbone_vars = module.bind(variables).resnet.unbind()
```

Here's what each step does:

- We extract the components of the `FlaxResNetForImageClassification` model provided by Hugging Face:
  - `module`: The full architecture definition
  - `variables`: The pretrained weights for all parts of the model
- To isolate just the ResNet backbone:
  - We `bind` the weights to the module.

— We access the `resnet` submodule, which excludes the classification head.

— We then `unbind` it, separating the backbone into its own callable Flax module with its own parameters.

This technique gives you direct access to intermediate model components—even when they aren't explicitly exposed. If you're unsure what submodules are available, you can inspect the bound module by listing its attributes or checking the Hugging Face docs.

Now, we can use the `backbone_module` just like any other Flax model by calling its `apply` method. For input, we pass the `backbone_vars` and a preprocessed image.

> Unlike many Hugging Face transformer models, Flax models typically expect input images in NHWC format (batch, height, width, channels). If the image is in NCHW format (batch, channels, height, width), use `transpose` or `moveaxis` to first reorder the dimensions.

```
outputs = backbone_module.apply(
    backbone_vars, jnp.transpose(inputs["pixel_values"], (0, 2, 3, 1))
)
last_hidden_state = outputs.last_hidden_state
last_hidden_state.shape
```

Output:

```
(1, 2048, 7, 7)
```

The output shape (1, 2048, 7, 7) represents:

- 2048: The number of feature channels output by the final ResNet block.

- 7, 7: The spatial dimensions after all convolutions and pooling layers have been applied. In other words, these operations have downsampled the image from 224 × 224 to 7 x 7 height and width.

At this stage, the feature maps are highly abstract, far removed from the original cat image. However, they still capture meaningful patterns learned by the model. These 2,048 feature maps would then be fed into the fully connected layers for the purpose of actually classifying the image. If we visualize them, we get something like Figure 5-14:

```
def display_feature_maps(feature_map, ncols=8):
    """Plot grid of the first 64 feature maps."""
    num_features = feature_map.shape[0]
    nrows = (num_features + ncols - 1) // ncols
    _, axes = plt.subplots(nrows, ncols, figsize=(ncols, nrows))
    axes = axes.flatten()
```

```
for i, (ax, feature) in enumerate(zip(axes, feature_map)):
    ax.imshow(feature, cmap="viridis")
    ax.axis("off")

plt.tight_layout()
plt.show()


display_feature_maps(last_hidden_state[0, 0:64, ...])
```

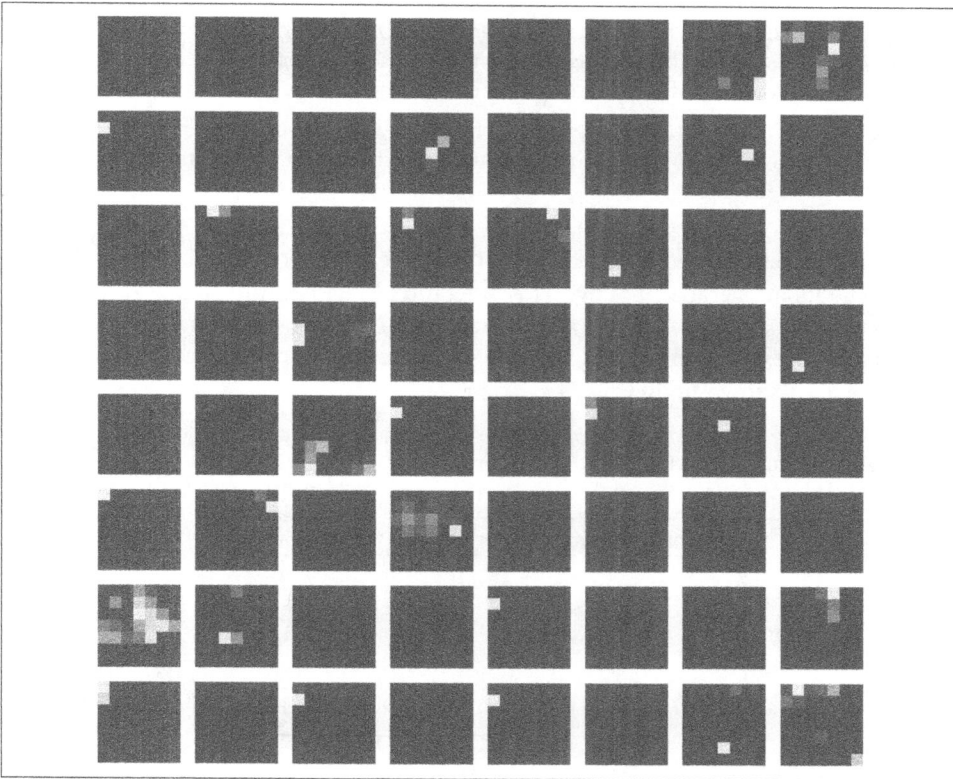

Figure 5-14. Feature maps from the final convolutional layer of ResNet50, illustrating how the model "sees" the cat image at that stage. This visualization shows activations from only the first 64 of 2048 learned filters. Each square represents the response of a different convolutional filter, with bright regions indicating areas of high activation. Many filters remain largely inactive, while others detect specific patterns or textures.

Some feature maps remain largely inactive (like those in the first row), while others highlight specific aspects of the cat image. At this late stage in the neural network, the representation of the cat image is highly abstract and no longer interpretable in a human-recognizable way. However, these learned patterns are still essential for the model's classification process.

> In this example, we simply examined the activations of late-stage convolutional filters in response to a cat image to demonstrate how to extract feature maps from a pretrained model. While useful as a sanity check, this doesn't reveal what each filter has actually learned.
>
> To dig deeper, try *feature visualization*: a technique where you iteratively modify an input image to maximize the activation of a specific filter. This reveals the kinds of patterns—like textures, edges, or object parts—that the filter responds to.
>
> For an excellent deep dive, check out the classic Distill blog post from 2017 (*https://oreil.ly/XpdSL*) on visualizing convolutional filters.

We can also access the model's final `pooler_output`, which represents the last hidden state after global average pooling. This operation collapses the spatial dimensions, leaving us with a single 2,048-dimensional feature vector—the final output of the model's feature extraction process. This vector serves as the condensed representation of the image, as shown in Figure 5-15:

```
_, ax = plt.subplots(1, 1, figsize=(6, 2))
ax.plot(np.squeeze(outputs.pooler_output), c="grey")
plt.title(f"Shape of outputs after pooling: {outputs.pooler_output.shape}");
```

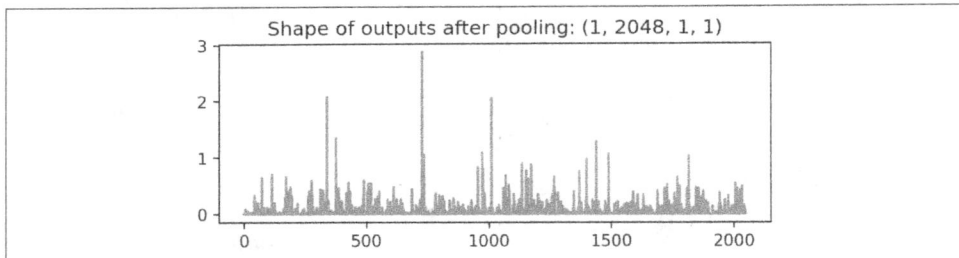

Figure 5-15. Visualization of the final pooled output of ResNet50 on the cat image input, also known as the last hidden state after global average pooling. This 2048-dimensional feature vector serves as the input to the classifier head.

All right, we now have a standalone ResNet module extracted from the full classifier. We've also inspected its final output—a 2,048-dimensional feature vector produced by global average pooling. This vector captures the distilled representation of the input image and will serve as the input to our custom classifier head.

With the backbone in place, we're now ready to construct the `SkinLesionClassifier Head`.

Building the SkinLesionClassifierHead

Now that we have extracted the pretrained ResNet50 backbone, we can add a new classification head tailored for skin lesion detection. This head will take the feature vector output from Figure 5-15 and produce logits for our target classes.

Although Hugging Face follows a naming convention like `FlaxResNetForSkinLesion Classification`, we will slightly simplify and call our head `SkinLesionClassifier Head`. This will perform the final classification by returning the logits for each skin class we want to predict. This is a small feedforward neural network consisting of sequential `Dense` layers with ReLU activations. Let's take a look at its implementation:

```python
class SkinLesionClassifierHead(nn.Module):
    """Skin lesion classification MLP head."""
    num_classes: int
    dropout_rate: float

    @nn.compact
    def __call__(self, x, is_training: bool):
        x = nn.Dense(256, kernel_init=nn.initializers.xavier_uniform())(x)
        x = nn.relu(x)
        x = nn.Dropout(rate=self.dropout_rate)(x, deterministic=not is_training)
        x = nn.Dense(128, kernel_init=nn.initializers.xavier_uniform())(x)
        x = nn.relu(x)
        x = nn.Dropout(rate=self.dropout_rate)(x, deterministic=not is_training)
        x = nn.Dense(
            self.num_classes, kernel_init=nn.initializers.xavier_uniform()
        )(x)
        return x
```

The final layer has as many neurons as the number of target classes, defined by the `num_classes` argument.

> In this chapter, we focus on multiclass classification, but the same code generalizes to binary classification. While a single `nn.Dense` neuron with sigmoid activation is perhaps more typical for binary tasks, we use a two-neuron output instead—one per class (melanoma or nonmelanoma)—producing logits that sum to 1 after softmax. This approach is slightly less efficient but more flexible, as it naturally extends to multiclass problems.

You will also notice that we specify `nn.initializers.xavier_uniform()` for our weight initialization. While the most naïve approach is to initialize all weights to zero, this leads to a major problem: all neurons in a layer would receive the same gradients during backpropagation and update identically, resulting in redundant feature learning. This is known as a *failure to break symmetry*, and it can severely limit the network's ability to learn.

Instead, we use *Xavier initialization* (also known as *Glorot initialization*), which assigns small, random values to each weight. This helps ensure that neurons start with diverse parameters and can learn different features. Xavier initialization is designed to maintain stable variance in both activations and gradients across layers, improving training dynamics.

> Most modern deep learning libraries—including Flax—automatically apply good initializers for standard layers. We specify `xavier_uniform()` here for clarity, but in practice, you can often rely on the library's defaults.

Building Our Models

We're now ready to begin the core modeling section of this chapter. We'll compare several architectures, gradually increasing in complexity. For the more promising ones, we'll also explore enhancements to improve performance. Let's begin with our baseline.

SimpleCnn model as a baseline

Our first model is a straightforward two-layer convolutional neural network trained from scratch. It doesn't use any pretrained features, and we won't spend time tuning it. The goal here is simple: to establish a reference point. How well can a minimal architecture perform on this task without any bells or whistles?

```python
class SimpleCnn(nn.Module):
    """Simple CNN model with small convolutional backbone and classifier head."""

    num_classes: int
    dropout_rate: float = 0.0

    def setup(self):
        """Initializes the CNN backbone and classification head."""
        self.backbone = CnnBackbone()
        self.head = SkinLesionClassifierHead(self.num_classes, self.dropout_rate)

    @nn.compact
    def __call__(self, x, is_training: bool = False):
        """Applies the backbone and classifier head to the input."""
        x = self.backbone(x)
        x = self.head(x, is_training=is_training)
        return x

    def create_train_state(
```

```
    self, rng: jax.Array, dummy_input, tx
) -> TrainStateWithBatchNorm:
    """Creates the training state with initialized parameters."""
    rng, rng_init, rng_dropout = jax.random.split(rng, 3)
    variables = self.init(rng_init, dummy_input)
    state = TrainStateWithBatchNorm.create(
        apply_fn=self.apply,
        tx=tx,
        params=variables["params"],
        batch_stats=None,
        key=rng_dropout,
    )
    return state
```

As you can see, it uses our shared classification head, `SkinLesionClassifierHead`, and a custom lightweight backbone:

```
class CnnBackbone(nn.Module):
    """Compact convolutional feature extractor for image data."""

    @nn.compact
    def __call__(self, x):
        """Applies two conv-pool blocks and a dense layer to the input."""
        x = nn.Conv(features=32, kernel_size=(3, 3))(x)
        x = nn.relu(x)
        x = nn.avg_pool(x, window_shape=(2, 2), strides=(2, 2))
        x = nn.Conv(features=64, kernel_size=(3, 3))(x)
        x = nn.relu(x)
        x = nn.avg_pool(x, window_shape=(2, 2), strides=(2, 2))
        x = x.reshape((x.shape[0], -1))  # flatten
        x = nn.Dense(features=256)(x)
        return x
```

This backbone is intentionally simple—no residual connections, no batch normalization. It's just a stack of convolutional layers followed by flattening. It won't win any benchmarks, but it should give us a useful lower bound on performance for this task.

Even though this model doesn't include batch normalization, we'll still use a training state that supports it—`TrainStateWithBatchNorm`—to keep our training loop compatible across models:

```
class TrainStateWithBatchNorm(train_state.TrainState):
    """Train state that tracks batch statistics and a PRNG key."""

    batch_stats: dict | None
    key: jax.Array
```

Using a unified training state lets us reuse the same training loop for both this simple CNN and the more advanced ResNet models we'll explore next.

Now let's move on to training our ResNet variants to better understand the impact of architecture and transfer learning on skin lesion classification. With the ResNet50 backbone we extracted earlier and our custom skin lesion classification head already in place, we now have all the components we need to assemble and train more advanced models.

We'll build three models, starting with the simplest: ResNet50 from scratch.

Building the ResNetFromScratch model

Our first ResNet variant is `ResNetFromScratch`, defined next. It uses the full ResNet50 architecture but does not load any pretrained weights—all parameters are initialized randomly and learned from scratch. While this model still benefits from the structure of a relatively deep, well-tested architecture, it doesn't directly inherit any knowledge from ImageNet or other large-scale datasets.

This gives us a useful comparison point: how far can we get with a good architecture alone, without any pretrained weights? Here's the code for the model:

```python
class ResNetFromScratch(nn.Module):
    """ResNet model initialized from scratch with a custom classification head."""

    num_classes: int
    layers: int = 50
    dropout_rate: float = 0.0

    def setup(self):
        """Initializes the backbone and classification head."""
        self.backbone = PRETRAINED_RESNETS[self.layers].module
        self.head = SkinLesionClassifierHead(self.num_classes, self.dropout_rate)

    def __call__(self, x, is_training: bool = False):
        """Runs a forward pass through the model."""
        x = self.backbone(x, deterministic=not is_training).pooler_output
        x = jnp.squeeze(x, axis=(2, 3))
        x = self.head(x, is_training=is_training)
        return x

    def create_train_state(
        self, rng: jax.Array, dummy_input, tx
    ) -> TrainStateWithBatchNorm:
        """Initializes model parameters and optimizer state."""
        rng, rng_init, rng_dropout = jax.random.split(rng, 3)
        variables = self.init(rng_init, dummy_input, is_training=False)
        variables = self.transfer_parameters(variables)
        tx = self.set_trainable_parameters(tx, variables)
        state = TrainStateWithBatchNorm.create(
            apply_fn=self.apply,
            tx=tx,
            params=variables["params"],
            batch_stats=variables["batch_stats"],
            key=rng_dropout,
        )
```

```
    return state

def transfer_parameters(_, variables):
    """Returns variables unchanged (no transfer learning)."""
    return variables

@staticmethod
def set_trainable_parameters(tx, _):
    """Returns optimizer configuration with all parameters trainable."""
    return tx
```

In the setup method, we instantiate both the backbone and the head. These are then used sequentially in the __call__ method to compute predictions. The backbone is accessed through the Hugging Face wrapper, so we reference it using .module to extract the underlying Flax-compatible model.

You will have noticed the the model takes a layer parameter. It picks the ResNet model with the required number of layers from the Hugging Face library of preloaded models:

```
from transformers import FlaxResNetModel
```

```
# Dictionary of pretrained ResNet models from Hugging Face.
PRETRAINED_RESNETS = {
    18: FlaxResNetModel.from_pretrained("microsoft/resnet-18"),
    34: FlaxResNetModel.from_pretrained("microsoft/resnet-34"),
    50: FlaxResNetModel.from_pretrained("microsoft/resnet-50"),
}
```

We have settled for a value of 50 to get the ResNet50 backbone, but the layers parameter could be conveniently explored as a hyper-parameter.

To prepare the model for training, we define a training state via the cre ate_train_state method. This initializes the parameters, optionally transfers pre-trained weights (not used here), and sets which parameters should remain trainable. While ResNetFromScratch doesn't apply any changes to the weights, this method is designed to be overridden by subclasses—as we'll see next with FinetunedResNet.

Even though ResNetFromScratch does not use any pretrained weights, it defines a transfer_parameters method (which is a no-op here) and a set_trainable_parame ters method (also a placeholder). These provide a consistent interface across models and make it easy to plug in transfer learning logic in downstream variants.

Since ResNet uses batch normalization, our training state must also track batch statistics—so we use the TrainStateWithBatchNorm class defined earlier.

Fully fine-tuning a pretrained model

The second model, `FinetunedResNet50`, is a subclass of `ResNet50FromScratch` that modifies just one key detail: it loads pretrained weights from the Hugging Face ResNet50 model.

This is done by overriding the `transfer_parameters` method. Unlike the base class (where this method is a no-op), here it copies over both the backbone's weights and its batch normalization statistics. The classification head remains randomly initialized, since it's specific to our skin lesion classification task:

```
class FinetunedResNet(ResNetFromScratch):
    """ResNet model with pretrained weights and full fine-tuning."""

    def transfer_parameters(self, variables):
        """Replaces model parameters with pretrained ResNet weights."""
        resnet_variables = PRETRAINED_RESNETS[self.layers].params
        variables["params"]["backbone"] = resnet_variables["params"]
        variables["batch_stats"]["backbone"] = resnet_variables["batch_stats"]
        return variables
```

Once the parameters are loaded, we will fine-tune `all` of them—both the pretrained ResNet backbone and the newly added classification head. This setup allows the model to fully adapt to the skin cancer classification task, and should in theory lead to stronger performance when the source (ImageNet) and target (skin) domains share similar visual features, such as textures, colors, and edges, which should be the case here.

Freezing the backbone with FinetunedHeadResNet

The third model, `FinetunedHeadResNet`, uses a different transfer learning strategy. Instead of fine-tuning all layers, it freezes the pretrained ResNet backbone and updates only the final classification head. This approach is common when working with small datasets. It allows us to leverage powerful pretrained features while reducing the number of trainable parameters—which can help avoid overfitting and reduce training time.

Flax and Optax make it easy to define separate optimizer behaviors for different parts of the model. Here, we apply our usual optimizer to the classification head and use a no-op optimizer (which performs no updates) for the frozen backbone, ensuring only the final layers are trained. This is implemented by overriding the `set_trainable_parameters` method.

This is implemented in the following model, which subclasses `FinetunedResNet` to inherit pre-trained weights but overrides the `set_trainable_parameters` method to freeze the backbone:

```
class FinetunedHeadResNet(FinetunedResNet):
    """ResNet model with a frozen backbone and trainable classification head."""

    @staticmethod
    def set_trainable_parameters(tx, variables):
        """Freezes backbone parameters and trains only the classification head."""
        return optax.multi_transform(
            transforms={"trainable": tx, "frozen": optax.set_to_zero()},
            param_labels=traverse_util.path_aware_map(
                lambda path, _: "frozen" if "backbone" in path else "trainable",
                variables["params"],
            ),
        )
```

By subclassing `FinetunedResNet`, we inherit all the logic for loading pretrained weights and constructing the model—and only need to override the part that controls which parameters are trainable. This modular design maximizes code reuse while making it easy to explore different fine-tuning strategies with minimal duplication.

More customization options

It's worth highlighting that fine-tuning doesn't have to be all or nothing—you're not limited to freezing the whole backbone or simply training all parameters. With tools like `optax.multi_transform`, we can easily assign different optimizer behaviors to different parts of the model.

For example, a common strategy is to freeze the earliest convolutional layers (which tend to learn general-purpose features like edges and textures) while fine-tuning later layers that capture more task-specific patterns. Some layers can even be updated using a reduced learning rate, allowing for more cautious adaptation.

The following is an example model, `PartiallyFinetunedResNet`, which demonstrates this idea. It applies:

- A standard learning rate to the classification head and late-stage backbone layers
- A reduced learning rate to intermediate layers (i.e., `stages/3`)
- A complete freezing of the early backbone layers:

```
class PartiallyFinetunedResNet(FinetunedResNet):
    """ResNet model with selective fine-tuning of deeper layers."""

    @staticmethod
    def set_trainable_parameters(tx, variables):
        """Freezes early layers, fine-tunes other layers at variable LR."""

        def label_fn(path, _):
            joined = "/".join(path)
            if "backbone" in path:
                if "stages/3" in joined:
```

```
      return "reduced_lr"
    return "frozen"
  return "trainable"

return optax.multi_transform(
  transforms={
    "trainable": tx,
    "reduced_lr": optax.adam(learning_rate=1e-5),
    "frozen": optax.set_to_zero(),
  },
  param_labels=traverse_util.path_aware_map(label_fn, variables["params"]),
)
```

We won't be training or evaluating this model in the rest of the chapter, but it's useful to be aware of this level of flexibility. Once your model is structured cleanly, these kinds of fine-tuning schemes are straightforward to implement and can make a big difference when adapting to smaller or more specialized datasets.

Training the Models

With our `DatasetBuilder` ready to generate data and all of our model variants defined, it's time to train them. We'll now implement a training loop that follows a familiar structure.

The Training Loop

The training loop follows the same approach as in previous chapters:

1. Initialize the model's training state.
2. Iterate over a defined number of steps.
3. At each step, fetch a batch of training data and update the model.
4. Periodically evaluate on a validation set to track progress.

In code:

```
@restorable
def train(
  state: TrainStateWithBatchNorm,
  rng: jax.Array,
  dataset_splits: dict[str, Dataset],
  num_steps: int,
  batch_size: int,
  preprocessor: Callable = crop,
  sampler: Callable = repeating_sampler,
  augmentor: Callable = None,
  eval_every: int = 10,
) -> tuple[TrainStateWithBatchNorm, dict]:
```

```
"""Trains a model using the provided dataset splits and logs metrics."""
# Set up with metrics logger and classes numbers.
num_classes = dataset_splits["train"].num_classes
metrics = MetricsLogger()

# Get train batch iterator from which to take batches.
rng, rng_train, rng_eval = jax.random.split(rng, 3)
train_batcher = BatchHandler(preprocessor, sampler, augmentor)
train_batches = train_batcher.get_batches(
    dataset_splits["train"], batch_size, rng_train
)

steps = tqdm(range(num_steps))  # Steps with progress bar.
for step in steps:
    steps.set_description(f"Step {step + 1}")

    rng, rng_dropout = jax.random.split(rng, 2)
    train_batch = next(train_batches)
    state, batch_metrics = train_step(
        state, train_batch, rng_dropout, num_classes
    )
    metrics.log_step(split="train", **batch_metrics)

    if step % eval_every == 0:
        for batch in BatchHandler(preprocessor).get_batches(
            dataset_splits["valid"], batch_size, rng_eval
        ):
            batch_metrics = eval_step(state, batch, num_classes)
            metrics.log_step(split="valid", **batch_metrics)
        metrics.flush(step=step)

    steps.set_postfix_str(metrics.latest(["loss"]))

return state, metrics.export()
```

At a high level, this function:

1. Sets up a metrics logger and prepares the data loaders

2. Iterates through the given number of training steps

3. Calls `train_step` to apply a gradient update on each training batch

4. Periodically runs `eval_step` on validation batches

5. Logs and flushes metrics so we can monitor progress

This design gives us a steady stream of training and validation feedback. The validation metrics are useful not just for model selection but also for deciding when to stop training or intervene (for example, if a model is doing horrendously poorly, we can just kill the hopeless run).

Next, let's take a look at what happens inside a single training step.

The training step

The core of our training process is the `train_step` function. This function wraps around a helper called `calculate_loss`, which handles forward propagation, loss computation, and gradient updates. This is a common design pattern in JAX code.

```python
@partial(jax.jit, static_argnums=(3,))
def train_step(
    state: TrainStateWithBatchNorm,
    batch: dict[str, list[Any]],
    rng_dropout: jax.Array,
    num_classes: int,
) -> tuple[TrainStateWithBatchNorm, dict[str, jax.Array]]:
    """Performs a single training step and returns updated state and metrics."""

    def calculate_loss(params, images, labels):
        variables, kwargs = {"params": params}, {"mutable": []}
        if state.batch_stats is not None:
            variables.update({"batch_stats": state.batch_stats})
            kwargs.update({"mutable": ["batch_stats"]})
        logits, updates = state.apply_fn(
            variables,
            x=images,
            is_training=True,
            rngs={"dropout": rng_dropout},
            **kwargs,
        )
        loss = optax.softmax_cross_entropy_with_integer_labels(
            logits, labels
        ).mean()
        return loss, (logits, updates)

    grad_fn = jax.value_and_grad(calculate_loss, has_aux=True)
    (loss, (logits, updates)), grads = grad_fn(
        state.params, batch["images"], batch["labels"]
    )
    state = state.apply_gradients(grads=grads)
    if state.batch_stats is not None:
        state = state.replace(batch_stats=updates["batch_stats"])

    metrics = {
        "loss": loss,
        **compute_metrics(batch["labels"], logits, num_classes),
    }

    return state, metrics
```

Inside `calculate_loss`:

- Images and labels are extracted from the batch.

- The model performs a forward pass to compute logits (unnormalized prediction scores for each class).

- The predicted logits are compared to ground truth labels using `optax.soft max_cross_entropy_with_integer_labels`.

- The mean cross-entropy loss is computed—averaged over the examples in the batch.

- Gradients are then calculated with respect to this loss and applied to update the model's parameters.

One important detail: if the model uses batch normalization (as ResNet does), we also need to update the running statistics tracked by the batch norm layers. The function `calculate_loss` handles both cases—whether batch norm is used or not —by conditionally including and updating the `batch_stats` entry in the variable collection.

This structure makes the training step general and reusable across all models we've defined, regardless of their architectural complexity or normalization layers.

The evaluation step

The evaluation step mirrors the training step but is simpler, as it does *not* update model weights. Its sole purpose is to compute performance metrics on a validation (or test) batch.

```
@partial(jax.jit, static_argnums=(2,))
def eval_step(
  state: TrainStateWithBatchNorm, batch: dict[str, Any], num_classes: int
):
  """Evaluates model performance on a batch and computes metrics."""
  variables, kwargs = {"params": state.params}, {}
  if state.batch_stats is not None:
    variables.update({"batch_stats": state.batch_stats})
    kwargs.update({"mutable": False})

  logits = state.apply_fn(
    variables, x=batch["images"], is_training=False, **kwargs
  )
  loss = optax.softmax_cross_entropy_with_integer_labels(
    logits, batch["labels"]
  ).mean()

  metrics = {
    "loss": loss,
    **compute_metrics(batch["labels"], logits, num_classes),
```

```
    }
    return metrics
```

Here's how it works:

- The model runs a forward pass with `is_training=False`, disabling training-time behavior like dropout and updating batch norm statistics.

- If batch normalization is used, the stored `batch_stats` are included in the variable collection—but they are *not* updated. Instead, we rely on the running averages (exponential moving averages of the mean and variance of activations) that were collected during training.

- Logits are compared to the ground truth labels using the same loss function as during training.

- In addition to computing the loss, the function returns performance metrics using `compute_metrics`, which includes weighted precision and recall (explained later).

By keeping evaluation stateless and side-effect-free—that is, avoiding any updates to model parameters or internal statistics—we ensure that validation scores remain consistent, reliable, and comparable across runs.

The evaluation metrics

During training, we typically monitor the loss, but for evaluation, we will also compute *precision* and *recall* to gain a more complete picture of the model's performance on the skin lesion classification task.

Let's first discuss how these metrics work in the binary case—for example, distinguishing melanoma from nonmelanoma—and then generalize to the multiclass setting we actually care about:

- *Precision* measures how many of the predicted positive cases (melanoma) are actually correct:

$$\text{Precision} = \frac{\text{True Positives}}{\text{True Positives} + \text{False Positives}}$$

  High precision means few false positives; so when the model flags a melanoma, it's likely to be correct. This is important in medical settings to avoid unnecessary follow-up procedures and patient anxiety.

- *Recall* (also called sensitivity) measures how many actual melanoma cases were correctly identified:

$$\text{Recall} = \frac{\text{True Positives}}{\text{True Positives} + \text{False Negatives}}$$

High recall means few missed melanoma cases: critical for ensuring early detection. In many clinical settings, recall is prioritized, since missing a true case (i.e., making a false negative prediction) is more harmful than incorrectly flagging a benign case.

In our case, we're working with multiple classes, not just melanoma versus nonmelanoma. To summarize performance across all classes, we use *weighted precision* and *weighted recall*. These metrics average per-class values while accounting for class imbalance—classes with more examples contribute more to the final score.

We'll be using weighted precision and recall as our main evaluation metrics, and we will plot them across training and validation splits as training proceeds. Here is our `compute_metrics` code:

```
@partial(jax.jit, static_argnums=(2,))
def compute_metrics(
    y_true: jax.Array, logits: jax.Array, n_labels: jax.Array
) -> dict[str, jax.Array]:
    """Computes weighted precision and recall metrics from logits and labels."""
    y_scores = jax.nn.softmax(logits)
    y_pred = jnp.argmax(y_scores, axis=1)
    metrics = {
        "recall_weighted": recall_score(
            y_true, y_pred, n_labels, average="weighted"
        ),
        "precision_weighted": precision_score(
            y_true, y_pred, n_labels, average="weighted"
        ),
    }
    return metrics
```

This function uses the predicted logits to compute softmax scores, takes the most likely class, and then calculates weighted precision and recall based on the true and predicted labels.

> Weighted metrics are concise and helpful but can mask problems—for instance, if the model performs poorly on a rare class, this might not show up clearly in the overall score. Always check *per-class metrics* as well. This can highlight weaknesses that might justify upweighting certain classes in the loss function or resampling your dataset to better balance class representation.

You may also encounter F1 scores in classification tasks. The F1 score is the harmonic mean of precision and recall—it provides a single number that balances both. Here, we'll stick to just plotting both precision and recall, as they are often easier metrics to interpret and let you diagnose whether your model struggles more with false positives or false negatives.

Why Not Use Accuracy to Track Performance?

Accuracy measures the overall percentage of correctly classified examples. It's simple and easy to understand, but it can break down in imbalanced datasets. For example, if 90% of the data is non-melanoma, a model that always predicts "non-melanoma" will achieve 90% accuracy—even though it fails to detect a single melanoma case.

That's why we use *precision* and *recall*, which provide a more nuanced view of model performance. These metrics distinguish between different types of errors. As mentioned, missing a melanoma (false negative) is often far worse than incorrectly flagging a benign case (false positive). Precision and recall allow us to focus on these critical trade-offs. In practice, you might, for example, track melanoma-specific recall alongside other metrics as you train.

What about *weighted* precision and recall? These average per-class values according to class frequency—giving more weight to common classes. This is often better than plain accuracy, but still not perfect: in many biological or clinical applications, rare classes can be the most important. A model might perform poorly on a rare but critical class, and the weighted metric won't reflect it clearly.

So, while weighted metrics are useful for summary plots, *per-class metrics* are essential. They help you spot blind spots and inform decisions about class weighting, sampling, or model design.

Faster evaluation metrics

You may have noticed that the `compute_metrics` function is jitted. Standard evaluation metrics—such as those from `sklearn` or `scipy`—are not typically JIT-compatible, which can become a bottleneck during training.

To address this, we've implemented JIT-friendly versions of precision and recall. These custom functions have been tested to match the outputs of their standard counterparts, but run significantly faster—especially during frequent batch-wise evaluation.

The metrics are computed on each batch and logged using the `MetricsLogger`, which aggregates results to provide full-dataset performance summaries.

In large-scale production models, the overhead from non-jitted metrics might be negligible. But for the relatively small models and fast iteration cycles we use here, the speedup from JIT compatibility is noticeable.

Creating the Multiclass Dataset

Finally, we can use our `DatasetBuilder` to construct the dataset that will be used to train and evaluate all models in this chapter. Using the same dataset setup across experiments allows us to make clean comparisons and attribute performance differences to model choices rather than data variation.

We also initialize and split a random seed up front, so we can reuse consistent seeds across model initializations and training runs—this helps ensure reproducibility.

```
from dlfb.cancer.dataset.preprocessors import skew

rng = jax.random.PRNGKey(seed=42)
rng, rng_dataset, rng_init, rng_train = jax.random.split(rng, num=4)

dataset_splits = DatasetBuilder(
  data_dir=assets("cancer/datasets/raw"),
).build(
  rng=rng_dataset,
  preprocessors=[skew, crop, resnet],
  image_size=(224, 224, 3),
  splits={"train": 0.70, "valid": 0.20, "test": 0.10},
)

num_classes = dataset_splits["train"].num_classes
```

There are three preprocessing options implemented:

crop
: Focuses on the central region of the image, where lesions are typically located. While simple and effective, it can discard potentially useful peripheral features such as hairs, texture, or pigment variation.

skew
: Retains as much of the lesion context as possible by resizing the entire image to a square without cropping—this means stretching or compressing the original aspect ratio. Since most of the input images are wider than they are tall, cropping to a square can cut out potentially useful surrounding tissue. Skewing avoids this by distorting the image to fit a square shape, preserving all pixels.

: While the image becomes slightly warped, it may still help the model capture broader contextual cues like skin texture or peripheral features.

resnet
: Applies standard preprocessing required for ResNet-based models, including resizing to the expected input shape and normalizing pixel values. Ensures compatibility with pretrained ResNet architectures.

These preprocessing steps are modular and can be swapped in and out easily. We can treat the choice between them as yet another tunable hyperparameter during experimentation.

Training the Baseline Model

Let's first train the baseline `SimpleCnn` model. There's nothing fancy here—but the point is to show that even a naive architecture can learn a surprising amount from the raw images. Here is the training setup:

```python
import optax

from dlfb.cancer.train.handlers.samplers import repeating_sampler
from dlfb.cancer.utils import decay_mask

learning_rate = 0.001
num_steps = 2000

state, metrics = train(
    state=SimpleCnn(num_classes=num_classes).create_train_state(
        rng=rng_init,
        dummy_input=dataset_splits["train"].get_dummy_input(),
        tx=optax.adamw(learning_rate, weight_decay=0.0, mask=decay_mask),
    ),
    rng=rng_train,
    dataset_splits=dataset_splits,
    num_steps=num_steps,
    batch_size=32,
    preprocessor=crop,
    sampler=repeating_sampler,
    augmentor=None,
    eval_every=100,
    store_path=assets("cancer/models/baseline"),
)
```

A few quick notes on this "no frills" setup:

- We use `optax.adamw` but set `weight_decay=0.0`, which makes it equivalent to plain `optax.adam`. This means no weight decay regularization is applied (we also do not need to worry about the `mask` parameter).

- The preprocessor is set to crop.

- We're using the `repeating_sampler` to cycle through the data. This does not resample images to balance class frequencies.

- No data augmentation is applied (`augmentor=None`).

Although we specify the `optax.adamw` optimizer, we initially leave weight decay unused. Why? Defining it uniformly across all of our models ensures consistency and makes hyperparameter tuning much simpler. Later on, we'll activate this placeholder to help reduce model overfitting.

Let's now evaluate the model's performance by plotting its metrics over the training steps. The results are shown in Figure 5-16.

```
from dlfb.cancer.inspect import plot_learning
```

```
plot_learning(metrics);
```

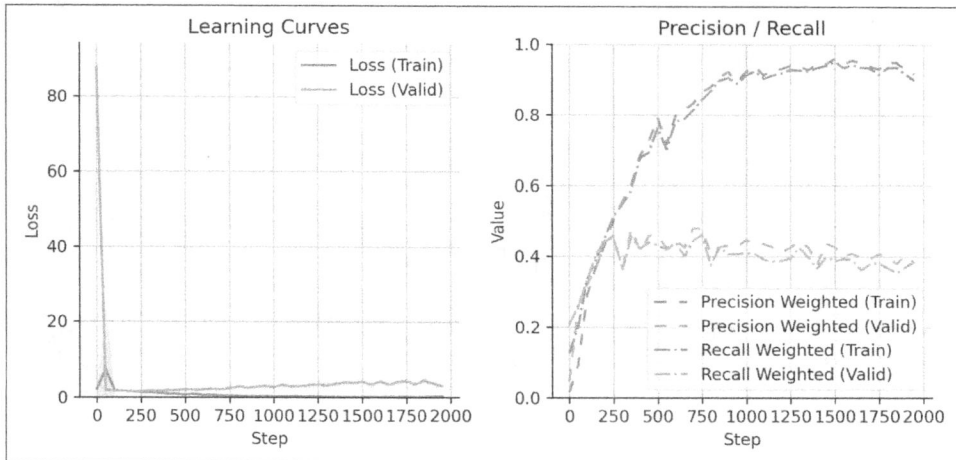

Figure 5-16. Evaluation metrics for the SimpleCnn model. The left plot shows the evolution of loss for the train and valid dataset splits, while the right plot tracks (weighted) precision and recall for the valid dataset.

We see that this baseline model reaches around 0.4 weighted precision and recall on the validation set—not bad for such a simple architecture. This gives us a reference point to compare against more advanced models.

Training metrics are much higher than validation metrics, indicating substantial overfitting. We also observe that validation loss increases over time, which is consistent with overfitting.

Another useful way to assess model performance is through the confusion matrix, shown in Figure 5-17. This gives a detailed view of how predictions align with ground truth labels—including which classes are commonly confused:

```
from dlfb.cancer.inspect import plot_classified_images
from dlfb.cancer.train import get_predictions
```

```
predictions = get_predictions(state, dataset_splits["valid"], resnet, 32)
plot_classified_images(
  predictions, dataset_splits["valid"], crop, max_images=8
);
```

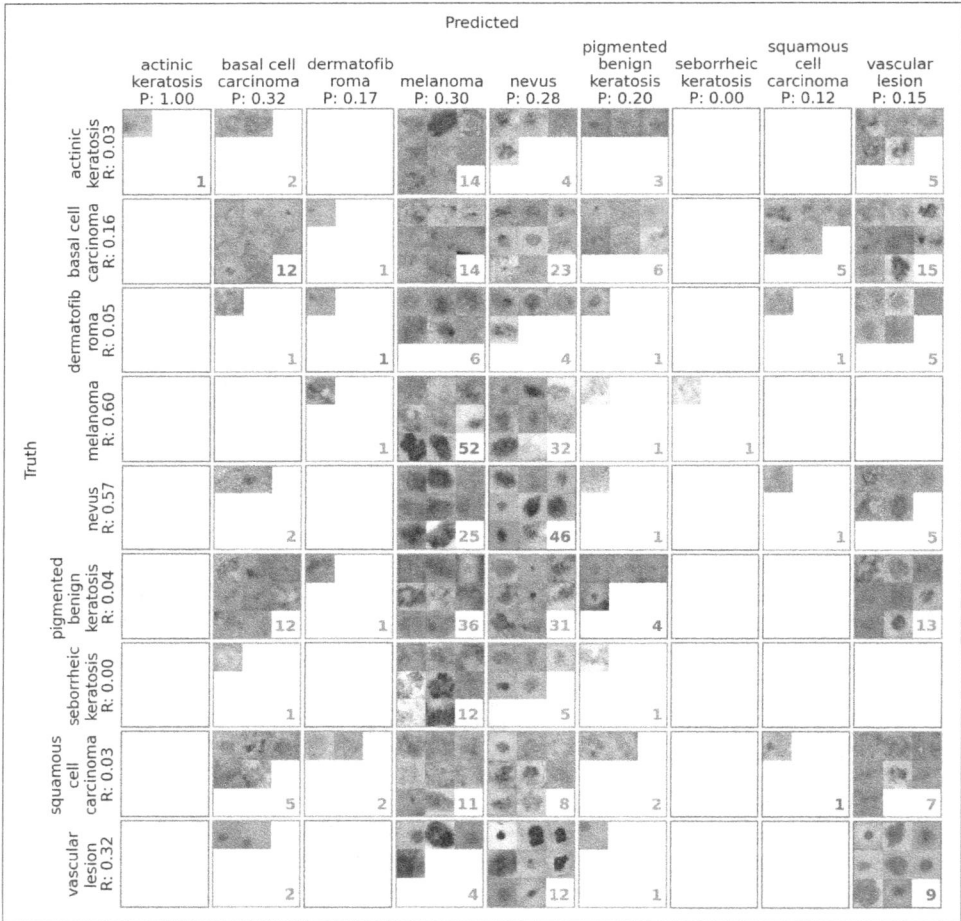

	actinic keratosis P: 1.00	basal cell carcinoma P: 0.32	dermatofibroma P: 0.17	melanoma P: 0.30	nevus P: 0.28	pigmented benign keratosis P: 0.20	seborrheic keratosis P: 0.00	squamous cell carcinoma P: 0.12	vascular lesion P: 0.15
actinic keratosis R: 0.03	1	2		14	4	3			5
basal cell carcinoma R: 0.16		12	1	14	23	6		5	15
dermatofibroma R: 0.05		1	1	6	4	1		1	5
melanoma R: 0.60			1	52	32	1	1		
nevus R: 0.57		2		25	46	1		1	5
pigmented benign keratosis R: 0.04		12	1	36	31	4			13
seborrheic keratosis R: 0.00		1		12	5	1			
squamous cell carcinoma R: 0.03		5	2	11	8	2		1	7
vascular lesion R: 0.32		2		4	12	1			9

Figure 5-17. Confusion matrix for the `SimpleCnn` *model on the nine-class skin lesion classification task. The rows represent the true labels, and the columns represent the predicted labels. Diagonal cells show correct predictions; off-diagonal cells indicate misclassifications. Each cell also includes example validation images for qualitative inspection, with the count of predictions shown in the lower-right. Precision (P:) and recall (R:) scores are provided per class.*

Some guidance on reading this plot—which we'll use throughout the chapter:

- Diagonal is good: These represent correctly classified images. An ideal model would have predictions only on the diagonal.

- Off-diagonal is bad: These are misclassifications.

- The number in the lower-right of each square indicates how many examples fall into that cell (in the validation set). Many confusion matrices simply report the counts here, but we also include actual images so that you can start to build intuition over what the different lesion types look like.

- Precision (`P:`) is shown above each column (per predicted class), and recall (`R:`) is shown along each row (per true class).

- The weighted precision and recall reported in the previous training plots are not simple averages of these values—they are class-weighted, meaning more common classes contribute more to the overall metric.

With that in mind, interpreting the confusion matrix in Figure 5-17 specifically, we can see that:

- This model is not that great—there are many off-diagonal entries.

- The model generally performs better on more common classes and struggles with rarer ones—we'll try resampling later to address this.

- It performs best on vascular lesions (bottom-right corner), which are visually distinct and easier to recognize, even to untrained eyes.

- The model also does relatively well on pigmented benign keratosis, nevus, and melanoma—though there's significant room for improvement.

- A key issue is that melanomas are often misclassified as nevi or other benign categories. This is particularly concerning given the clinical importance of detecting melanoma. We'll keep a close eye on whether more advanced models reduce these errors.

With this baseline in mind, let's turn our attention to the `ResNetFromScratch` model.

Training the ResNetFromScratch Model

Recall that this model follows the ResNet architecture but starts from randomly initialized weights—meaning it will probably require the most training time to reach strong performance compared to models that makes use of pretrained weights.

Before training, we need to set up both the dataset and model instance. We are aware of the class imbalance and small dataset size, and we have already prepared techniques like `balanced_sampler` and `rich_augmentor` to mitigate these issues. However, to better understand their individual impact, we will first run training without any augmentation or class balancing.

For this model, we do not need to apply `resnet` preprocessing to the images, since we are training from scratch and don't require compatibility with pretrained ResNet weights.

Now, let's train our model:

```
state, metrics = train(
    state=ResNetFromScratch(num_classes=num_classes).create_train_state(
        rng=rng_init,
        dummy_input=dataset_splits["train"].get_dummy_input(),
        tx=optax.adamw(learning_rate, weight_decay=0.0, mask=decay_mask),
    ),
    rng=rng_train,
    dataset_splits=dataset_splits,
    num_steps=num_steps,
    batch_size=32,
    preprocessor=crop,   # No preprocessing.
    sampler=repeating_sampler,   # No balancing.
    augmentor=None,   # No augmentation.
    eval_every=100,
    store_path=assets("cancer/models/resnet_from_scratch"),
)
```

As before, let's evaluate the model's performance by plotting its metrics over the training steps. The results are shown in Figure 5-18:

```
plot_learning(metrics);
```

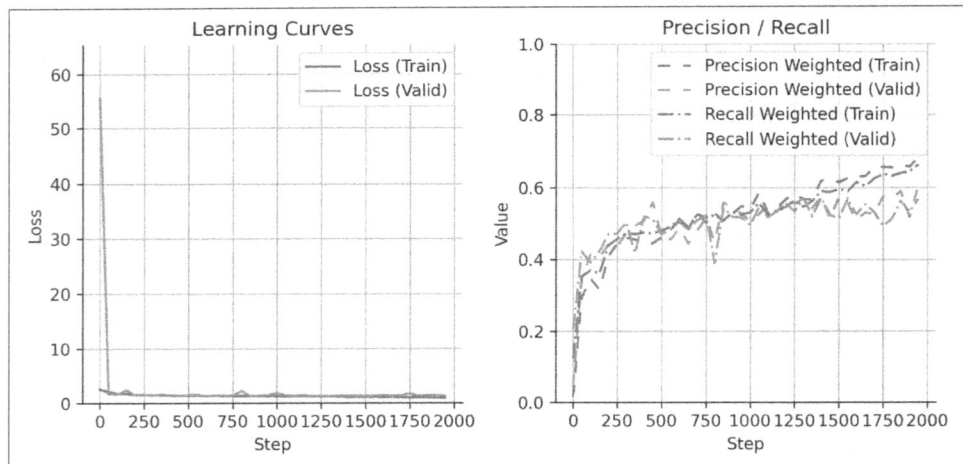

Figure 5-18. Model performance over training and validation steps for the `ResNetFrom Scratch` *multiclass skin lesion classification model.*

The results are promising and already show an improvement over the baseline—precision and recall metrics are significantly higher. This makes sense, as the ResNet architecture is generally strong for image tasks.

Importantly, both train and validation metrics increase steadily and remain closely aligned. Likewise, the train and validation losses both decrease over time, suggesting the model is not significantly overfitting. In fact, the training curves indicate that the model is still learning—so with more training steps, we might be able to achieve even better results.

To see which classes benefit most from this improved model, let's look again at the confusion matrix in Figure 5-19:

```
predictions = get_predictions(state, dataset_splits["valid"], resnet, 32)
plot_classified_images(
  predictions, dataset_splits["valid"], crop, max_images=8
);
```

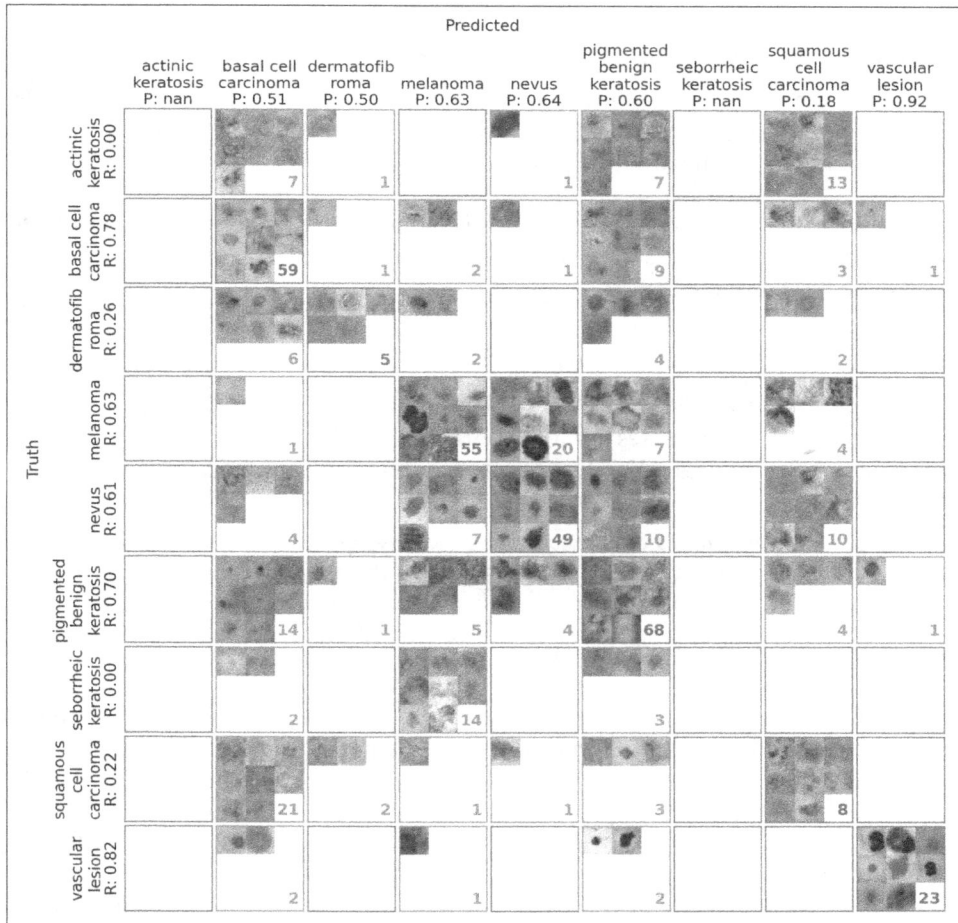

Truth \ Predicted	actinic keratosis P: nan	basal cell carcinoma P: 0.51	dermatofibroma P: 0.50	melanoma P: 0.63	nevus P: 0.64	pigmented benign keratosis P: 0.60	seborrheic keratosis P: nan	squamous cell carcinoma P: 0.18	vascular lesion P: 0.92
actinic keratosis R: 0.00		7	1		1	7		13	
basal cell carcinoma R: 0.78		59	1	2	1	9		3	1
dermatofibroma R: 0.26		6	5	2		4		2	
melanoma R: 0.63		1		55	20	7		4	
nevus R: 0.61		4		7	49	10		10	
pigmented benign keratosis R: 0.70		14	1	5	4	68		4	1
seborrheic keratosis R: 0.00		2		14		3			
squamous cell carcinoma R: 0.22		21	2	1	1	3		8	
vascular lesion R: 0.82		2		1		2			23

Figure 5-19. Confusion matrix for the `ResNetFromScratch` model on the nine-class skin lesion classification task.

Some observations:

- Overall improvement: We see better recall and precision across most categories compared to the baseline model. The model is generally more confident and more accurate. The diagonal is visibly stronger overall, indicating more correct predictions.

- Vascular lesions are now being classified almost perfectly (bottom-right corner). These lesions are visually distinctive, and the model is clearly picking up on that.

- Nevus and melanoma both show solid improvements in recall. However, there is still some confusion between the two, as seen in the off-diagonal cells between those classes.

- Pigmented benign keratosis shows a marked improvement in both precision and recall, with a notable jump from the baseline.

- Actinic keratosis and seborrheic keratosis are now never predicted by the model—resulting in `NaN` precision values for those classes. This is odd and is something to revisit with future models.

We now move on to the `FinetunedHeadResNet`, which incorporates pretrained ImageNet features. This should give us a much stronger model, especially given the small size of our training dataset.

Training the FinetunedHeadResNet Model

This time around, we need to be careful about image preprocessing. Since we're using pretrained weights from a ResNet model originally trained on ImageNet, our input images must be preprocessed in the same way as those used for ImageNet.

> If the images deviate too much—for example, if they're scaled differently or normalized inconsistently—the pretrained features may no longer be as applicable. This can force the model to overcorrect during training, reducing the benefits of transfer learning.

To handle this, we apply the `resnet` preprocessing function, which ensures that input images are resized and normalized in a way that matches the expectations of the pretrained ResNet backbone:

```
IMAGE_PROCESSOR = AutoImageProcessor.from_pretrained("microsoft/resnet-50")
```

```
def resnet(image: jax.Array) -> jax.Array:
    """Preprocess from pretrained model with transpose for compatibility."""
    image = IMAGE_PROCESSOR(image, return_tensors="jax", do_rescale=True)
    image = image["pixel_values"]
```

```
image = convert_nchw_to_nhwc(image)
return image
```

We are now ready to train the `FinetunedHeadResNet` model, which freezes the pre-trained backbone and trains only the classification head. To isolate the effects of transfer learning, we keep the setup minimal—no regularization, augmentation, or class rebalancing is applied at this stage.

```
state, metrics = train(
    state=FinetunedHeadResNet(num_classes=num_classes).create_train_state(
        rng=rng_init,
        dummy_input=dataset_splits["train"].get_dummy_input(),
        tx=optax.adamw(learning_rate, weight_decay=0.0, mask=decay_mask),
    ),
    rng=rng_train,
    dataset_splits=dataset_splits,
    num_steps=num_steps,
    batch_size=32,
    preprocessor=resnet,
    sampler=repeating_sampler,
    augmentor=None,
    eval_every=100,
    store_path=assets("cancer/models/resnet_just_head"),
)
```

Let's examine the model's performance in Figure 5-20.

```
plot_learning(metrics);
```

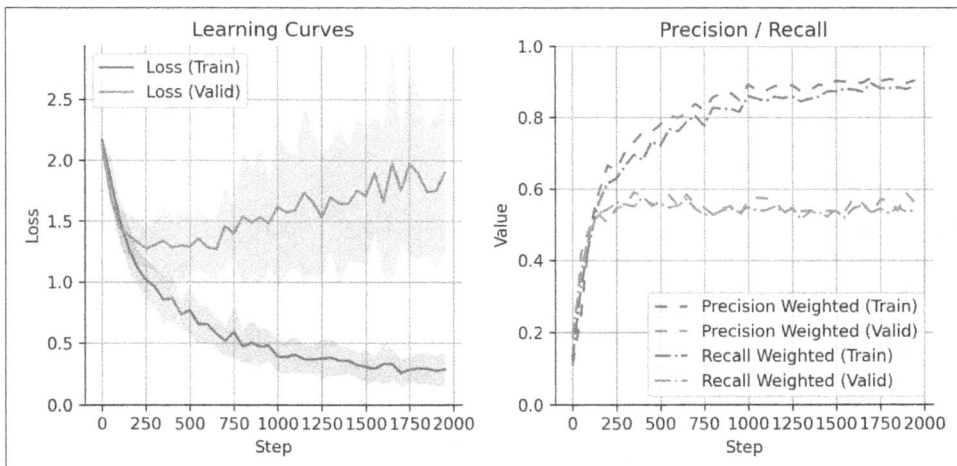

Figure 5-20. Model performance over training and validation steps for the `Finetuned HeadResNet` multiclass skin lesion classification model.

Some observations:

- Loss curves: The training loss steadily decreases, as expected, but the validation loss begins to rise again after around step 250. This is a classic sign of overfitting—the model is becoming more confident on the training set, but its generalization to new data begins to degrade.

- Train vs. validation gap: There is still a large gap between training and validation metrics, though it's smaller than what we saw with the `SimpleCnn` baseline.

- Precision and recall: Interestingly, while the validation loss increases, the validation precision and recall metrics remain relatively stable. This suggests that although the model's softmax confidence may be worsening, its actual classification decisions are not deteriorating significantly as training proceeds—at least not yet.

- Plateauing metrics: Validation performance levels off around step 250. This likely reflects the fact that we are only training the classification head—the majority of the model (the ResNet backbone) remains frozen. That limited flexibility may be capping performance, despite early gains.

It's worth noting that validation performance here is not noticeably better than what we achieved by training ResNet from scratch (though notice how much faster `FinetunedHeadResNet` trained). This may suggest that training only the classification head doesn't provide enough flexibility for the model to properly adapt to our dataset. The much higher training performance relative to validation performance also points to overfitting. While we could try to address this with regularization or augmentation, we'll instead move on to what we expect will be a significantly stronger approach—fine-tuning the entire network.

Training the FinetunedResNet Model

Next, we'll train a model that updates all the pretrained weights—not just the classification head. This approach, known as full fine-tuning, allows the model to gradually adapt its internal feature representations to better match our dataset:

```
state, metrics = train(
  state=FinetunedResNet(num_classes=num_classes).create_train_state(
    rng=rng_init,
    dummy_input=dataset_splits["train"].get_dummy_input(),
    tx=optax.adamw(learning_rate, weight_decay=0.0, mask=decay_mask),
  ),
  rng=rng_train,
  dataset_splits=dataset_splits,
  num_steps=num_steps,
  batch_size=32,
  preprocessor=resnet,
  sampler=repeating_sampler,
```

```
    augmentor=None,
    eval_every=100,
    store_path=assets("cancer/models/resnet50_basic"),
)
```

Let's examine the model's performance in Figure 5-21:

```
plot_learning(metrics);
```

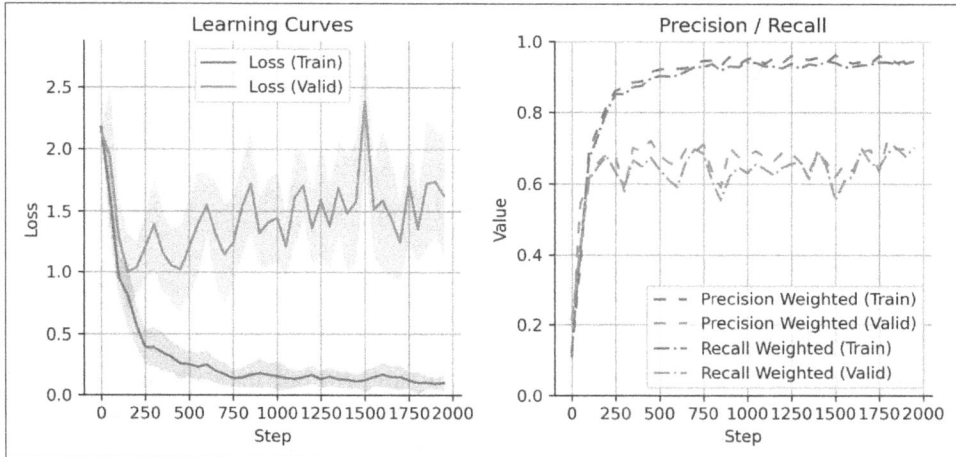

Figure 5-21. *Model performance over training and validation steps for the initial (non-optimized)* FinetunedResNet *multiclass skin lesion classification model.*

There are a few takeaways from these plots:

- Comparison to previous models: This model clearly outperforms the baseline SimpleCnn, the ResNetFromScratch, and the FinetunedHeadResNet models in terms of validation metrics.

- Precision and recall: Validation metrics are consistently higher than in previous models—with weighted precision and recall reaching around 0.65–0.7 by the end of training. However, there is still a noticeable gap between training and validation performance, indicating that overfitting remains an issue we'll need to address.

- Loss curves: The training loss steadily drops to near-zero, while the validation loss stays high and rises gradually. This is consistent with overfitting, although it's worth remembering that rising softmax loss doesn't always reflect a drop in classification quality (and indeed, here the validation metrics stay stable while validation loss rises).

In summary, this is our best-performing model so far—confirming the benefit of full fine-tuning when using pretrained backbones. Before examining the confusion matrix of the fully fine-tuned model, let's first apply a few targeted optimizations to further improve headline metric performance.

Optimizing the FinetunedResNet Model

We'll apply four key modifications:

- A warmup plus cosine decay learning rate schedule
- Data augmentation (as introduced in "Augmenting the dataset" on page 262)
- High-rate dropout to reduce overfitting (set to 0.7)
- Weight decay using the adamw optimizer (set to 1e-4)

We'll walk through each of these in turn.

Learning rate schedule

Fine-tuning a pretrained model requires care to preserve the valuable low-level features it learned from large datasets like ImageNet. A common strategy is to use a *warmup learning rate schedule*, which starts training with a small learning rate that gradually increases—allowing the model to adapt gently—before decaying smoothly over time. This helps:

- Stabilize training in the early stages
- Prevent abrupt updates to pretrained weights
- Allow gradual adaptation to our new task

We'll use a warmup and cosine decay schedule where the learning rate will:

- Warm up for the first 20% of steps, gradually increasing the learning rate
- Peak at a learning rate of 0.001 (1e-3)
- Smooth decay to 0.00001 (1e-5) by the end of training

Let's visualize how the learning rate evolves in Figure 5-22:

```
import optax

learning_rate = 0.001
num_steps = 2000

warmup_cosine_decay_scheduler = optax.warmup_cosine_decay_schedule(
    init_value=0.0001,
    peak_value=learning_rate,
```

```
    end_value=0.00001,
    warmup_steps=int(num_steps * 0.2),
    decay_steps=num_steps,
)

lrs = [warmup_cosine_decay_scheduler(i) for i in range(num_steps)]

plt.scatter(range(num_steps), lrs)
plt.title("Learning Rate over Steps")
plt.ylabel("Learning Rate")
plt.xlabel("Step");
```

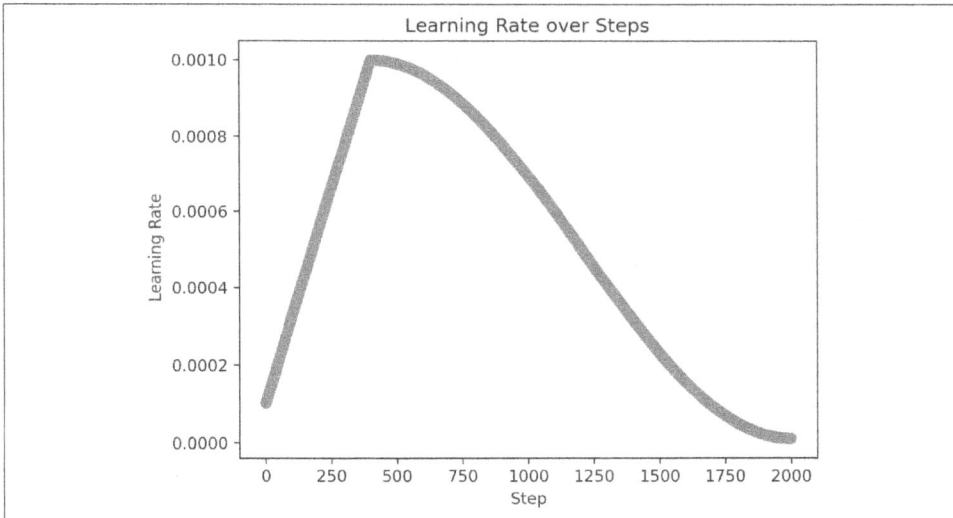

Figure 5-22. Learning rate evolution during training, following a warmup-cosine decay schedule. The learning rate starts small, gradually increases over the first 20% of training steps (warmup), then peaks and decays smoothly.

Data augmentation

We previously observed overfitting even in our best model. One way to mitigate this is to apply data augmentation, which increases the effective diversity of the training set by introducing small, label-preserving transformations (e.g., flips, brightness jitter, crops). As discussed earlier in this chapter, augmentation helps the model generalize more robustly by preventing it from memorizing the training images.

We'll apply these augmentation strategies by using the rich_augmentor defined earlier in the chapter.

Regularization via dropout and adamw

In addition to data augmentation, we'll also apply explicit regularization using two methods:

Dropout

Randomly disables neurons during training, encouraging the network to rely on distributed representations and reducing co-adaptation. We set a high dropout rate of 0.7 to strongly combat overfitting.

Weight decay

Discourages overly large weights by penalizing them in the loss function. We apply this using `optax.adamw`, which is designed to work correctly with L2 regularization.

> You cannot safely use L2 regularization with the standard Adam optimizer—it won't apply weight decay in the intended way. This is exactly why `adamw` was introduced, and it should be used when adding weight decay.

It's also standard practice *not* to apply weight decay to:

Bias parameters

These are typically small and few in number and don't contribute substantially to model complexity. Regularizing them doesn't tend to help generalization and can sometimes hurt performance.

Batch normalization parameters

These include the learned scale (`gamma`) and shift (`beta`) terms, as well as running statistics. Applying weight decay to these can destabilize training, especially in fine-tuning scenarios, as they control the distribution of activations rather than model capacity.

Instead, weight decay is usually applied only to the main weights of convolutional or linear layers, where it can help prevent overfitting by discouraging overly complex solutions. To enforce this, we'll use a parameter mask when constructing the optimizer—applying decay only to the relevant parameters.

Training the Optimized FinetunedResNet Model

Now that we've introduced a learning rate schedule, data augmentation, dropout, and weight decay—it's time to put everything together. We can train the optimized `FinetunedResNet` model and see whether these changes help reduce overfitting and improve validation performance:

```
state, metrics = train(
  state=FinetunedResNet(
    num_classes=num_classes, dropout_rate=0.7
  ).create_train_state(
    rng=rng_init,
    dummy_input=dataset_splits["train"].get_dummy_input(),
    tx=optax.adamw(
      warmup_cosine_decay_scheduler, weight_decay=1e-4, mask=decay_mask
    ),
  ),
  rng=rng_train,
  dataset_splits=dataset_splits,
  num_steps=num_steps,
  batch_size=32,
  preprocessor=resnet,
  sampler=repeating_sampler,
  augmentor=rich_augmentor,
  eval_every=100,
  store_path=assets("cancer/models/resnet50_optimized"),
)
```

Now, let's analyze the model's performance over the training steps, as shown in Figure 5-23:

```
plot_learning(metrics);
```

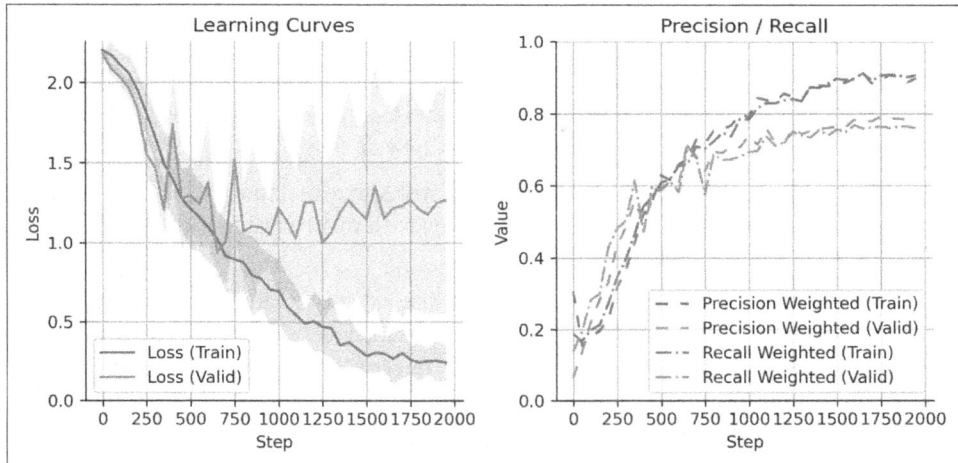

Figure 5-23. Model performance over training and validation steps for the optimized FinetunedResNet multiclass skin lesion classification model.

This training run demonstrates the cumulative benefit of regularization and optimization strategies. Some observations:

- Precision and recall: Both metrics improve steadily on both the training and validation sets. Most notably, the validation precision and recall reach nearly 0.75 by the end of training—our best performance so far. The gap between training and validation metrics is also now narrower and more stable.

- Loss curves: While the validation loss remains higher than training loss, it's still lower than before and appears to not impact the validation metrics—a major improvement.

In short, we've trained a model that generalizes much better and achieves strong, stable performance across all metrics. This serves as a great foundation for further experimentation.

To better understand this model's behavior, let's plot the confusion matrix as Figure 5-24:

```
predictions = get_predictions(state, dataset_splits["valid"], resnet, 32)
plot_classified_images(
    predictions, dataset_splits["valid"], crop, max_images=8
);
```

Some observations about the model's behavior:

- Most predictions lie along the diagonal: This indicates strong overall performance and class-specific accuracy. Nearly all classes show high recall, with vascular lesions and basal cell carcinoma especially well classified.

- Melanoma versus nevus: While there are still some melanoma cases being misclassified as nevi, the number is much lower than in earlier models and melanoma recall is the highest level yet.

- Rare classes: Performance on rare classes like actinic keratosis, dermatofibroma, and squamous cell carcinoma has improved. These categories are now being correctly identified more consistently, with fewer scattered misclassifications.

- Emptier off-diagonal cells: Many potential misclassification combinations are completely empty. This indicates that the model is no longer making widespread or erratic mistakes—it's more selective and confident.

- Visual inspection: The image thumbnails in each cell help confirm that correct predictions tend to look visually consistent, and many of the remaining errors involve subtle or understandable confusions.

Altogether, this confusion matrix reinforces the earlier metrics: the optimized model is significantly more accurate, more reliable, and ultimately would be more clinically useful than earlier versions.

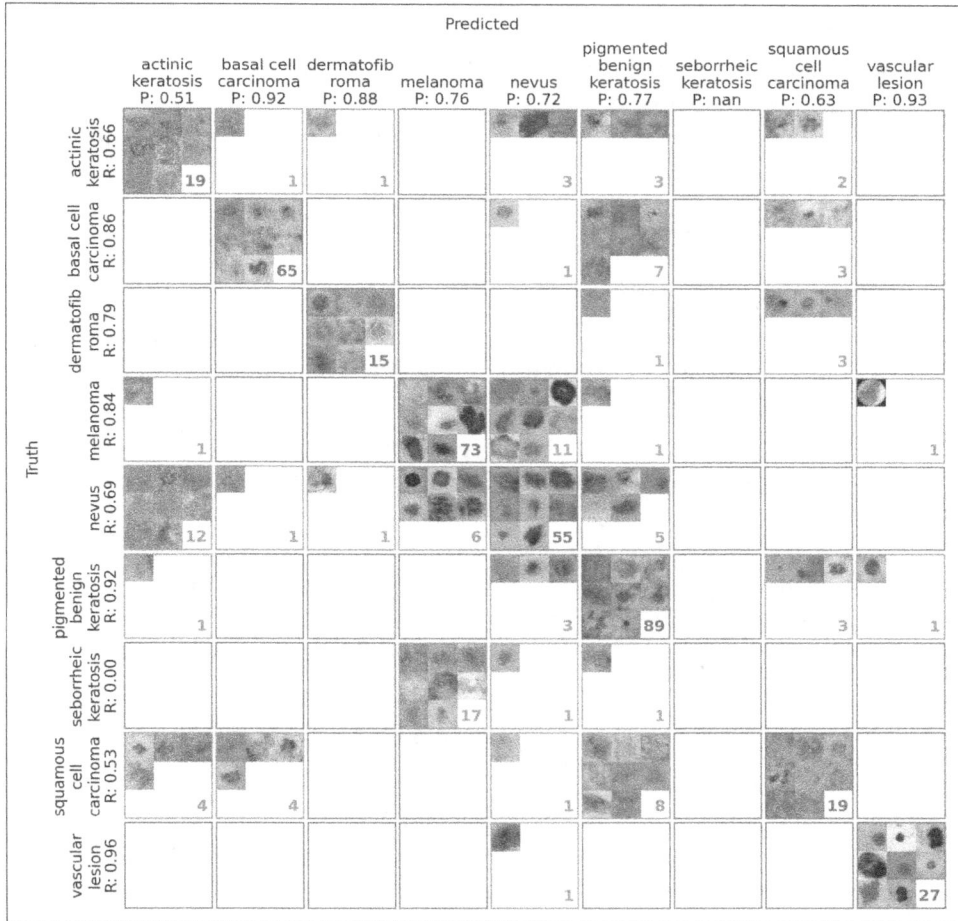

Figure 5-24. Confusion matrix for the final optimized `FinetunedResNet` *model on the nine-class classification task.*

Confusion matrices can reveal a lot—but they won't turn us into dermatologists. Interpreting model errors often requires deep domain knowledge. Collaborating with subject matter experts, like dermatologists in this case, can provide critical insights into misclassifications, uncover hidden biases in the data, and suggest clinically meaningful improvements. For any real-world machine learning project, involving domain experts early and often is one of the most effective ways to build useful, trustworthy models.

Further Improving the Model

We'll stop optimizing this model for now, but it's far from perfect—for example, it completely fails to predict seborrheic keratosis. Here are several ideas for pushing it further, roughly ordered by expected bang for your buck:

Longer training
> Metrics were still improving slightly at the end of 2000 steps. Training for longer could yield further gains, especially since we are using a warmup schedule (which starts learning slowly to protect pretrained weights).

Ensembling
> Combine predictions from multiple trained models to reduce variance and improve robustness:
>
> - The most common ensembling approach here would be to average the softmax probabilities across models. Alternatively, you could use majority voting on predicted labels (less smooth, but sometimes effective).
>
> - You can ensemble models trained with different random seeds (e.g., 2, 4, or 8), or even ensemble structurally different models—such as the various ResNet models we've explored.

Exploring augmentation strategies
> There's still plenty of room to be more creative with data augmentation. Try playing around with our augmentation setup (e.g., the specific transformations we encoded) and check their effect on model learning.

Class re-weighting
> Increase the loss contribution from rare or clinically important classes like melanoma. This can be done by computing inverse class frequencies and passing per-example weights into the loss.

Hard example mining
> Focus on examples that are consistently misclassified (e.g., melanoma confused with nevus). You could maintain a pool of high-loss samples and upsample them in training batches, or alternatively, increase loss weights for these types of samples.

Stronger regularization
> Tune `dropout_rate` and `weight_decay`, or explore additional regularization methods such as:

Label smoothing
> Instead of using hard one-hot labels, you assign most of the probability to the true class (e.g., 0.9) and distribute the rest evenly across other classes. This reduces model overconfidence and can improve generalization.

Mixup or CutMix

Blend or patch together two images and their labels. These techniques regularize the model and encourage the model to learn more robust decision boundaries between classes.

Multimodal data

Dig into the original dataset to see if you can incorporate metadata such as patient age or sex, anatomical site of the lesion, or dermoscopic features. These features can be embedded (e.g., as one-hot or learned embeddings) and concatenated with image features before classification. Metadata often provides key disambiguating context that pixels alone can't capture.

It can be hard to know where to start when trying to improve a model—and changes can interact in unexpected ways. Progress is rarely linear, so think of it as an exploration process. Use intuition, literature, and diagnostics to guide you—and have fun getting to know your model.

Summary

Classifying skin lesions is a genuinely difficult task—if it weren't, early detection would be routine, and fewer cases would be missed. Still, the models we built in this chapter demonstrate how machine learning can meaningfully assist in this challenge, offering scalable tools to support (not replace) clinical expertise.

This project came with plenty of constraints, especially the limitations of the dataset. We had to adapt carefully—testing model variants, countering overfitting, and making the most of each part of the training setup. While more powerful models and larger datasets exist, we hope the mindset and practical tools introduced here help you tackle similar problems—particularly when working with limited or imperfect data.

The techniques we explored—class balancing, data augmentation, fine-tuning pretrained models, diagnosing overfitting, and incrementally improving architectures—are widely applicable. Whether you're detecting lung disease in chest X-rays, identifying tumors in MRIs, or analyzing retinal images for diabetic retinopathy, the same principles apply: adapt to the data, learn from errors, and iterate thoughtfully.

Now let's move on. We've seen more than enough skin lesion images for one chapter. But before we go, a final reminder: keep an eye on your own skin. If something looks unusual or changes over time, don't hesitate to seek medical advice. Early detection truly does save lives.

## CHAPTER 6

Learning Spatial Organization Patterns Within Cells

In this chapter, we shift focus from classifying high-level cell states—such as distinguishing cancerous from healthy tissue—to something more low level and foundational: understanding the *spatial organization inside individual cells*. Specifically, we'll train a deep learning model to analyze microscopy images and learn where exactly in the cell different proteins are located, a task known as *protein localization*.

Protein localization plays a crucial role in cell biology. A protein's position within the cell—for example, whether it's in the nucleus or the mitochondria—often determines its function. Mislocalization of proteins is implicated in many diseases, even when the protein's structure is normal (i.e., not mutated or altered). Thanks to modern fluorescence microscopy, we can observe a protein's location in a cell directly, but the resulting images are often high dimensional, noisy, and hard to interpret at scale.

Unlike earlier chapters, the goal here isn't to strictly optimize a metric like accuracy, recall, or precision on a specific classification or regression task. Instead, we'll train a model to learn a *latent representation* of protein localization directly from raw microscopy images. You can think of a *latent space* as the model's internal map—a compressed representation where proteins with similar localization patterns are grouped together, even without explicit labels. This approach falls under *representation learning*: the goal is to uncover meaningful structure in the data that reflects biological patterns.

Why focus on representation learning instead of just training a classifier for a specific task of interest? In many biological settings, and especially in protein localization, we don't have clean, comprehensive labels. Instead of forcing the model to solve a narrow, predefined task, we want it to learn a *rich internal representation* of the data that captures spatial patterns and similarities between proteins. These representations can be reused for clustering, visualization, identifying unknown cellular compartments, or understanding how protein localization changes across cell types or conditions. This is analogous to how large language models learn general-purpose representations of words or sentences, which can then be reused for many downstream tasks.

Our approach to modeling protein localization in this chapter is based on *cytoself*, a self-supervised deep learning method published in *Nature Methods* in 2022.[1] The model combines image reconstruction (rebuilding the microscopy input image) with a secondary task—predicting the protein's identity—to learn a rich and interpretable embedding space that reflects biological localization patterns.

Unlike in previous chapters, the model's primary output isn't a classification label or regression score. Instead, for a given microscopy image of a fluorescently tagged protein, it produces an *embedding*—a position in the latent space—that captures the spatial characteristics of the protein's localization. These embeddings can then be visualized, clustered, or compared to known annotations.

This is the most advanced chapter in the book. You'll work with a large, real-world microscopy dataset and implement a custom vector-quantized variational autoencoder (VQ-VAE) from scratch. The chapter takes you deep into self-supervised learning, large-scale image processing, and the spatial organization of cells—and reproduces core results from a recent deep learning biology paper.

> More than in any previous chapter, we strongly recommend keeping the companion Colab notebook open as you read. You may need to scale down the model to fit within memory limits, but actively running the code will solidify your understanding and give you room to explore.
>
> To run the full-scale model, we recommend using a powerful GPU such as an A100. These are available through platforms like Google Colab Pro, Kaggle Notebooks (with upgrades), Google Cloud Platform (GCP), or AWS EC2.

Biology Primer

The cell was first observed in 1665 by the British scientist Robert Hooke (*https://oreil.ly/6JmDd*), who used a microscope to describe its structure in cork tissue. Since then, microscopy has become one of biology's most essential tools. Modern microscopes allow researchers to visualize living cells in astonishing detail—and increasingly, to capture this data at massive scale.

---

1 Kobayashi, H., Cheveralls, K. C., Leonetti, M. D., & Royer, L. A. (2022). Self-supervised deep learning encodes high-resolution features of protein subcellular localization. *Nature Methods*, 19(8), 995–1003. https://doi.org/10.1038/s41592-022-01541-z

Microscopy is now central to many areas of biomedical research. For example, pharmaceutical companies routinely screen the effects of drug candidates by imaging thousands of treated cells and then use machine learning models to assess cellular responses in an automated way. Does a cell look alive or dead? Does its observable structure change in response to a particular drug compound? Do the cells divide more or less rapidly as a result of a treatment?

Despite its power to capture biological detail, microscopy data can be challenging to analyze. Cells vary naturally in size, shape, and appearance, and the imaging and sample preparation process itself introduces noise and artifacts. Furthermore, unlike genomes, which can to some extent be compared to a universal "reference" genome, cells don't come with a "baseline cell" or a standardized coordinate system, which can make analysis difficult. And microscopy can challenge computational resources: high-resolution microscopy produces large volumes of data that can strain both memory and compute.

Traditionally, microscopy image analysis relied on manually defined, hand-engineered features. For example, in drug screening experiments, scientists might measure whether a compound causes cells to shrink, swell, or change shape—signs that the drug is affecting cell health or behavior. To do so, researchers would extract properties such as cell size, shape, brightness, or texture using classical image processing techniques, including thresholding, edge detection, and morphological operations (which manipulate shapes in the image to clean up noise, fill gaps, or separate touching cells). These features were then fed into relatively simple models such as logistic regression or decision trees to predict cellular outcomes.

This earlier approach to microscopy image analysis required domain expertise to decide which features mattered, and they often missed subtle patterns not obvious to the human eye. Deep learning changed this: modern convolutional neural networks (CNNs) can learn to extract meaningful features directly from raw pixels, capturing complex visual cues without relying on handcrafted rules.

These advances in image analysis have opened the door to exploring deeper biological questions—including how proteins are spatially organized within cells.

Spatial Organization Within the Cell

Cells aren't just sacks of molecules; they're intricately organized. Subcellular compartments evolved billions of years ago and represent one of the most fundamental principles of biology: *specialization through spatial organization.* Instead of letting everything float freely in an undifferentiated soup, cells developed internal structures that separate and coordinate different functions.

All three domains of life—bacteria, archaea, and eukaryotes—show signs of this spatial complexity. In eukaryotic cells (the kinds of cells found in humans, plants, fungi, and more), this organization is especially pronounced. These cells contain membrane-bound compartments called *organelles*, each with a specialized job, as shown in Figure 6-1.

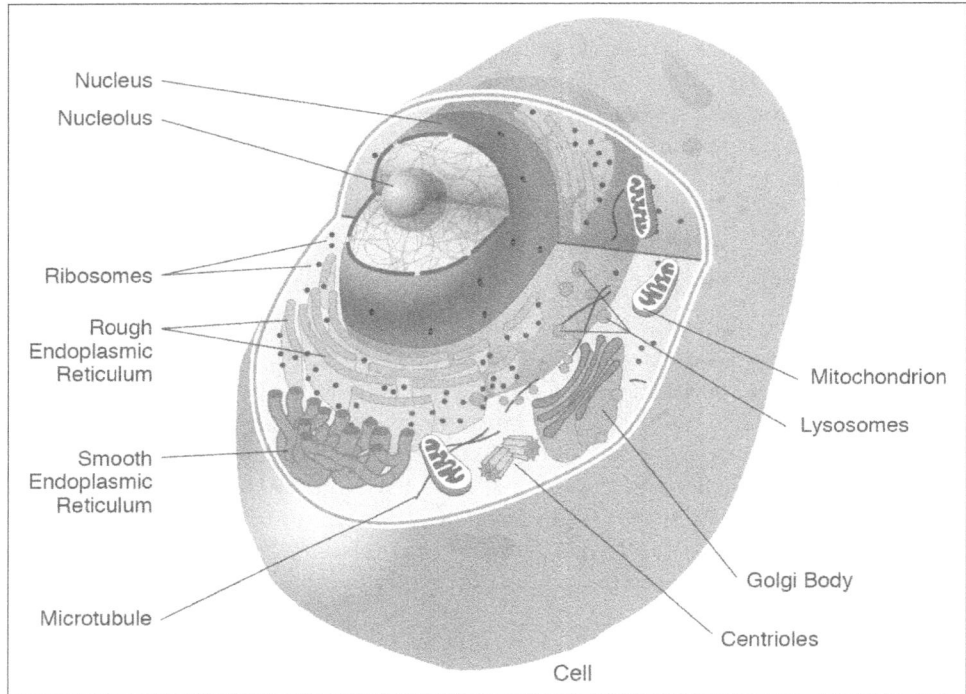

Figure 6-1. A visual representation of the complex organization within a eukaryotic cell. The different organelles compartmentalize the cell into regions where different specialized functions are performed (illustration from the National Institutes of Health).

We won't dive into full organelle flashcards here (no need to relive high school biology trauma), but here's a quick recap of the structures that will be most relevant in this chapter:

Plasma membrane

> A lipid bilayer that wraps the cell, protecting its internal environment and controlling what enters and exits.

Nucleus

> The command center of the cell. It's enclosed in its own membrane and houses the DNA—the genetic blueprint.

Cytoplasm

The gel-like substance that fills the cell. This is where most cellular activity happens, and it's home to many other organelles.

Mitochondria

Often called the "powerhouses" of the cell. They generate energy in the form of ATP. Evolutionarily, they were once free-living bacteria that formed a mutually beneficial partnership with early cells.

Endoplasmic reticulum (ER)

Comes in two flavors. The *rough ER* is studded with ribosomes (protein-making factories) and helps synthesize proteins. The *smooth ER* handles lipids and detoxification.

Golgi apparatus

The cell's shipping center. It modifies, packages, and routes proteins and lipids to their final destinations.

With that background on cellular compartments in place, let's turn to protein localization.

Protein Localization

Every protein begins its life in the same way: it's transcribed from a gene into messenger RNA (mRNA) in the nucleus, and then it is translated into a chain of amino acids by ribosomes in the cytoplasm. But from there, things get much more spatially complex.

Once a protein is made, it doesn't just float around randomly. It gets shipped to a specific destination inside the cell, guided by a sort of molecular postal code. Some proteins are sent to the nucleus, others to the mitochondria, the cell membrane, or the ER. This process is known as *protein localization*, and it's critical for proper cellular function.

A protein's location is often just as important as its molecular structure. Even a perfectly formed protein can't function if it's in the wrong place. For example, the protein DNA polymerase needs to be inside the nucleus to carry out its function of replicating DNA. If it ends up in the cytoplasm, it's effectively useless. On the other hand, some proteins aren't confined to a specific location. For instance, the protein actin is found throughout the cell, where it helps maintain its shape and enable movement.

Mislocalization—when a protein ends up in the wrong compartment—is a key driver of disease. For example, in amyotrophic lateral sclerosis (ALS), the protein TDP-43 accumulates in the cytoplasm of neurons, even though its normal location is in the nucleus.[2] The protein isn't mutated, misfolded, or present in abnormal quantities—it's simply in the wrong place. And that alone is enough to disrupt cellular function and trigger disease.

Another example is the tumor suppressor BRCA1, which normally helps repair DNA in the nucleus. In some breast and ovarian cancers, BRCA1 is mislocalized to the cytoplasm, where it can no longer perform its repair function, even though the protein itself may be entirely structurally normal.[3]

Understanding Protein Localization

Once a protein is made, how does it end up in the right part of the cell? The answer lies in short amino acid sequences, which are like molecular zip codes embedded within the protein. These address tags are recognized by the cell's transport machinery and guide proteins to their proper destinations: the nucleus, mitochondria, plasma membrane, and so on.

The rules of this system are still only partially understood. Many proteins localize to multiple compartments. Some are misdirected under stress or disease conditions. And some just seem to defy categorization.

Even in healthy cells, we don't yet have a complete map of where all human proteins go—or how those localizations change across cell types, developmental stages, or environmental conditions.

Mapping the protein localization landscape in detail is one of the big open challenges in cell biology. Understanding protein localization at scale could:

Reveal new, previously unknown subcellular compartments

It may be surprising, but we're still discovering fundamental structures inside cells. In recent years, researchers have identified the *exclusome* (a cytoplasmic DNA compartment in mammalian cells), *paraspeckles* and *nuclear speckles* (membraneless nuclear bodies involved in RNA processing), and even new entire organelles. The most recent example is the *nitroplast*, a nitrogen-fixing organelle discovered in marine algae as recently as 2024.[4] These findings show how much more there is to uncover about the cell's internal architecture.

---

2 Suk, T. R. (2020). The Role of TDP-43 Mislocalization in Amyotrophic Lateral Sclerosis. *Molecular Neurode-Generation*, 15(1), 45.

3 Rodriguez, J. A., Au, W. W., & Henderson, B. R. (2003). Cytoplasmic mislocalization of BRCA1 caused by cancer-associated mutations in the BRCT domain. *Experimental Cell Research*, 293(1), 14–21. https://doi.org/10.1016/j.yexcr.2003.09.027

4 Coale, T. H. et al. (2024). Nitrogen-fixing organelle in a marine alga. *Science*, 384(6692), 217–222. https://doi.org/10.1126/science.adk1075

Help assign functions to poorly annotated proteins

If a protein consistently localizes to mitochondria, you could hypothesize that it plays a role in energy metabolism or apoptosis (programmed cell death), which are key functions of the mitochondria. For example, the cytoself model we study in this chapter grouped several previously uncharacterized proteins with known mitochondrial proteins, leading researchers to propose their involvement in oxidative phosphorylation—the process by which cells generate ATP within the mitochondrial matrix.

Detect early cellular changes that mark disease

Shifts in protein localization can serve as early warning signs of various diseases. For example, we previously mentioned that in ALS, the protein TDP-43 moves from the nucleus to the cytoplasm, but remarkably, this change has been observed in presymptomatic individuals carrying disease-linked mutations.[5] More broadly, large-scale profiling of localization patterns could help detect early cellular dysfunction across a wide range of conditions.

Therapies that correct protein mislocalization

In diseases where a protein is physically functional but simply ends up in the wrong place, one therapeutic strategy is to restore its proper localization using engineered localization signals. For example, researchers have used nuclear localization signals to redirect tumor suppressors like p53 or BRCA1 back to the nucleus, where they can resume their normal function.

Better therapeutic targeting

Another approach is to guide drugs, proteins, or nanoparticles to specific subcellular compartments, such as the lysosome or mitochondria, to maximize their effectiveness and minimize side effects. This strategy is used in emerging nanomedicine platforms.[6]

Imagine being able to say not just *what* a protein does but *where* it does it—and how its journey changes as a cell divides, differentiates, or begins to break down in disease.

Ideally, that gives you a sense of how exciting the protein localization space is. Now let's dive into how machine learning can help us explore it.

---

5 Paré, B., et al. (2015). Early detection of structural abnormalities and cytoplasmic accumulation of TDP-43 in tissue-engineered skins derived from ALS patients (*https://doi.org/10.1186/s40478-014-0181-z*). *Acta Neuropathologica Communications*, 3(1).

6 Ye, D., et al. (2023). Recent Advances in Nanomedicine Design Strategies for Targeting Subcellular Structures,. *iScience*, 28(1), 111597.

Machine Learning Primer

The model we'll be building in this chapter—the cytoself from the Kobayashi paper mentioned earlier—is based on a type of neural network called a *vector-quantized variational autoencoder (VQ-VAE)*. If you haven't seen this type of model before, don't worry: we'll walk through the key ideas that lead up to this model so you understand not just *what* we're building but *why* it works.

Why use a VQ-VAE? Unlike most models, which map images into a continuous feature space, a VQ-VAE forces the model to describe each image using a limited set of learned visual patterns, called a *codebook*. You can think of this like a tile set or a visual vocabulary, where each pattern gets reused across many inputs. This encourages the model to represent each protein image using discrete building blocks, making it easier to group proteins by similar localization and uncover shared visual motifs. In other words, instead of inventing new coordinates for every input, the model says, "This protein looks like a mix of tile #7 and tile #241."

This kind of representation is especially useful in biology, where we're trying to *discover* structure in the data, not just reconstruct it. The discreteness also makes the model more interpretable and allows downstream tools (like clustering or dimensionality reduction) to work more effectively.

Autoencoders (AEs)

Let's start by unpacking the last and arguably most important part of the VQ-VAE acronym: the AE, *autoencoder*.

An autoencoder is a type of neural network trained to reconstruct its input. It does this in two steps:

1. An *encoder* compresses the input into a lower-dimensional representation.
2. A *decoder* then attempts to reconstruct the original input from that compressed version.

In the simplest case, both the encoder and decoder might consist of fully connected linear layers. The internal representation, known as the bottleneck, typically has fewer neurons than the input, forcing the model to compress the data, as illustrated in Figure 6-2 (diagram taken from "Introduction to Autoencoders").[7]

---

7 Jordan, J. (2018, March 19). *Introduction to autoencoders* (*https://oreil.ly/iLKwc*). Jeremy Jordan.

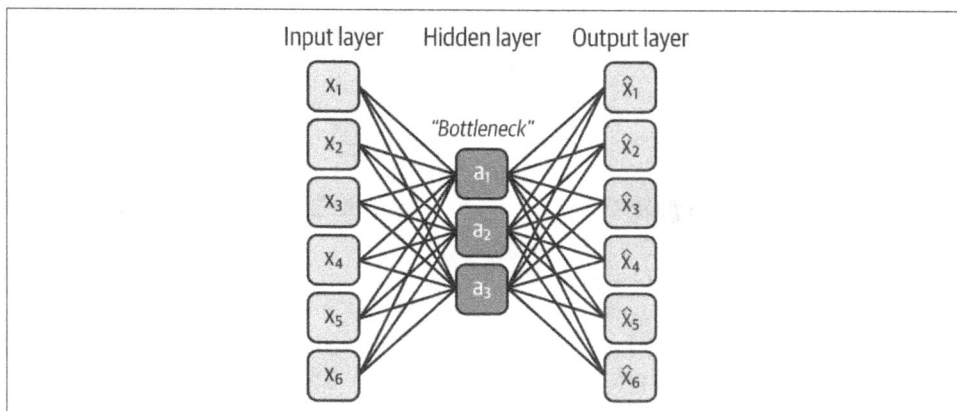

Figure 6-2. An autoencoder learns a compressed internal representation of its input by forcing it through a low-dimensional "bottleneck." The network is trained to reconstruct the original input as closely as possible from this bottleneck representation.

The bottleneck forces the model to distill the most important patterns in the data, while discarding irrelevant details. This is a form of dimensionality reduction, similar in spirit to techniques like principal component analysis (PCA). But unlike PCA, autoencoders can learn nonlinear transformations and scale to large, complex datasets, making them especially powerful for data like microscopy images.

One important detail: in a standard autoencoder, the internal representation—the activations of the neurons in the bottleneck layer—are *continuous*. That means each neuron in the bottleneck can take on any real number (like 1.27, –3.14, etc.), so the representation of each input lives in a continuous space. This gives the model a lot of flexibility, but it can also make the latent space less structured and harder to interpret. Two inputs that look quite similar might map to distant points in the latent space, and it can be difficult to understand what each dimension represents.

Later, we'll see how VQ-VAEs address this by introducing a fixed set of allowed *codes*. Instead of letting the encoder output arbitrary values, it's forced to pick from a dictionary of discrete latent vectors—a design that makes the representation more structured, compressible, and interpretable. Of course, this comes with a trade-off: the model gives up some of the expressivity of continuous representations in exchange for a more constrained and interpretable latent space.

Cytoself uses a ResNet-style CNN as its encoder. If you'd like a refresher on CNNs and ResNets, see the previous chapter's explanation of these topics.

Variational Autoencoders (VAEs)

Now that we've introduced regular autoencoders, let's look at a variation that adds a twist—and opens the door to powerful generative capabilities. The *V* in *VAE* stands for *variational*. In a *variational autoencoder*, we no longer compress each input (like a microscopy image) into a single point in the latent space. Instead, the model learns to represent each input as a *probability distribution* over the latent space. That's a bit of a mouthful, so let's break it down step-by-step.

In a standard autoencoder:

- The encoder takes an input and produces a fixed set of numbers: one activation for each neuron in the bottleneck layer we described earlier.
- These numbers describe a single point in latent space.
- The decoder then uses that point to reconstruct the input.

In a variational autoencoder:

- The encoder outputs two numbers per latent dimension: a *mean* and a *standard deviation*.
- These define a normal distribution—a bell curve—for each coordinate in the latent space.
- Instead of feeding a fixed number to the decoder, the model samples a value from each distribution.

To make this concrete:

- In a standard autoencoder, a given input might be compressed to a single point in the latent space, for example, `[1.3, -0.7, 1.9]` for a bottleneck layer with three neurons. This exact vector is passed *directly* to the decoder to reconstruct the input.
- In a variational autoencoder, the encoder instead outputs two separate vectors:
  — One for the means (e.g., `[1.3, -0.7, 1.9]`)
  — And one for the standard deviations (e.g., `[0.2, 0.5, 0.1]`)

This is often done using a shared hidden layer followed by two parallel linear layers: one predicts the means, the other the standard deviations. So, in this example, the encoder outputs six values in total: three means and three standard deviations.

Together, these describe a 3D Gaussian distribution: not a single point, but a cloud of likely values. During training, the model randomly samples a vector from this distribution; for example:

- Something like 1.1 from a distribution centered at 1.3 ± 0.2
- -0.5 from -0.7 ± 0.5
- 1.7 from 1.9 ± 0.1

The sampled vector—in this case, [1.1, -0.5, 1.7]—is what actually gets passed to the decoder for reconstruction.

Why add randomness?

At first, introducing randomness might sound like unnecessary fuzziness: why not just stick with a fixed encoding? But this design choice has several powerful benefits:

It smooths the latent space.
  Because the model samples slightly different encodings for the same input, it learns to decode nearby points into similar outputs. This forces the latent space to be smooth, continuous, and meaningful—with small movements in the space leading to small, realistic changes in the output.

It groups similar inputs together.
  Inputs that are alike produce similar distributions, so their samples overlap in the latent space. This naturally pulls similar data points closer together, helping the model learn structure in the dataset.

It prevents overfitting.
  By introducing controlled randomness during training, the model can't just memorize exact input-output pairs. It has to learn patterns that hold across perturbations.

It enables generation.
  Once trained, the model can generate new, realistic-looking outputs by simply sampling from the latent space, even in regions that weren't seen during training. This makes VAEs useful not just for reconstruction, but for creative or exploratory tasks.

Continuous latent space

We can visualize this difference in how latent spaces are structured in VAEs versus standard autoencoders using the diagram in Figure 6-3 (based on a figure by Saul Dobilas).[8]

---

8 Dobilas, S. (2025, January 22). VAE: Variational Autoencoders – How to employ neural networks to generate new images (*https://oreil.ly/0SB-Q*). *Towards Data Science.*

Data mapped as points in the latent space
(e.g., undercomplete AE): Not suitable for generating "meaningful" new data

The point that has not been explicitly mapped is "meaningless"

Data mapped as distributions
(e.g., VAE) with the latent space being continuous and regularized

The point that has not been explicitly mapped is "meaningless"

Points from a regularized latent space produce new "meaningful" shapes that are close to the original ones

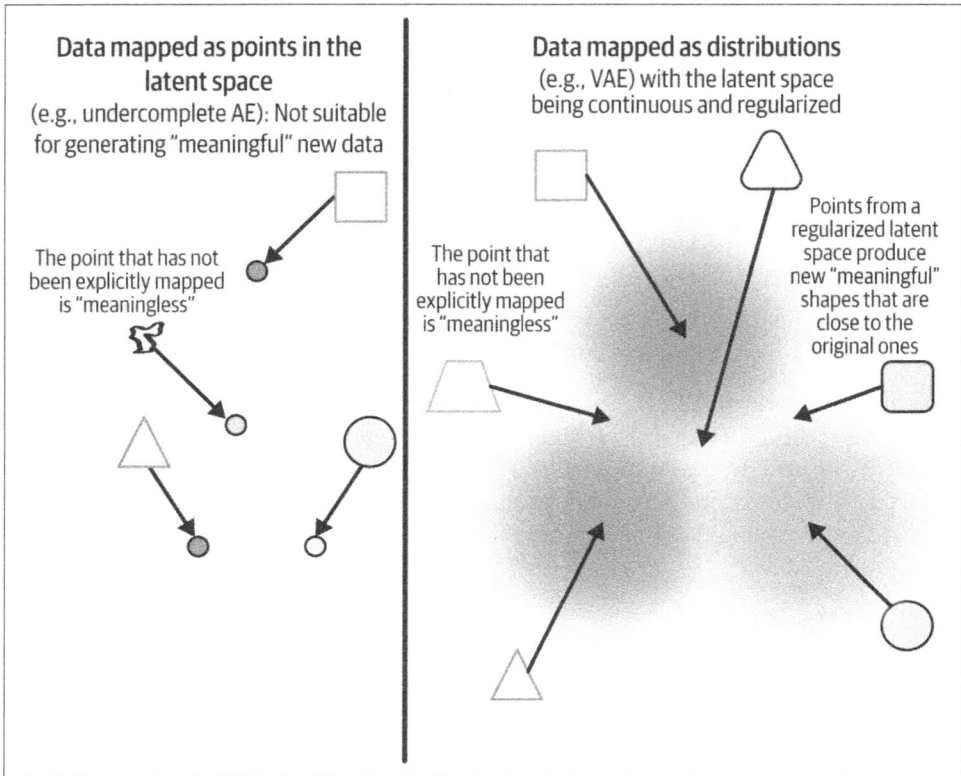

Figure 6-3. An intuitive way to think about regularized continuous latent spaces. In a standard autoencoder, points are mapped discretely and may not generalize meaningfully. In a VAE, points are sampled from smooth distributions, enabling meaningful interpolation and generative sampling.

This diagram highlights how VAEs encourage a smooth and continuous latent space, enabling interpolation and generation, a key distinction from regular autoencoders.

You might wonder: can't you use a regular autoencoder for grouping similar inputs together and for generative sampling?

In theory, yes; similar inputs often *do* end up close together in the latent space, and you *can* try sampling from it. But there's no guarantee that the space will be smooth, continuous, or meaningful. Some regions may decode into nonsense, and small changes in latent space might lead to big, unpredictable jumps in the output.

VAEs solve this by explicitly *shaping the latent space*. They use probability distributions and regularization to encourage the model to use the space in a more structured and consistent way.

There's just one more concept to cover before we can understand this chapter's architecture: *vector quantization.*

Vector-Quantized Variational Autoencoders (VQ-VAEs)

The *VQ* in *VQ-VAE* stands for *vector quantization,* a classical technique borrowed from signal processing and data compression. At its core, vector quantization means taking a continuous input, such as a floating-point vector, and snapping it to the nearest match from a fixed set of allowed vectors. This set is called a *codebook.*

To make this concrete:

- Imagine our codebook contains just two vectors: [2, 0.5] and [1, -3].
- Now suppose the encoder outputs [1.8, 0.3].
- Instead of passing this continuous vector to the decoder, the VQ-VAE finds the nearest codebook entry—in this case, [2, 0.5] —and *replaces* the encoder output with that vector.

This "snapping" process is called *quantization.* You can think of it like rounding a continuous input to the closest available option. The decoder never sees the raw encoder output, only the snapped codebook vector.

This makes VQ-VAEs different from variational autoencoders or standard autoencoders: rather than learning a continuous latent space, the model learns a *discrete vocabulary of embeddings* and uses that to represent everything it sees. By compressing the encoder output to a finite set of discrete vectors, the model encourages robustness, interpretability, and reusability in its representations, all of which are particularly helpful for downstream tasks like clustering or biological discovery.

Where does the codebook come from?

The codebook in a VQ-VAE is *learned during training.* Just like the weights in the encoder and decoder, the vectors in the codebook start out random and are gradually refined through backpropagation. Over time, the vectors adapt to represent recurring, meaningful patterns in the data, so the model gets better at snapping encoder outputs to useful representations.

How large should the codebook be?

There's no one-size-fits-all answer: the optimal number of vectors depends on your data and goals. A *larger codebook* (e.g., 1,024+ entries) allows finer distinctions between inputs. A *smaller codebook* (e.g., 64–128) forces the model to reuse patterns more often, which can help with generalization and interpretability.

In biological imaging tasks like protein localization, codebook sizes typically range from 128 to 512, depending on the number of proteins, the resolution of localization patterns, and the desired balance between expressiveness and interpretability. Cytoself, for instance, uses two codebooks of 2,048 entries, each made up of 64D vectors—one for the global representation and one for the local—giving the model rich capacity to represent complex spatial patterns.

Dissecting a VQ-VAE Diagram

Now that we've covered the key building blocks—autoencoders, variational inference, and vector quantization—we're ready to interpret the original VQ-VAE diagram, shown in Figure 6-4.[9]

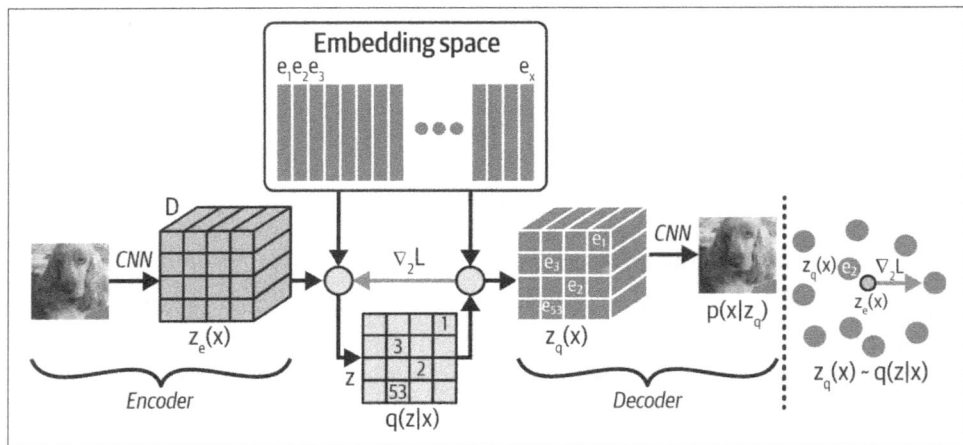

Figure 6-4. Illustration of the major components in a VQ-VAE, based on van den Oord et al. (2017).

Here's a step-by-step breakdown of what's happening in the above VQ-VAE diagram:

1. Input image (dog photo on the far left)

The process begins with a raw input image; here, the input is a photo of a dog. In our case, this would be a microscopy image showing protein localization. The image is passed into a CNN encoder, which extracts meaningful features and transforms the image into a compressed latent representation.

This representation is shown as the *cube labeled $z_e(x)$* (the variable z is often used to denote embeddings). It's a 3D tensor because CNNs process images into pseudoimages called feature maps, which highlight specific parts of the input.

---

9 Van Den Oord, Aaron, Oriol Vinyals, and Koray Kavukcuoglu. 2017. "Neural Discrete Representation Learning" (*https://oreil.ly/D-Rst*). arXiv.Org. November 2, 2017.

This cube holds the distilled, continuous version of the input, and it's what will be quantized next using the codebook.

2. Quantization step (dot to the right of the left cube)

Each vector in the cube $z_e(x)$—the continuous encoder output—is passed to the *vector quantization step*. This step compares each D-dimensional vector (D = 64 for cytoself) in the grid to every entry in a learned *codebook* (e_1, e_2, ..., e_k), shown at the top of the diagram.

The model then *replaces* that vector with the nearest codebook vector, snapping it to the closest match. This creates a new tensor called $z_q(x)$, the quantized version of the encoder output, where every original vector has been replaced by a discrete one from the codebook.

3. Decoder (cube on the right)

Once quantized, the decoder input $z_q(x)$ (right cube) is passed into the *decoder* ("CNN" arrow), which is another CNN. Its job is to take the snapped, discrete representation and *reconstruct the original image* as closely as possible. In this diagram, the decoder output is shown as a second image of the dog, visually similar to the input. For us in this chapter, this would be a reconstructed microscopy image.

The closer the reconstruction is to the original (i.e., the lower the reconstruction loss), the better the model has learned to encode meaningful information in its quantized representation.

4. Codebook embedding space (far right)

The panel on the far right shows a *zoomed-in view of the embedding space*, the set of vectors in the codebook. The central dot (with border and arrow) represents the encoder's original output vector, $z_e(x)$, which sits somewhere in continuous space. The model snaps it to the closest codebook vector, here shown as e_2. That snapped value becomes the quantized output $z_q(x)$.

The arrow indicates the *training signal*: during learning, the encoder is nudged to produce outputs that are closer to the codebook entries they get snapped to. This improves the efficiency of quantization over time and ensures that codebook entries are actually used.

This is starting to get into VQ-VAE training strategy, so let's now cover this topic explicitly.

Training a VQ-VAE

Now that we've broadly seen how a VQ-VAE is structured, let's look at how it's trained.

Training a VQ-VAE involves optimizing more than just the *reconstruction loss*, that is, how closely the output image matches the original input. The model also needs to *refine the codebook vectors* over time and ensure that the encoder learns representations that snap cleanly to those vectors. This creates an interesting balance between three components:

The quality of the encoder's representations
How well does the encoder capture the meaningful signals in the input data?

The usefulness and coverage of the codebook
Does the codebook provide enough diversity to represent a wide range of inputs, and are all entries being used effectively?

The decoder's ability to reconstruct from quantized codes
How accurately can the decoder recover the original input using only the compressed, discrete representations?

There are a few unique challenges in training VQ-VAEs to be aware of:

Quantization error
If the nearest codebook vector is a poor match, some detail is lost during quantization, and the reconstruction suffers.

The key is to make sure your codebook is large enough and your encoder is flexible enough to produce embeddings that land near useful entries.

Codebook collapse
If only a few codebook vectors are used repeatedly while others are ignored, the model wastes capacity. Preventing this requires some extra care during training:

Commitment loss
This loss term encourages the encoder to commit to a chosen codebook vector by penalizing large differences between the encoder output and its nearest codebook entry. It keeps the encoder from drifting too far away from the codebook.

Entropy penalty
This encourages the model to use a wider range of codebook entries more evenly, increasing the diversity of representations and reducing collapse.

Despite these challenges, the VQ-VAE is a powerful architecture, especially when you want a model that produces discrete, interpretable representations while still learning directly from data.

> If this feels a bit complex, you're not alone—engaging with modern deep learning models can be challenging at first. But hands-on practice is the best way to build intuition and fluency. In the next section, you'll start building and training a VQ-VAE yourself, applying everything we've covered so far.

Constructing the Dataset

All models need good data. In this section, we'll take a closer look at how to prepare a cell imaging dataset to learn about protein localization.

Data Requirements

A high-quality dataset is crucial for successfully applying deep learning. Fortunately, there are several excellent protein localization resources available. One of the best is OpenCell (*https://oreil.ly/km_v9*), which provides standardized, high-resolution images of human proteins inside cells, such as that shown in Figure 6-5.

Because all images were taken using a single, consistent imaging pipeline, they are particularly well suited for machine learning. This consistency ensures that the model focuses on learning biological variation in protein localization, rather than being distracted by irrelevant differences in imaging conditions or processing pipelines.

> Want to explore the dataset yourself? Head to OpenCell (*https://opencell.sf.czbiohub.org*), which has a beautifully designed interface that makes browsing the data fast and intuitive.

The model we'll build in this chapter relies entirely on imaging data; it doesn't use any labels or annotations during training. This is a major strength: it learns localization patterns directly from raw microscopy images, without any manual supervision. While OpenCell does provide localization annotations, we'll only use them at the end to sanity-check our results. Overreliance on curated labels can introduce bias and limit scalability, as manual annotation is both expensive and subjective. This is where self-supervised learning shines—it enables models to uncover meaningful patterns without needing predefined labels.

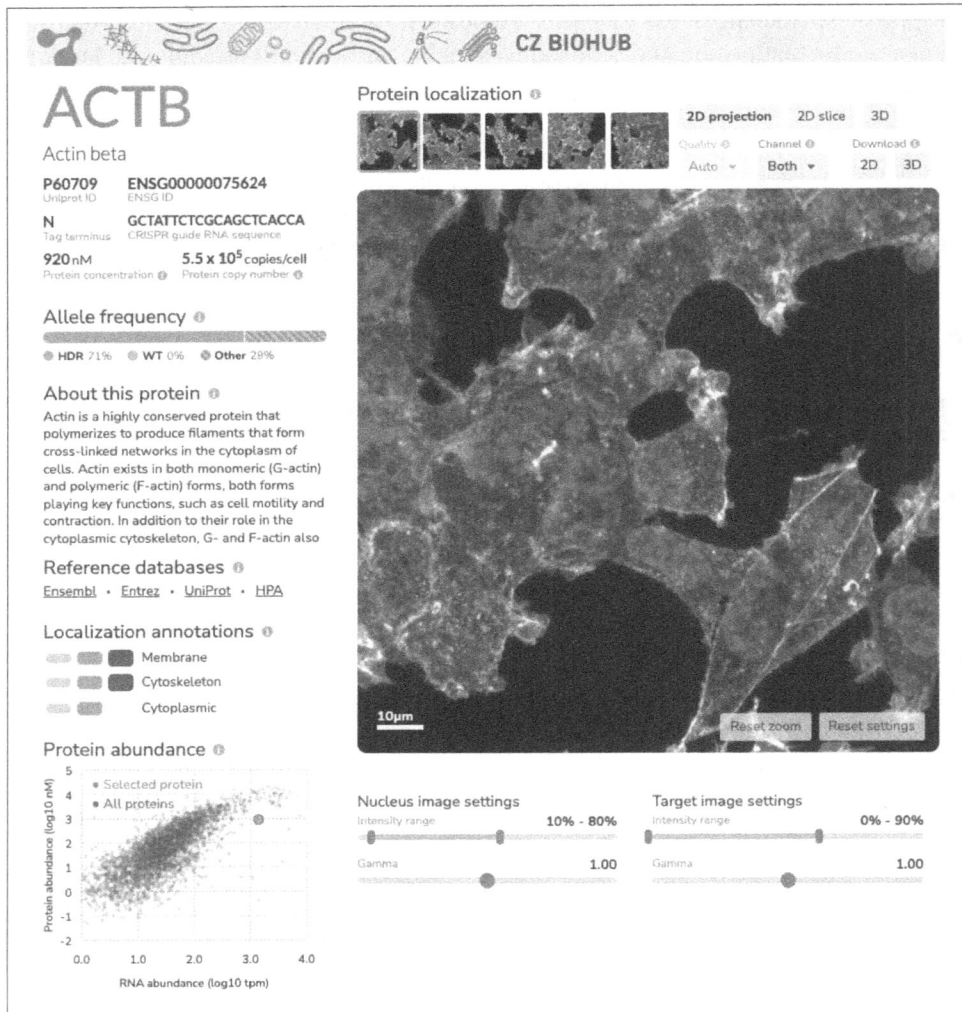

Figure 6-5. Example entry for the protein ACTB (beta-actin) from the OpenCell database. Each protein has a detailed summary page that includes its identifiers (e.g., UniProt ID), sequence information, expression levels (how many copies of the gene and protein are present in the cell), and subcellular localization annotations. On the right, you can see a fluorescence microscopy image showing ACTB localization (gray) alongside nuclear staining (blue). These multichannel images, together with protein-specific metadata, form the basis of the dataset used in this chapter.

Sourcing the Data

Preparing images for deep learning can be a challenge in itself, especially when it comes to format, consistency, and scale. We saw this previously in Chapter 5, where careful preprocessing was essential. Fortunately, the cytoself authors have released a preprocessed version of the dataset they used in their paper, so we don't have to start from scratch here.

The dataset contains 1,311 fluorescently tagged proteins, each imaged in roughly 18 different fields of view. From each field of view, about 45 individual crops are extracted, each typically containing around three cells. This yields roughly 800 image crops per protein, totaling more than 1,048,800 images in the full dataset. Each image includes two channels:

- The green *protein channel*, showing the location of the fluorescently tagged protein of interest. This is the only channel used for training in this chapter.
- A *nuclear stain* using Hoechst, a blue fluorescent dye that binds strongly to DNA. This highlights the nuclei of the cells and provides spatial context, for example, helping to identify whether a protein is nuclear, cytoplasmic, or membrane bound.

From the nuclear stain, a third representation is computed: the *distance-to-nucleus map*. This is not a separate imaging channel, but a spatial map derived from the Hoechst signal that gives the model more information about relative positioning:

- Pixels inside the nucleus are assigned positive values, representing the shortest distance to the nuclear boundary.
- Pixels outside the nucleus are assigned negative values, also based on shortest distance to the nuclear edge.

To simplify the task, we will use only the tagged protein channel for model training. The other channel (Hoechst) and the derived distance map provide useful spatial context but are not required for this self-supervised learning setup.

In addition to the image data, the dataset includes:

- Protein annotations: Such as gene names and unique IDs
- Curated localization labels: Manually assigned organelle or compartment categories

These curated labels are *not* used during training. Instead, they're only used afterward to evaluate whether the model has learned biologically meaningful representations on its own.

> This dataset is significantly larger than those used in previous chapters, introducing new challenges in terms of memory usage, data loading speed, and training efficiency.
>
> Make sure you have enough disk space and access to a capable GPU. For example, use an A100 available through Colab Pro+, or be prepared to scale down the model or batch size to make training feasible.

Getting a Glimpse of the Dataset

For more complex datasets, it's common to wrap the data loading and preprocessing logic in a custom `Dataset` class, as we have done in earlier chapters in this book. This helps make the data easier to explore and use in training.

Let's first use the utility function `get_dataset` to start inspecting a few of the image frames in the dataset; see Figure 6-6. Sampling a handful of frames quickly reveals just how much variation there is in patterns and intensity across the dataset:

```python
import jax
from dlfb.localization.dataset.utils import get_dataset
from dlfb.utils.context import assets

rng = jax.random.PRNGKey(42)
rng, rng_frames = jax.random.split(rng, 2)

dataset = get_dataset(data_path=assets("localization/datasets"))
n_frames = 16
dataset.plot_random_frames(n=n_frames, rng=rng_frames);
```

This diversity is expected: each image corresponds to a different tagged protein, and many proteins aren't restricted to a single cellular compartment. Instead, they localize to multiple regions of the cell, often with varying abundance. This adds another layer of complexity to the learning task.

Interestingly, this heterogeneity isn't just seen across different proteins. It also appears *within the same protein*, across different cells. Even when the protein and the imaging setup are held constant, localization patterns can vary from cell to cell.

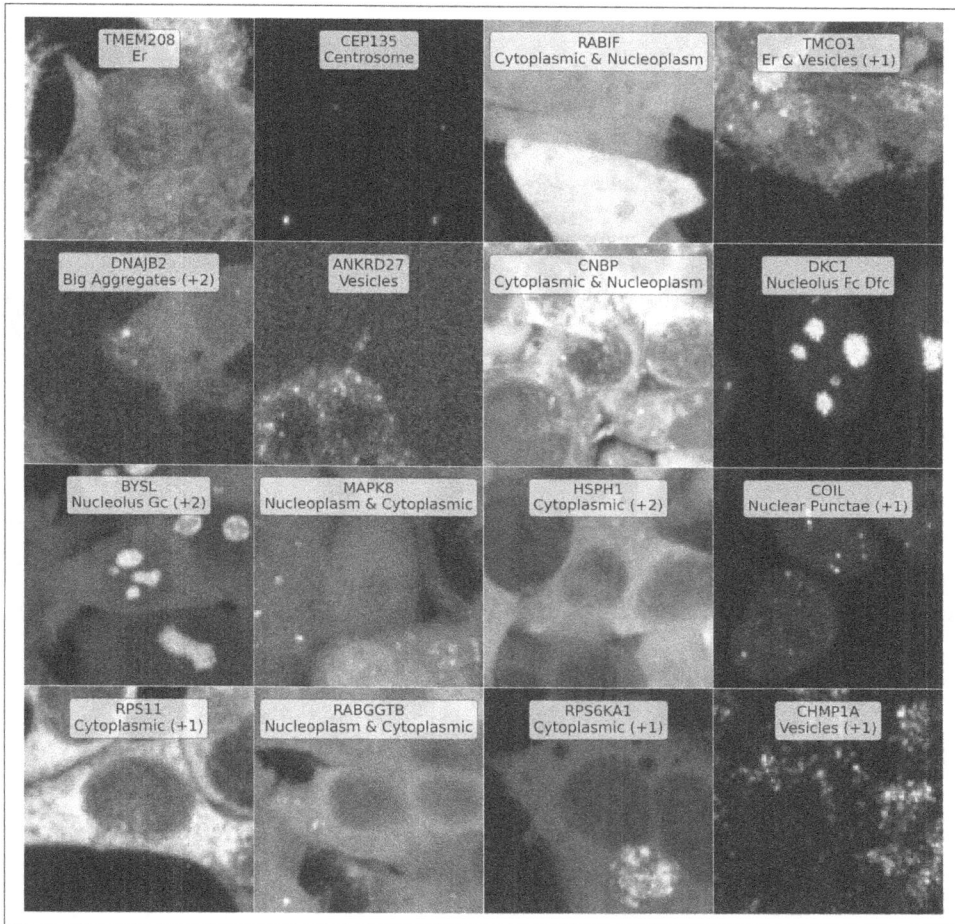

Figure 6-6. Plot of a random subset of frames. The protein symbols and their primary localization(s) are given for each frame (number indicates additional secondary localizations measured).

To illustrate this, we'll load a dataset containing all image frames for a single protein, ACTB (beta-actin), and plot a few random samples. In Figure 6-7, you'll see that even for this single protein, there is substantial variation in signal strength, shape, and localization simply due to the nature of biological variability—remember that all frames were captured under identical experimental conditions.

```
selected_protein = "ACTB"
dataset.plot_random_frames(
    n=n_frames, with_labels=False, rng=rng_frames, gene_symbols=[selected_protein]
);
```

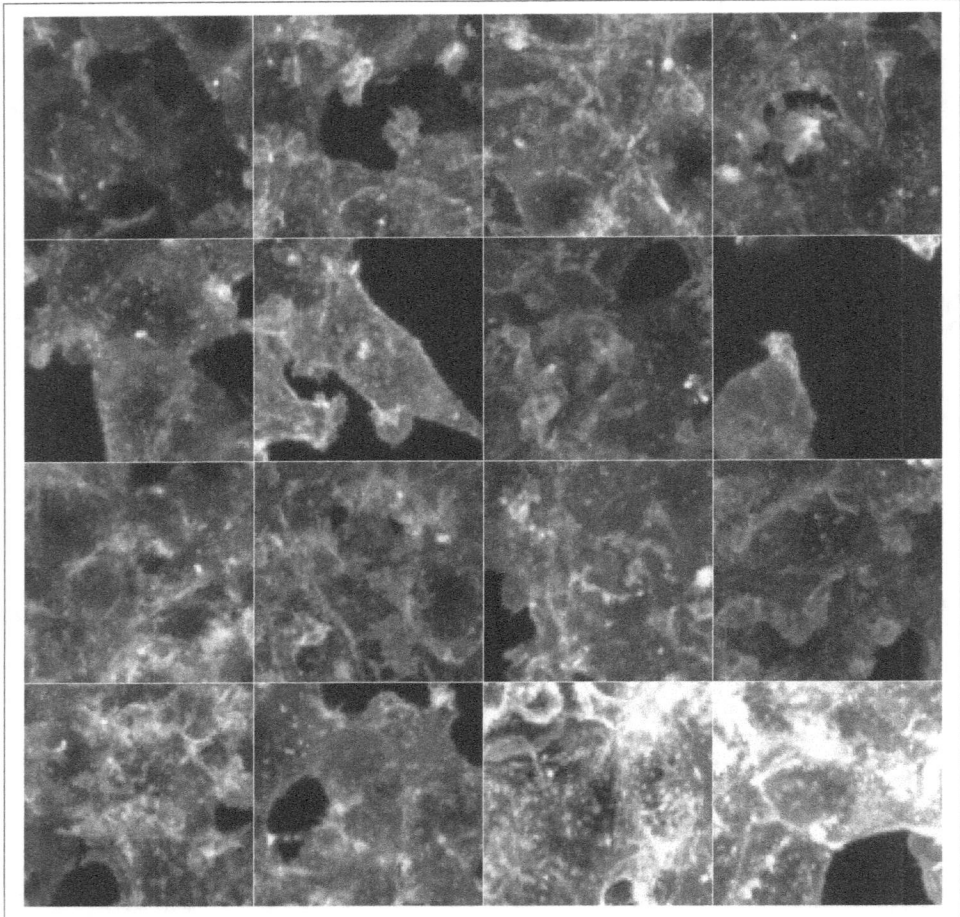

Figure 6-7. Random subset of 16 image frames showing the localization of the protein ACTB (beta-actin), a cytoskeletal protein annotated as localizing to the membrane, cytoskeleton, and cytoplasm. Even though these frames all represent the same protein under identical experimental conditions, they display considerable variation in shape, brightness, and localization pattern, highlighting the inherent biological variability across cells.

Let's now take a closer look at just a single frame in Figure 6-8, which will give us a better sense of what the model will be processing during training. The figure is deliberately shown large so that you can see the individual pixels clearly. Try to think of the image not just as a picture, but as a matrix of numbers: white pixels have values close to 255, black pixels are near 0, and all the shades of gray are values in between:

```
dataset.plot_random_frames(n=1, with_labels=False, rng=rng_frames);
```

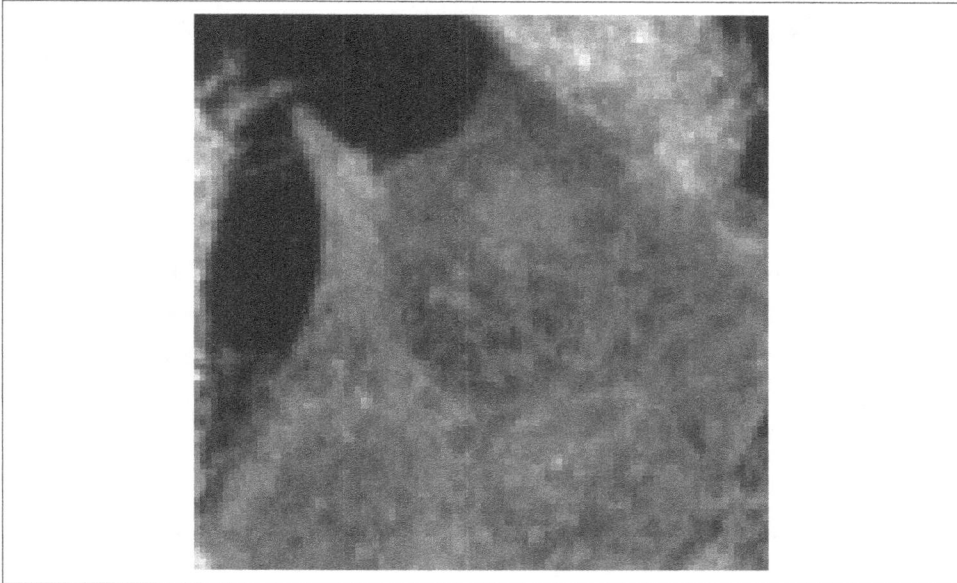

Figure 6-8. Close-up view of a single image frame for the protein ACTB. The pixelated appearance reflects the raw input that the model will see: a 2D matrix of intensity values. Bright regions correspond to high fluorescence signal (values near 255), and dark regions correspond to low signal (values near 0). This is the actual numerical input used to learn spatial localization patterns. Despite the images looking noisy or unclear to the human eye, the model must learn to extract consistent patterns from them.

Implementing a DatasetBuilder Class

Here, we've implemented a helper class called `DatasetBuilder`, which handles dataset setup and caching behind the scenes:

```python
class DatasetBuilder:
    """Builds a dataset with splits for learning."""

    def __init__(self, data_path: str, force_recreate: bool = False):
        self.images = ImageLoader(data_path).load(force_recreate=force_recreate)
        self.labels = LabelLoader(data_path).load(force_recreate=force_recreate)

    def build(
        self,
        rng: jax.Array,
        splits: dict[str, float],
        exclusive_by: str = "fov_id",
        n_proteins: int | None = None,
        max_frames: int | None = None,
    ) -> dict[str, Dataset]:
        """Retrieve a dataset of proteins split into learning sets."""
        validate_splits(splits)

        if not n_proteins:
            n_proteins = self.labels.get_n_proteins()

        # Sample frames from chosen proteins.
        rng, rng_proteins = jax.random.split(rng, num=2)
        frames = self.labels.get_frames_of_random_proteins(rng_proteins, n_proteins)

        n_frames = frames.shape[0]
        if max_frames is not None and n_frames > max_frames:
            # Limit number of frames used.
            frames = frames.head(max_frames)
            n_frames = max_frames

        # Get random entities to exclusively be assigned across splits
        rng, rng_perm = jax.random.split(rng, 2)
        set_ids = jnp.array(frames[exclusive_by].to_numpy(np.int32))
        shuffled_set_ids = jax.random.permutation(rng_perm, jnp.unique(set_ids))

        # Assign consecutive ids to proteins across all frames
        frame_ids = jnp.array(frames["frame_id"].to_numpy(np.int32))
        lookup_with_protein_encoding = self._encode_proteins_across_frames(
            self.labels.lookup.iloc[frame_ids.tolist()]
        )

        # Assemble the dataset by splits considering exclusive sets
        dataset_splits, start = {}, 0
        for name, size in self._get_split_sizes(
            splits, n_sets=len(shuffled_set_ids)
        ):
```

```
        mask = jnp.isin(set_ids, shuffled_set_ids[start : (start + size)])
        dataset_splits[name] = Dataset(
          images=self.images,
          labels=Labels(
            lookup=lookup_with_protein_encoding.loc[
              frame_ids[mask].tolist()
            ].reset_index(drop=True)
          ),
        )
        start += size

    return dataset_splits

  def _get_split_sizes(self, splits, n_sets):
    """Convert split fractional sizes to absolute counts."""
    names = list(splits.keys())
    sizes = [int(n_sets * splits[name]) for name in names[:-1]]
    sizes.append(n_sets - sum(sizes))  # Ensure total adds up
    for name, size in zip(names, sizes):
      yield name, size

  def _encode_proteins_across_frames(self, lookup) -> pd.DataFrame:
    """Encode protein labels across dataset to consecutive integers."""
    protein_ids_in_frames = lookup["protein_id"].to_list()
    unique_protein_ids = sorted(set(protein_ids_in_frames))
    mapping = pd.DataFrame(
      [
        {"protein_id": id_, "code": idx}
        for idx, id_ in enumerate(unique_protein_ids)
      ]
    )
    return lookup.merge(mapping, how="left", on="protein_id").set_index(
      "frame_id", drop=False
    )
```

During initialization, we provide the path to where the data is stored. This allows the DatasetBuilder to create linked instances of Images and Labels. You can also see that the DatasetBuilder has a single public method, .build, which returns a dict[str, Dataset] where every key is a dataset split and the value corresponds to a dataset. This method gives you the option to subset the dataset to a limited number of randomly selected proteins (n_proteins) and/or a maximum number of frames (max_frames).

It also splits the dataset into different sets for the learning stage. The sets can be provided with the splits parameter where the names and fractional sizes can be requested. Finally, it is possible to request with the exclusive_by parameter that across splits, fields of view or proteins are never shared (with fov_id or protein_id, respectively). The former ensures that during learning, we do not inadvertently bleed information between splits, as two frames could otherwise capture an overlapping zone in a field of view. Not having the same proteins in the training and evaluation sets ensures that more general representations are learned, as was done in the original paper. Here we use fov_id, as we will restrict our dataset to a smaller number of proteins for faster training. Implementation-wise, we first randomly split the dataset by unique fields of view, and then, as we build the splits, we mask frames not to appear across splits. You will also have noticed that we are encoding the original protein IDs to consecutive integers, which, as you will see later, is required by the loss function during the training loop.

Building a first dataset instance

We'll now look at using the DatasetBuilder to create instances of Dataset splits:

```python
from dlfb.localization.dataset import Dataset
from dlfb.localization.dataset.builder import DatasetBuilder
from dlfb.utils.context import assets

builder = DatasetBuilder(data_path=assets("localization/datasets"))

rng, rng_dataset = jax.random.split(rng, 2)
dataset: dict[str, Dataset] = builder.build(
    rng=jax.random.PRNGKey(42),
    splits={"train": 0.80, "valid": 0.10, "test": 0.10},
    exclusive_by="fov_id",
    n_proteins=50,
)
```

We now have a dataset. Having a dataset builder provides flexibility that makes it easy to explore, debug, and prototype with small slices of the dataset before scaling up to the full training set.

Accessing the dataset internals

Behind the scenes, the Dataset class contains two main components: Images and Labels. These handle the raw microscopy image data and the metadata annotations, respectively.

You don't need to dive into their implementations to follow along in this chapter, but if you're curious, the full source code is available. The lower-level methods let you load specific proteins, subsample data, or query for localization annotations—helpful tools for deeper biological exploration or interactive experimentation.

If you read the earlier chapters, this modular data structure should feel familiar. We assemble data from different blocks, use memory mapping for large arrays, and wrap everything into Python classes for convenience and speed. There are of course a lot of improvements that you can make to allow for even faster access to larger datasets, and TensorStore (*https://oreil.ly/m29ho*) certainly ticks a lot of boxes here.

Now that we've examined how the data is structured and loaded, let's move on to training our first model.

Building a Prototype Model

In this section, we'll build a simplified version of the cytoself model we introduced earlier. The aim is to distill the core architectural ideas into a somewhat compressed prototype that is easy to understand, train, and modify. The code implementation here is adapted in part from the official Haiku VQ-VAE repository (*https://oreil.ly/KF0A0*) and the VQ-VAE Flax implementation (*https://oreil.ly/z-ZXk*) from Arnaud Aillaud.

We intentionally omit some of the more complex features from the original work, such as split quantization, hierarchical vector quantization, and multiresolution training, in order to keep the code accessible and focused on the key mechanisms. This also makes it easier to tinker and explore your own architectural experiments. Check out the final section of this chapter for more information on possible extensions to the model.

Defining the LocalizationModel

The following code defines the core model we'll use throughout this chapter:

```
class LocalizationModel(nn.Module):
    """VQ-VAE model with a fully connected output head."""

    embedding_dim: int
    num_embeddings: int
    commitment_cost: float
    num_classes: int | None
    dropout_rate: float
    classification_head_layers: int

    def setup(self):
        """Builds the encoder, decoder, quantizer, and output head."""
        self.encoder = Encoder(latent_dim=self.embedding_dim)
        self.vector_quantizer = VectorQuantizer(
            num_embeddings=self.num_embeddings,
            embedding_dim=self.embedding_dim,
            commitment_cost=self.commitment_cost,
        )
        self.decoder = Decoder(latent_dim=self.embedding_dim)
        self.classification_head = ClassificationHead(
            num_classes=self.num_classes,
            dropout_rate=self.dropout_rate,
            layers=self.classification_head_layers,
        )

    def __call__(self, x: jax.Array, is_training: bool):
```

```
"""Runs a forward pass."""
ze = self.encoder(x)
zq, perplexity, codebook_loss, commitment_loss = self.vector_quantizer(ze)
decoded = self.decoder(zq)
logits = self.classification_head(
    zq.reshape((zq.shape[0], -1)), is_training
)
return decoded, perplexity, codebook_loss, commitment_loss, logits

def create_train_state(
    self, rng: jax.Array, dummy_input: jax.Array, tx
) -> TrainState:
    """Initializes training state."""
    rng, rng_init, rng_dropout = jax.random.split(rng, 3)
    variables = self.init(rng_init, dummy_input, is_training=False)
    return TrainState.create(
        apply_fn=self.apply, params=variables["params"], tx=tx, key=rng_dropout
    )

def get_encoding_indices(self, x: jax.Array) -> jax.Array:
    """Returns nearest codebook indices for input."""
    ze = self.encoder(x)
    encoding_indices = self.vector_quantizer.get_closest_codebook_indices(ze)
    return encoding_indices
```

This model is defined with the main components we already introduced:

Encoder
Maps the input image into a latent space

Vector quantizer
Discretizes this latent space using a learned codebook

Decoder
Reconstructs the input image from the quantized representation

These are all wired together in the model's setup method. You'll also notice an additional component, the ClassificationHead, which we'll come back to later. For now, just know that it's used to mitigate the previously mentioned codebook collapse, a failure mode where the model uses only a small fraction of the codebook entries, reducing the representational power of the latent space.

In the following sections, we'll walk through each part of the model, starting with the Encoder, which is responsible for extracting latent features from the input microscopy images.

The Encoder: Processing Input Images

The Encoder is the first part of our model and is responsible for converting raw input images into a continuous latent representation. Its main job is to process the input

---

into a form that the `VectorQuantizer` can work with—a spatial feature map with rich, expressive features per pixel. This latent representation will later be discretized by the `VectorQuantizer`, so it needs to be of the right shape and dimensionality.

The encoder is built from three convolutional layers followed by two residual blocks:

- The first two conv layers downsample the image by a factor of 4 overall (each has a stride length of 2), reducing spatial resolution while increasing feature dimensionality.
- The third conv layer preserves spatial resolution but deepens the feature map.
- The two `ResnetBlocks` further refine the features with normalization, nonlinearity, and skip connections.

Together, this forms the pipeline that turns a 100×100 grayscale image into a smaller grid of `latent_dim`-dimensional feature vectors, ready for quantization:

```python
class Encoder(nn.Module):
    """Convolutional encoder producing latent feature maps."""

    latent_dim: int

    def setup(self):
        """Initializes convolutional and residual layers."""
        self.conv1 = nn.Conv(
            self.latent_dim // 2, kernel_size=(4, 4), strides=(2, 2), padding=1
        )
        self.conv2 = nn.Conv(
            self.latent_dim, kernel_size=(4, 4), strides=(2, 2), padding=1
        )
        self.conv3 = nn.Conv(
            self.latent_dim, kernel_size=(3, 3), strides=(1, 1), padding=1
        )
        self.res_block1 = ResnetBlock(self.latent_dim)
        self.res_block2 = ResnetBlock(self.latent_dim)

    def __call__(self, x):
        """Forward pass applying convolution and residual blocks to input."""
        x = self.conv1(x)
        x = nn.relu(x)
        x = self.conv2(x)
        x = nn.relu(x)
        x = self.conv3(x)
        x = self.res_block1(x)
        x = self.res_block2(x)
        return x


class ResnetBlock(nn.Module):
    """Residual convolutional block with GroupNorm and Swish activation."""
```

```
latent_dim: int

def setup(self):
  """Initializes normalization and convolutional layers."""
  self.norm1 = nn.GroupNorm()
  self.conv1 = nn.Conv(
    self.latent_dim, kernel_size=(3, 3), strides=(1, 1), padding=1
  )
  self.norm2 = nn.GroupNorm()
  self.conv2 = nn.Conv(
    self.latent_dim, kernel_size=(3, 3), strides=(1, 1), padding=1
  )

def __call__(self, x):
  """Applies two conv layers with Swish activation and skip connection."""
  h = nn.swish(self.norm1(x))
  h = self.conv1(h)
  h = nn.swish(self.norm2(h))
  h = self.conv2(h)
  return x + h
```

As you can see, the only parameter we are passing into the Encoder is latent_dim. This controls how many channels the network will output; in other words, how expressive each spatial location in the latent space is. In the context of our full model, this is set to embedding_dim, which ensures that the output from the encoder matches the dimensionality expected by the VectorQuantizer.

In the context of a LocalizationModel model instantiation, it is set to self.embed ding_dim, since the Encoder has to encode raw input images into a form that is compatible with the quantization step afterwards; in other words, it needs to have a compatible shape. In the context of the VQ-VAE, this is directly related to the quantized latent space dimensions, as it will need to return encoded images that are shaped correctly for the quantization process.

The ResnetBlock uses *group normalization* and the *swish* activation function, a choice inspired by more recent diffusion models (like *stable diffusion*). As a quick reminder, the key idea with residual blocks is that the output is refined by residual learning: instead of trying to learn a full transformation of the input from scratch, the network learns a correction on top of the input. This typically helps with gradient flow and generalization.

At the end of the encoder, we are left with a spatial feature map with shape [batch_size, height, width, latent_dim], which will then be passed into the VectorQuantizer. This completes the encoding stage of the VQ-VAE architecture.

You can change up various parts of this encoder—convolution kernels, number of layers, activation functions, stride length, and so on. For simplicity, we don't expose them as parameters to Encoder here, but these are all hyperparameters that you can try tuning to optimize performance.

The VectorQuantizer: Discretizing the Embeddings

The VectorQuantizer is the heart of a VQ-VAE—it's where we turn continuous embeddings into discrete codes. This step forces the model to commit to a limited vocabulary of learned feature vectors, improving compression and encouraging meaningful, reusable representations.

Let's break down the key components of the VectorQuantizer module:

```python
class VectorQuantizer(nn.Module):
    """Vector quantization module for VQ-VAE."""

    num_embeddings: int
    embedding_dim: int
    commitment_cost: float

    def setup(self):
        """Initializes the codebook as trainable parameters."""
        self.codebook = self.param(
            "codebook",
            nn.initializers.lecun_uniform(),
            (self.embedding_dim, self.num_embeddings),
        )

    def __call__(self, inputs: jax.Array):
        """Applies quantization and returns outputs with losses and perplexity."""
        quantized, encoding_indices = self.quantize(inputs)
        codebook_loss, commitment_loss = self.compute_losses(inputs, quantized)
        perplexity = self.calculate_perplexity(encoding_indices)
        ste = self.get_straight_through_estimator(quantized, inputs)
        return ste, perplexity, codebook_loss, commitment_loss

    def quantize(self, inputs: jax.Array):
        """Snaps inputs to nearest codebook entries."""
        encoding_indices = self.get_closest_codebook_indices(inputs)
        flat_quantized = jnp.take(self.codebook, encoding_indices, axis=1).swapaxes(
            1, 0
        )
        quantized = jnp.reshape(flat_quantized, inputs.shape)
        return quantized, encoding_indices

    def get_closest_codebook_indices(self, inputs: jax.Array) -> jax.Array:
        """Returns indices of closest codebook vectors."""
        distances = self.calculate_distances(inputs)
        return jnp.argmin(distances, 1)
```

```python
    def calculate_distances(self, inputs: jax.Array) -> jax.Array:
        """Computes Euclidean distances between inputs and codebook vectors."""
        flat_inputs = jnp.reshape(inputs, (-1, self.embedding_dim))
        distances = (
            jnp.sum(jnp.square(flat_inputs), 1, keepdims=True)
            - 2 * jnp.matmul(flat_inputs, self.codebook)
            + jnp.sum(jnp.square(self.codebook), 0, keepdims=True)
        )
        return distances

    def compute_losses(self, inputs: jax.Array, quantized: jax.Array):
        """Computes codebook and commitment losses."""
        codebook_loss = jnp.mean(jnp.square(quantized - lax.stop_gradient(inputs)))
        commitment_loss = self.commitment_cost * jnp.mean(
            jnp.square(lax.stop_gradient(quantized) - inputs)
        )
        return codebook_loss, commitment_loss

    def calculate_perplexity(self, encoding_indices: jax.Array) -> jax.Array:
        """Computes codebook usage perplexity."""
        encodings = jax.nn.one_hot(
            encoding_indices,
            self.num_embeddings,
        )
        avg_probs = jnp.mean(encodings, 0)
        perplexity = jnp.exp(-jnp.sum(avg_probs * jnp.log(avg_probs + 1e-10)))
        return perplexity

    @staticmethod
    def get_straight_through_estimator(
        quantized: jax.Array, inputs: jax.Array
    ) -> jax.Array:
        """Applies straight-through estimator to pass gradients through
        quantization.
        """

        ste = inputs + lax.stop_gradient(quantized - inputs)
        return ste
```

Following are the key parts of the VectorQuantizer class:

Codebook

A learnable matrix of shape (embedding_dim, num_embeddings), initialized with lecun_uniform. Each column is a codebook vector—essentially, a prototype that the model can match against. This matrix is what defines the discrete latent space.

Quantization

The `quantize()` function replaces each encoded input vector with the nearest codebook vector, based on Euclidean distance. This forces the model to express its understanding of each input frame using a fixed vocabulary of learned visual patterns.

Losses

`codebook_loss`

Encourages codebook vectors to move toward the encoder output. This updates the codebook.

`commitment_loss`

Encourages the encoder output to commit to a chosen codebook vector rather than fluctuate. This updates the encoder. The two losses are combined to maintain a balance between encoder stability and codebook usage.

Straight-through estimator (STE)

Quantization is nondifferentiable: you can't backpropagate through a hard lookup operation. The STE solves this by copying the quantized vector for the forward pass, but passing gradients as if the quantization didn't happen. It's a standard trick that allows training to proceed via approximate gradients.

Perplexity

A metric that tells us how many codebook vectors the model is using. High perplexity (close to `num_embeddings`) means the model is spreading its attention across many entries. Low perplexity means collapse—only a few vectors are being used, which limits the model's capacity.

You can see that the parameters for this module are `num_embeddings`, `embedding_dim`, and `commitment_cost`, which define the structure of the discrete latent space:

Number of embeddings

Determines how many discrete vectors the model has to choose from. A higher number allows for more fine-grained and diverse representations.

Embedding dimension (also called latent dimension)

Defines the richness of each vector. For example, if it's 64, each codebook entry has 64 values. Higher-dimensional embeddings can capture more nuanced patterns, but they require more data and computation to train effectively.

Commitment cost

A weighting factor used in the loss function to penalize encoder outputs that deviate from their selected codebook vectors. If this value is too low, the encoder may ignore the codebook. If it's too high, the encoder may be overly constrained and learn less expressive representations.

Here is the main work the `VectorQuantizer` does during the forward pass, via its `__call__` method:

```
def __call__(self, inputs: jax.Array):
    """Applies quantization and returns outputs with losses and perplexity."""
    quantized, encoding_indices = self.quantize(inputs)
    codebook_loss, commitment_loss = self.compute_losses(inputs, quantized)
    perplexity = self.calculate_perplexity(encoding_indices)
    ste = self.get_straight_through_estimator(quantized, inputs)
    return ste, perplexity, codebook_loss, commitment_loss
```

This function performs four key operations:

- Quantization: It quantizes the input by replacing each encoded vector with the nearest entry from the codebook.

- Loss computation: It calculates two loss terms: one to pull codebook entries toward encoder outputs (`codebook_loss`) and another to encourage the encoder to stay near a codebook entry (`commitment_loss`).

- Perplexity: It calculates the codebook perplexity to assess how well the model is utilizing the full range of codebook entries.

- STE: It enables backpropagation through the nondifferentiable quantization step.

Let's walk through the quantization step in more detail:

```
def quantize(self, inputs: jax.Array):
    """Snaps inputs to nearest codebook entries."""
    encoding_indices = self.get_closest_codebook_indices(inputs)
    flat_quantized = jnp.take(self.codebook, encoding_indices, axis=1).swapaxes(
        1, 0
    )
    quantized = jnp.reshape(flat_quantized, inputs.shape)
    return quantized, encoding_indices


def calculate_distances(self, inputs: jax.Array) -> jax.Array:
    """Computes Euclidean distances between inputs and codebook vectors."""
    flat_inputs = jnp.reshape(inputs, (-1, self.embedding_dim))
    distances = (
        jnp.sum(jnp.square(flat_inputs), 1, keepdims=True)
        - 2 * jnp.matmul(flat_inputs, self.codebook)
        + jnp.sum(jnp.square(self.codebook), 0, keepdims=True)
    )
    return distances
```

In `quantize`, the goal is to replace each input vector with its nearest neighbor in the codebook. This is done in a few steps:

- Flatten the input so that we can process all spatial positions as a list of vectors.
- Compute distances between each input vector and all codebook vectors using Euclidean distance.
- Find the nearest codebook vector for each input location (`argmin` over distances).
- Gather the corresponding codebook entries using the indices.
- Reshape the quantized result back to the original input shape.

The distances are computed using `calculate_distances`, which implements the squared Euclidean distance between a flattened input vector x and codebook vector y. This is based on the identity:

$$\| x - y \|^2 = \| x \|^2 - 2\langle x,y \rangle + \| y \|^2$$

This formulation efficiently computes the distances using matrix operations.

In summary, during the forward pass, the `VectorQuantizer` finds the best-matching codebook vector for each encoded input, replaces it, and enables gradients to flow using the STE. The result is a discretized latent representation that is both more structured and interpretable.

Calculating VQ-VAE–specific losses

With the quantized embeddings in hand, most of the hard work is done. But we still need to evaluate how well the original inputs are captured. In other words, we need to calculate the *VQ-VAE–specific* losses:

```
def compute_losses(self, inputs: jax.Array, quantized: jax.Array):
    """Computes codebook and commitment losses."""
    codebook_loss = jnp.mean(jnp.square(quantized - lax.stop_gradient(inputs)))
    commitment_loss = self.commitment_cost * jnp.mean(
      jnp.square(lax.stop_gradient(quantized) - inputs)
    )
    return codebook_loss, commitment_loss
```

There are two components here:

Codebook loss

This term measures how far the quantized vectors are from the original encoder outputs. We want this difference to be small. Ideally, the quantized version should be nearly identical to what the encoder originally produced. The key detail is that `inputs` are wrapped in `lax.stop_gradient`. This prevents gradients from flowing back into the encoder so that only the codebook is updated to better match the encoder output.

Commitment loss

This encourages the encoder to produce outputs that are close to some entry in the codebook. It helps avoid drifting too far from quantized values.

Here, `quantized` is wrapped in `lax.stop_gradient`. The gradient flows only to the encoder and not the codebook, encouraging it to "commit" to one of the existing codebook vectors. The `self.commitment_cost` parameter scales this loss to control how strongly the encoder is pulled toward existing codebook entries.

These two losses play complementary roles: one pulls the codebook toward the encoder outputs and the other pulls the encoder toward the codebook. Together, they ensure that both parts of the model co-adapt and stabilize over time, leading to a high-quality quantized latent space.

Using perplexity to measure codebook use

After quantization, we want to understand how well the model is using its codebook. Is it relying on just a few entries, or is it spreading its attention across many? That's what the *perplexity* metric captures:

```python
def calculate_perplexity(self, encoding_indices: jax.Array) -> jax.Array:
    """Computes codebook usage perplexity."""
    encodings = jax.nn.one_hot(
        encoding_indices,
        self.num_embeddings,
    )
    avg_probs = jnp.mean(encodings, 0)
    perplexity = jnp.exp(-jnp.sum(avg_probs * jnp.log(avg_probs + 1e-10)))
    return perplexity
```

Let's break this down:

- The `encoding_indices` gives the index of the selected codebook entry for each input.

- We one-hot encode these indices to count how often each codebook vector is used.

- Taking the average of this one-hot matrix gives a frequency distribution over the codebook entries.
- We then compute the entropy of this distribution, and we exponentiate it to get the perplexity.

The result is a number between 1 and `num_embeddings`, indicating how many codebook entries are effectively in use.

Different Meanings of Perplexity

The term *perplexity* is used across different areas of machine learning, where it reflects the entropy of a probability distribution.

Entropy is a measure of uncertainty or unpredictability. A distribution with *high entropy* spreads its weight relatively evenly across many options, meaning there's more uncertainty about which option will be chosen. In contrast, *low entropy* means the distribution is concentrated on a few options, indicating higher certainty.

In a VQ-VAE, high entropy in codebook usage (and thus high perplexity) is desirable. It means the model is using many different codebook vectors rather than relying on just a few.

In language modeling, perplexity is also used, but with a slightly different interpretation: lower perplexity is better. It is computed over the predicted next token (or a masked token), and lower values indicate that the model assigns higher probability to the correct word, meaning it's more confident and accurate in its predictions.

Using the straight-through estimator

The final operation in the `__call__` method of the `VectorQuantizer` is the computation of the STE:

```python
@staticmethod
def get_straight_through_estimator(
    quantized: jax.Array, inputs: jax.Array
) -> jax.Array:
    """Applies straight-through estimator to pass gradients through
    quantization.
    """
    ste = inputs + lax.stop_gradient(quantized - inputs)
    return ste
```

In a typical neural network, gradients are computed via backpropagation through a series of differentiable operations. However, *quantization isn't differentiable*. It involves snapping continuous values to the nearest discrete codebook entry, and you can't take a gradient through that. This poses a challenge for gradient-based optimization.

To solve this, we use the STE trick. It works like this:

Forward pass
The output of this expression is quantized, so the decoder receives the discrete codebook entries.

Backward pass
Because `quantized - inputs` is wrapped with `stop_gradient`, it *has no effect on the gradients*. During backpropagation, only `inputs` contributes to the gradient. In other words, the gradient of the STE output with respect to the encoder input is treated as if the quantization step never happened.

This means that while the model behaves as if quantized during inference and reconstruction, the encoder receives useful gradients during training—treating quantization as if it were an identity function. Using the STE is essential for training models with discrete bottlenecks, like VQ-VAEs. It allows us to maintain the representational advantages of a discrete latent space, while still optimizing with gradient descent, the foundation of modern deep learning.

JAX's stop_gradient

JAX's `stop_gradient` is a function that blocks the flow of gradients during backpropagation. It acts like an identity operation during the forward pass, but during the backward pass, it tells the model: "Don't compute gradients through this expression."

You'll see `stop_gradient` used in two main situations:

- *To handle nondifferentiable operations*, such as vector quantization, where we want to bypass the operation during gradient computation while still using it in the forward pass.

- *To freeze part of a model* during fine-tuning; for example, running a pretrained backbone without updating its weights. Wrapping the backbone's output in `stop_gradient` ensures that gradients only flow into newly added layers, like classification heads.

In earlier chapters, we froze parameters using optimizer-level tricks like `optax.multi_transform` and `set_to_zero`. These prevent *parameter updates*, but they still allow *gradients to be computed* and flow through the model.

In contrast, `stop_gradient` blocks gradient flow entirely. It's like cutting the gradient circuit at a specific point in the computation. Both approaches are valid and often interchangeable. Your choice just depends on the level of control you need. If you want to avoid updates but still propagate gradients (e.g., for analysis, visualization, or mechanisms that depend on gradient information), use the optimizer route. If you want to shut down all gradients past a point, use `stop_gradient`. It's the more "nuclear" option.

This covers the core of the VQ-VAE and the most complex aspect of this chapter. All that's left now is the decoder—the final piece that turns the quantized latent codes back into a reconstructed image.

Decoder: Decoding the Discretized Embeddings Back to Images

The `Decoder` is the final stage of our VQ-VAE model. Conceptually, it mirrors the `Encoder`, but instead of compressing the image, it transforms the quantized latent representation back into the original image space. Its job is to turn the discrete, low-resolution feature map back into a full-resolution grayscale image.

Here's the code:

```python
class Decoder(nn.Module):
    """Decoder module for reconstructing input from quantized representations."""

    latent_dim: int

    def setup(self) -> None:
        """Initializes residual blocks and upsampling layers."""
        self.res_block1 = ResnetBlock(self.latent_dim)
        self.res_block2 = ResnetBlock(self.latent_dim)
        self.upsample1 = Upsample(latent_dim=self.latent_dim // 2, upfactor=2)
        self.upsample2 = Upsample(latent_dim=1, upfactor=2)

    def __call__(self, x: jax.Array) -> jax.Array:
        """Applies the decoder to input and returns the reconstructed output."""
        x = self.res_block1(x)
        x = self.res_block2(x)
        x = self.upsample1(x)
        x = nn.relu(x)
        x = self.upsample2(x)
        return x


class Upsample(nn.Module):
    """Upsampling block using bilinear interpolation followed by convolution."""

    latent_dim: int
    upfactor: int
```

```
def setup(self) -> None:
    """Initializes the convolutional layer for post-interpolation refinement."""
    self.conv = nn.Conv(
        self.latent_dim, kernel_size=(3, 3), strides=(1, 1), padding=1
    )

def __call__(self, x: jax.Array) -> jax.Array:
    """Upsamples input using bilinear interpolation and applies convolution."""
    batch, height, width, channels = x.shape
    hidden_states = jax.image.resize(
        x,
        shape=(
            batch,
            height * self.upfactor,
            width * self.upfactor,
            channels,
        ),
        method="bilinear",
    )
    x = self.conv(hidden_states)
    return x
```

And here is the structure:

- Two `ResnetBlocks` refine the latent representation.
- Two `Upsample` layers then increase the spatial resolution step-by-step.
- The final output has shape [`batch_size, height, width, 1`], a single-channel image.

The `Upsample` module works like this:

- It uses bilinear interpolation to resize the feature maps to a larger spatial size (e.g., doubling width and height).
- Then, it applies a 3 × 3 convolution to learn a transformation of the upsampled features.

As a reminder, *bilinear interpolation* resizes images by estimating new pixel values based on the four nearest neighbors. It performs linear interpolation twice: first along one axis (e.g., left to right) and then along the other (top to bottom). For example, when upsampling a 2 × 2 image to 3 × 3, the new center pixel is computed as a weighted average of the four corner values, creating smooth transitions. This avoids the blocky appearance of nearest-neighbor resizing, which simply assigns each new pixel the value of the single closest original pixel, leading to sharp, jagged edges.

ClassificationHead: A Simple but Crucial Module

There is one final component in our model that we need to discuss: the `Classifica tionHead`. This module performs a seemingly simple task—predicting protein IDs—but it turns out to be one of the most important parts of the architecture. The original Kobayashi paper discovered that this model block, which they called *FcBlock*, was actually crucial in guiding the model to learn general protein localization patterns.

In essence, the `ClassificationHead` takes the quantized embeddings and tries to classify which protein the input microscopy image contained. This is implemented as a small, fully connected network (hence, "Fc"), with one or two dense layers, ReLU activations, and dropout:

```python
class ClassificationHead(nn.Module):
    """Fully connected MLP head with optional dropout."""

    num_classes: int
    dropout_rate: float
    layers: int

    @nn.compact
    def __call__(self, x: jax.Array, is_training: bool) -> jax.Array:
        for i in range(self.layers - 1):
            x = nn.Dense(features=1000)(x)
            x = nn.relu(x)
            x = nn.Dropout(rate=self.dropout_rate)(x, deterministic=not is_training)
```

```
x = nn.Dense(features=self.num_classes)(x)
return x
```

The parameters include:

- `num_classes`: The number of protein IDs in the dataset
- `dropout_rate`: Helps regularize the model and prevent overfitting
- `layers`: Whether to use a one- or two-layer classifier

Despite its simplicity, this block had the largest impact on performance in the original cytoself paper. It acts as a form of auxiliary task: the model is explicitly asked to predict the protein identity from its embedding; a task that is possible only if the embeddings encode useful spatial features.

This changes the role of the embeddings: instead of only trying to minimize reconstruction loss, the model is now encouraged to organize the latent space in a way that helps with protein discrimination. This helps prevent codebook collapse and leads to much better localization-specific features.

> Interestingly, this auxiliary protein classification task is hard for the model—and that makes intuitive sense. Many proteins are part of the same complex and share the same localization, making their image frames visually indistinguishable. But that's the point: by attempting this difficult task, the model is pushed to extract subtle, generalizable cues related to localization, even if it doesn't achieve perfect classification.

Later in the chapter, you'll see the difference when this component is removed. It also demonstrates a broader lesson: *adding the right auxiliary task can transform a model's ability to learn*. You can control how strongly this auxiliary task influences training via the `classification_weight` parameter.

We now have a model. Let's train it.

Setting Up Model Training

We will now train the `LocalizationModel` model we've built. To begin, we'll use a smaller number of image frames to allow for faster iteration and debugging.

The main training loop is defined in the `train` function. It sets up the training state, splits the data into batches, and iterates over the dataset for a number of epochs:

```python
@restorable
def train(
    state: TrainState,
    rng: jax.Array,
    dataset_splits: dict[str, Dataset],
    num_epochs: int,
    batch_size: int,
    classification_weight: float,
    eval_every: int = 10,
) -> tuple[TrainState, dict[str, dict[str, list[dict[str, float]]]]]:
    """Train the VQ-VAE model with optional classification."""
    # Setup metrics logging
    metrics = MetricsLogger()

    epochs = tqdm(range(num_epochs))
    for epoch in epochs:
        epochs.set_description(f"Epoch {epoch + 1}")
        rng, rng_batch = jax.random.split(rng, 2)

        # Perform a training step on a batch of train data and log metrics.
        for batch in dataset_splits["train"].get_batches(
            rng_batch, batch_size=batch_size
        ):
            rng, rng_dropout = jax.random.split(rng, 2)
            state, batch_metrics = train_step(
                state, batch, rng_dropout, classification_weight
            )
            metrics.log_step(split="train", **batch_metrics)

        # Evaluate on the validation split
        if epoch % eval_every == 0:
            rng, rng_batch = jax.random.split(rng, 2)
            for batch in dataset_splits["valid"].get_batches(
                rng_batch, batch_size=batch_size
            ):
                batch_metrics = eval_step(state, batch, classification_weight)
                metrics.log_step(split="valid", **batch_metrics)

        metrics.flush(epoch=epoch)
        epochs.set_postfix_str(metrics.latest(["total_loss"]))

    return state, metrics.export()
```

The first thing to notice is that training proceeds in epochs, which form the main loop. As a reminder, an *epoch* is a full pass through the entire training set; every training example is seen once. Before we enter the loop for the first time, we initialize the training state so that we have a starting point. Then, the training begins with the first epoch.

Before diving into the training logic, let's briefly look at how the dataset is fed into the model. A key part of this is the `Dataset.get_batches` method, which handles how image examples are served during training:

```python
def get_batches(
    self,
    rng: jax.Array,
    batch_size: int,
):
    """Yields batches of image and label data for training or evaluation."""
    frame_ids = self.labels.get_frame_ids()

    n_frames = len(frame_ids)
    batches_per_epoch = n_frames // batch_size

    # Shuffle data.
    _, rng_perm = jax.random.split(rng, num=2)
    shuffled_idx = jax.random.permutation(rng_perm, n_frames)

    # The model has a softmax layer and expects consecutive integers.
    all_labels = self.labels.lookup[["frame_id", "code"]].set_index("frame_id")

    for idx_set in shuffled_idx[: batches_per_epoch * batch_size].reshape(
      (batches_per_epoch, batch_size)
    ):
      frame_id_set = frame_ids[idx_set]
      yield {
        "frame_ids": frame_id_set,
        "images": self.images.frames[frame_id_set],
        "labels": all_labels.loc[frame_id_set]["code"].to_numpy(dtype=int),
      }
```

You can see that we first select either the *training* or the *test* set of image frames, split them into batches of a preset size, and shuffle their indices. The protein labels for each frame are then encoded as integers so that they can be used with the `optax.soft max_cross_entropy_with_integer_labels` loss function. Finally, each batch yields the image data, the integer-encoded protein labels, and the corresponding frame IDs (which can be useful for analysis or visualization).

Once the data is batched, it's passed into two key functions: `train_step` for updating the model and `eval_step` for monitoring performance. Let's take a closer look at `train_step`:

```python
@jax.jit
def train_step(
  state: TrainState,
  batch: dict[str, jax.Array],
  rng_dropout: jax.Array,
  classification_weight: float,
) -> tuple[TrainState, dict[str, float]]:
  """Train for a single step."""
```

```python
def calculate_loss(params: dict) -> tuple[jax.Array, dict[str, float]]:
    """Forward pass and loss computation."""
    (
        x_recon,
        perplexity,
        codebook_loss,
        commitment_loss,
        logits,
    ) = state.apply_fn(
        {"params": params},
        batch["images"],
        is_training=True,
        rngs={"dropout": rng_dropout},
    )

    loss_components = {
        "recon_loss": optax.squared_error(
            predictions=x_recon, targets=batch["images"]
        ).mean(),
        "codebook_loss": codebook_loss,
        "commitment_loss": commitment_loss,
        "classification_loss": classification_weight
        * optax.softmax_cross_entropy_with_integer_labels(
            logits=logits, labels=batch["labels"]
        ).mean(),
    }

    metrics = {
        "total_loss": sum_loss_components(**loss_components),
        "perplexity": perplexity,
        "accuracy": accuracy_score(batch["labels"], y_pred=logits.argmax(-1)),
        **loss_components,
    }
    return metrics["total_loss"], metrics

# Compute gradients and apply update.
grad_fn = jax.value_and_grad(calculate_loss, has_aux=True)
(_, metrics), grads = grad_fn(state.params)
state = state.apply_gradients(grads=grads)
return state, metrics
```

Within the train_step function, the inner calculate_loss defines how the model's loss is computed. This function is the core of the training step. It determines how well the model is performing and guides the weight updates to minimize the loss.

First, we call state.apply_fn, which runs the __call__ method of the Localization Model model. This returns the reconstruction (x_recon), the codebook-related losses (codebook_loss and commitment_loss), the classification logits, and the perplexity of the quantization.

We then compute two additional losses:

- `recon_loss`: How different the reconstructed image is from the original, using squared error
- `classification_loss`: How well the model predicts the protein ID from the image embedding, using cross-entropy

These losses are assembled into `loss_components`, and then they are combined into a `total_loss` that drives training. Notably, the classification loss is multiplied by a `classification_weight`, allowing us to control how much it contributes to learning. Setting it to zero *ablates* (removes) the classification task, something we will test later in the chapter.

Finally, we calculate evaluation metrics to track how training is progressing. These include:

- `perplexity`: How effectively the model is using the codebook
- `accuracy`: How effectively the model is predicting the protein IDs

All of this is used to compute gradients via `jax.value_and_grad` and then update the model with `state.apply_gradients`. This design cleanly separates different objectives (reconstruction, quantization, classification) and lets you experiment with different trade-offs by adjusting loss weights.

After each epoch, we store the metrics collected across batches. These include reconstruction loss, codebook and commitment losses, classification accuracy, and perplexity. The metrics are averaged across batches to give a summary per epoch, allowing us to monitor model progress and convergence.

The `eval_step` is essentially the same as the `train_step`, with one key difference: it does not update the model weights. Instead, it runs the model in inference mode and is used to assess how well the current model performs on a held-out test set. This gives us an unbiased signal of generalization performance.

We've now covered the model, dataset, and training logic. It's finally time to give it a spin and see what it can learn.

Training with a Small Image Set

Let's see the model in action. We'll start by training it on a small subset of the data: 50 proteins, split into 80% training data, 10% validation data, and 10% test data, using a fixed random seed for reproducibility.

We define our model architecture by setting the `embedding_dim`, `num_embeddings`, `commitment_cost`, `dropout_rate`, and `classification_head_layers`. Then, we specify the training parameters: number of epochs, batch_size, learning_rate, and classification_weight.

```
from dlfb.localization.dataset.utils import count_unique_proteins
from dlfb.localization.model import LocalizationModel
from dlfb.localization.train import train

model = LocalizationModel(
  num_classes=count_unique_proteins(dataset_splits),
  embedding_dim=64,
  num_embeddings=512,
  commitment_cost=0.25,
  dropout_rate=0.45,
  classification_head_layers=2,
)
```

Now we can start training:

```
rng, rng_init, rng_train = jax.random.split(rng, 3)

state, metrics = train(
  state=model.create_train_state(
    rng=rng_init,
    dummy_input=dataset_splits["train"].get_dummy_input(),
    tx=optax.adam(0.001),
  ),
  rng=rng_train,
  dataset_splits=dataset_splits,
  num_epochs=10,
  batch_size=256,
  classification_weight=1,
  eval_every=1,
  store_path=assets("localization/models/small"),
)
```

After training, we now have a LocalizationModel model that has learned to compress, quantize, and reconstruct protein localization patterns, while also performing auxiliary classification. But how well has it actually learned? Let's find out. We'll start by visually inspecting its reconstructions.

Inspecting Image Reconstruction

Before diving deeper, let's do a quick sanity check: can the LocalizationModel model we just trained reconstruct the input images at all? If it has learned any meaningful representation of the input data, its reconstructions should at least vaguely resemble the original frames. We will evaluate this on the validation set of the dataset, that is, on frames that were never seen during training.

Indeed, the model captures some of the structural features present in the inputs (see Figure 6-9). The reconstructions are far from perfect—blurry and low resolution—

but that's OK. Remember, our goal isn't to generate photorealistic images, it's to learn discrete representations that encode spatial localization patterns. Reconstruction is just a training objective to help guide that process.

```
from dlfb.localization.inspect.reconstruction import show_reconstruction

show_reconstruction(dataset["valid"], state, n=8, rng=rng_frames);
```

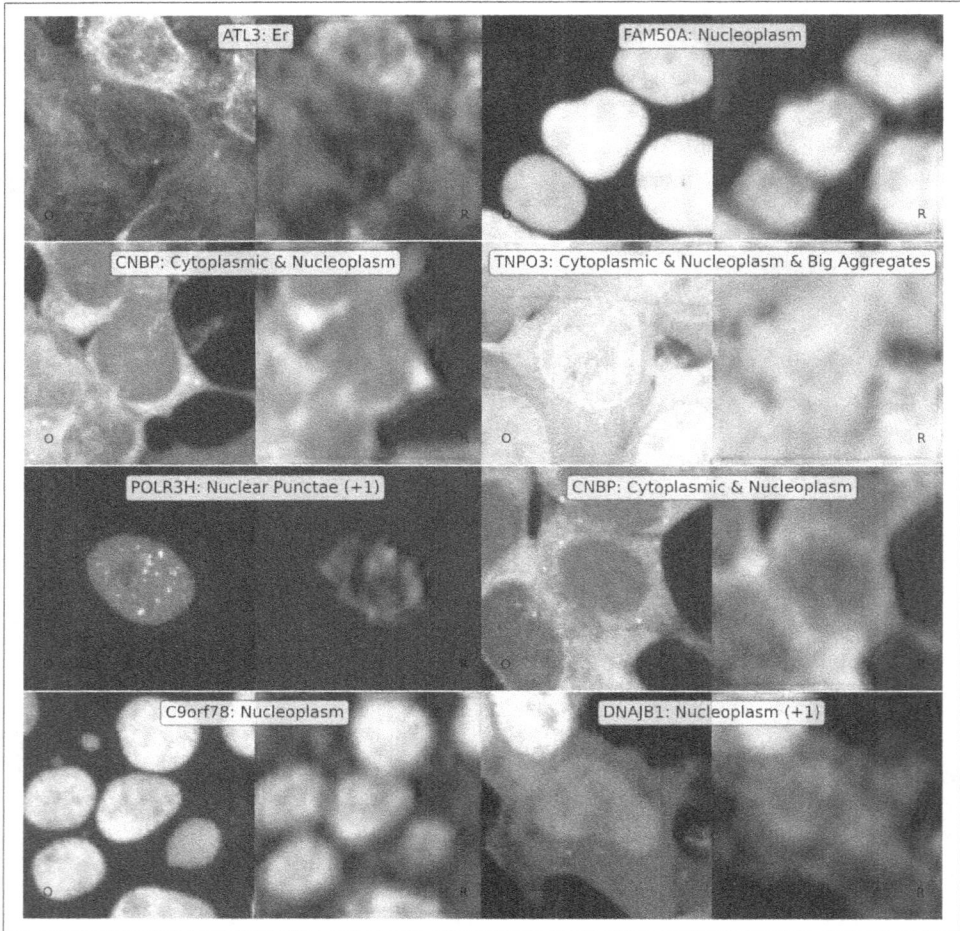

Figure 6-9. Reconstructed images of random proteins from the small-scale model. Each pair shows an input image (left) and its reconstruction (right). While blurry, the reconstructions often capture core structural features—a sign that the model is learning to encode localization-relevant information. Protein is indicated per panel (number indicates additional secondary localizations measured).

Examining Evaluation Metrics Over Epochs

Let's now look more closely at how training progressed by examining the loss curves (see Figure 6-10). The left panel shows the four individual loss components used during training, while the right panel displays total training and test loss across epochs.

```
from dlfb.localization.inspect.metrics import plot_losses

plot_losses(metrics);
```

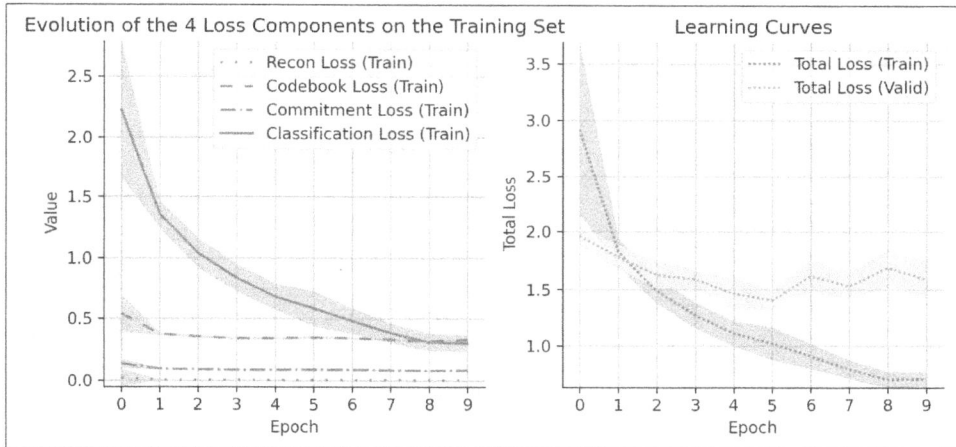

Figure 6-10. Training dynamics represented by individual loss components on the training set (left) and total training versus validation loss over epochs (right). All loss components decrease over time. While training loss continues to improve steadily, validation loss plateaus early, suggesting the model generalizes reasonably well but gains from further training may be limited.

As expected, all loss components steadily decrease over time—especially the classification loss, which dominates the total due to its larger magnitude. This makes sense: distinguishing between 50 proteins based solely on their localization patterns is a challenging task. Remember that we can always adjust the classification loss's relative weight using the `classification_weight` parameter later.

Meanwhile, validation loss closely tracks training loss throughout, with only a slight and stable gap between them. There's no clear sign of overfitting, which is encouraging given the small dataset.

Next, let's inspect how the codebook is being used—by looking at the evolution of its *perplexity* over time. In VQ-VAEs, perplexity measures how many codebook entries are effectively being used during vector quantization. If the model relies on just a handful of embeddings (e.g., 10 out of 512), the perplexity will be low. If it spreads usage more evenly across many entries, the perplexity rises, approaching the total number of codebook vectors available. In Figure 6-11 we see the evolution of perplexity:

```
from dlfb.localization.inspect.metrics import plot_perplexity

plot_perplexity(metrics);
```

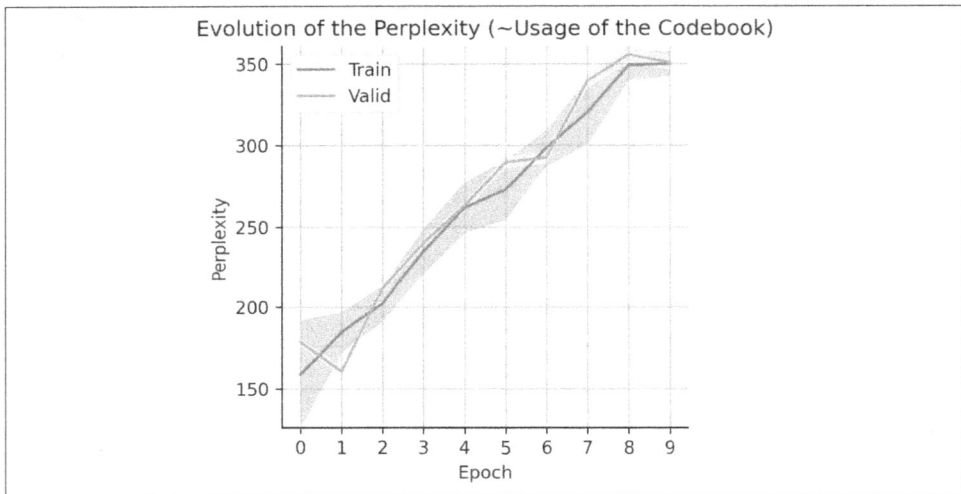

Figure 6-11. Perplexity increases over epochs on both the training and validation sets. A rising perplexity indicates that the model is using more of the available codebook entries, rather than collapsing onto a small subset. The close alignment between training and validation perplexity suggests that this richer, more diverse representation generalizes well beyond the training data.

As you can see in Figure 6-11, perplexity steadily increases across epochs for both the training and validation sets. This is a strong signal that the model is learning to encode diverse spatial patterns using a broad vocabulary of codebook entries. Instead of relying on a narrow set of common features, it's finding more nuanced ways to represent the variation in protein localization—effectively condensing complex microscopy images into compact, expressive codes.

Training a Model Without a Classification Task

Next, we'll train the exact same model on the same data, but this time with the classification weight set to 0. This simulates removing the auxiliary protein ID classification task, in other words, disabling the `ClassificationHead`. Note that we are using the same random seed to provide the closest possible comparison. This lets us see how much that component helps guide the model's learning.

> As a general design principle, it's helpful to structure your code so that model components can be ablated (i.e., effectively removed) by setting their weight to zero via a config or flag—rather than rewriting or commenting out parts of the architecture. This makes it much easier to test hypotheses, compare model variants, and run controlled experiments. It's a simple practice that promotes modular, reproducible research.

Let's give it a go:

```
state_alt, metrics_alt = train(
  state=model.create_train_state(
    rng=rng_init,
    dummy_input=dataset_splits["train"].get_dummy_input(),
    tx=optax.adam(0.001),
  ),
  rng=rng_train,
  dataset_splits=dataset_splits,
  num_epochs=10,
  batch_size=256,
  classification_weight=0,  # i.e. the protein id are ignored
  eval_every=1,
  store_path=assets("localization/models/small_alt"),
)
```

This time around, the model performs much worse, and its perplexity collapses (down to ~30, previously ~350), as shown in Figure 6-12.

```
plot_perplexity(metrics_alt);
```

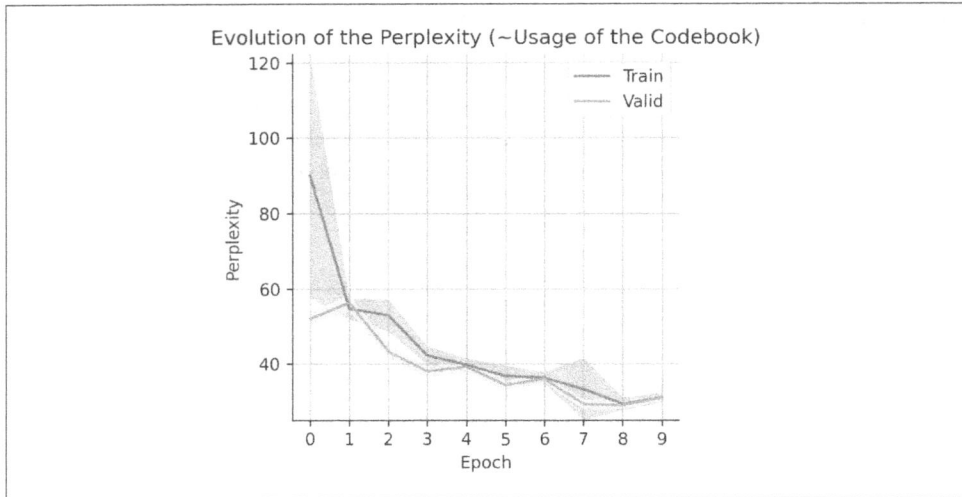

Figure 6-12. Perplexity over epochs collapses without the `ClassificationHead` in the model architecture, suggesting that the auxiliary protein identification task plays a critical role in encouraging a diverse and informative representation.

Perhaps counterintuitively, this model's reconstructed microscopy images actually look a bit *better* in Figure 6-13.

```
show_reconstruction(dataset["valid"], state_alt, n=8, rng=rng_frames);
```

At first glance, this might seem like an improvement, but it actually reveals a critical trade-off in the model's objectives.

Without the auxiliary classification task (i.e., without the `ClassificationHead`), the model can focus more of its efforts on minimizing the reconstruction loss. This encourages it to memorize the input data as precisely as possible, often by collapsing to a small number of frequently used codebook entries. That's why the reconstructions look sharper: the model has overfit to pixel-level detail rather than learning generalizable representations.

But VQ-VAEs aren't just about pretty reconstructions. They're about learning discrete, structured representations of the input. With the classification task enabled, the model is forced to organize its internal representations in a way that's useful for predicting protein identity. This encourages it to capture high-level biological features, such as localization patterns, at the cost of slightly blurrier reconstructions.

This same observation was highlighted in the original cytoself paper: adding classification-like objectives improves the quality of the latent space by pushing the model to encode meaningful, discriminative features. In fact, cytoself showed that models trained only to reconstruct images were far less effective at clustering localizations or identifying complexes, even when their reconstructions looked visually fine.

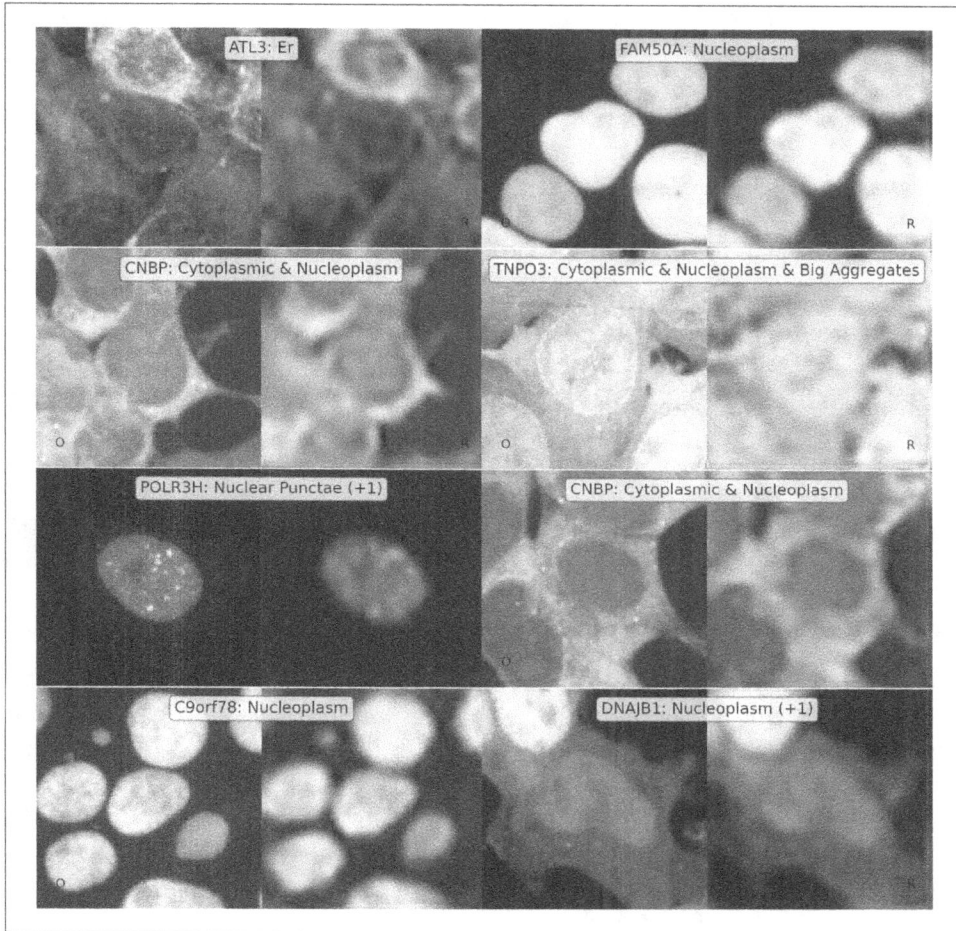

Figure 6-13. Reconstructed images appear slightly more faithful to the input when the ClassificationHead *is removed—likely because the model can focus its efforts on the reconstruction task, resulting in less blurring.*

How Auxiliary Tasks Can Help

An *auxiliary task* is a secondary prediction task that helps the model learn better internal representations, and they can be very helpful to machine learning models. Adding a related task guides the model to focus on meaningful structure in the data. You implement this by adding additional heads to your model and literally summing their loss values during training to get the total loss. You can control how much each task contributes by adjusting the loss weights.

For example, for the chapter on skin cancer, you could implement such auxiliary tasks as predicting *tissue type*, *cell density*, or even *tumor grade*. One popular approach in the cancer imaging space is to add a segmentation task (pixel-wise labeling), such as detecting cell nuclei or highlighting tumor regions, to help the model learn more localized, interpretable features.

These auxiliary heads can be safely discarded after training, or they can be kept if they're useful. Even when thrown away, they shape the learned representations in ways that often improve generalization, robustness, and interpretability.

Understanding the Model

What has the model actually learned about spatial organization within cells? To answer this question, we need to take a closer look at the model's *latent space*: the internal representation it builds from the input images. In particular, since our model uses a codebook, we can analyze *which* entries get used and *how often* for each protein. This gives us a kind of *feature spectrum* or a summary of how each protein maps onto the learned visual vocabulary.

Understanding Localization Clustering

A natural question is whether the model can distinguish between different subcellular compartments. One way to test this is to apply the uniform manifold approximation and projection (UMAP) dimensionality reduction technique to the learned embeddings. UMAP projects high-dimensional data into two dimensions while preserving local structure, making it easier to visualize complex relationships.

> There are other dimensionality reduction techniques, such as *t-SNE* and *PCA*, but *UMAP* is especially well suited for this task. Unlike PCA, it can capture complex nonlinear relationships. UMAP is similar to t-SNE in many ways, but with a key advantage: it tends to preserve both local *and* global structure more effectively. This makes it especially useful for visualizing patterns across diverse datasets like this one.
>
> UMAP has become a go-to tool in fields like proteomics and genomics—not just for its performance, but also because it works well out of the box, with minimal hyperparameter tuning.

If the model has learned meaningful spatial features, we'd expect image frames from similar cellular compartments—like the mitochondria, nucleus, or ER—to cluster together in the UMAP space. To visualize this, we need to extract the model's internal representation of each image.

The function `get_frame_encoding_index_histogram` does this by calculating a code-book usage histogram for each frame:

```python
def get_frame_embeddings(
    state: TrainState,
    dataset_split: Dataset,
    batch_size: int = 256,
) -> dict[str, np.ndarray]:
    """Returns per-frame histograms of codebook encoding indices."""
    num_embeddings = get_num_embeddings(state)
    frame_ids, frame_histograms = [], []

    rng = jax.random.PRNGKey(42)
    for batch in dataset_split.get_batches(rng, batch_size):
        frame_ids.append(batch["frame_ids"])
        encoding_indices = pluck_encodings(state, batch)

        # Reshape and count codebook usage per frame.
        frame_histograms.append(
            np.apply_along_axis(
                lambda x: np.histogram(x, bins=np.arange(0, num_embeddings + 0.5))[0],
                axis=1,
                arr=jnp.reshape(encoding_indices, (batch_size, -1)),
            )
        )

    return {
        "frame_ids": np.concatenate(frame_ids),
        "frame_histograms": np.concatenate(frame_histograms, axis=0),
    }
```

For each batch of images, we use the model to obtain the encoding indices: the discrete codebook entries selected for each spatial patch in each frame. These are returned by the `pluck_encodings` helper function. Each frame's output is then flattened and passed to `np.histogram`, which counts how often each codebook entry is used.

The result is a histogram vector per frame—one value per codebook entry—describing how frequently that entry was activated. These vectors can be thought of as discrete fingerprints of each image, which we can then reduce to 2D with UMAP for visualization.

> Why proceed in batches? Batching during the embedding extraction—just like during training—helps manage memory and computation. It ensures that we don't load the entire dataset into memory at once. Note that the batch size used here doesn't have to match the training batch size.

We'll generate projections for two versions of our model: one with the `Classifica tionHead` enabled and one without. By plotting both side-by-side, we can observe how the presence of the `ClassificationHead` affects the structure of the learned embedding space. We continue using the validation dataset to evaluate our model. The resulting visualization is shown in Figure 6-14:

```python
from dlfb.localization.inspect.embeddings.clustering import (
    calculate_projection,
    plot_projection,
)
from dlfb.localization.inspect.embeddings.utils import get_frame_embeddings

frame_embeddings = {}
for name, s in zip(["no_head", "with_head"], [state_alt, state]):
    frame_embeddings[name] = get_frame_embeddings(s, dataset_splits["valid"])

projection = calculate_projection(frame_embeddings)
plot_projection(
    projection,
    dataset_splits["valid"],
    titles=["No ClassificationHead", "With ClassificationHead"],
);
```

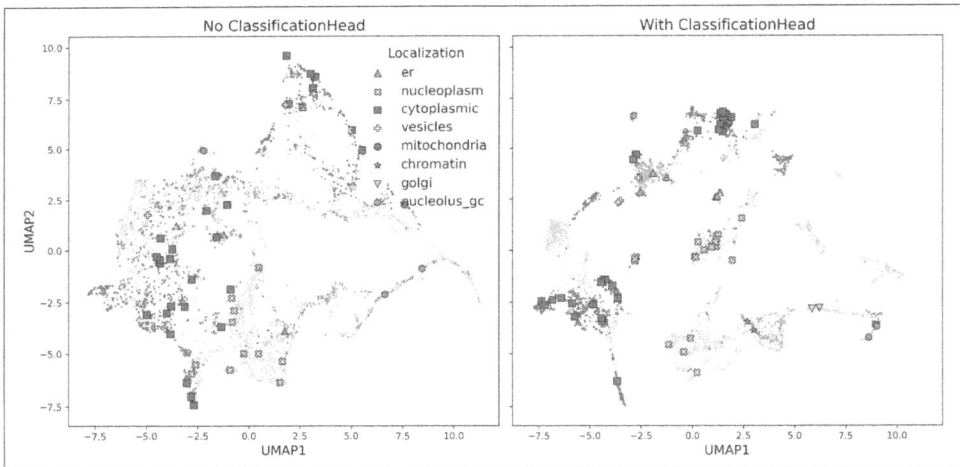

Figure 6-14. UMAP projections of the model with (left) and without (right) the `Classi ficationHead`. The model on the left learns clearer, more meaningful structure—with frames from the same subcellular compartment (e.g., vesicles, chromatin, nucleolus) forming tighter clusters. Without the `ClassificationHead`, the structure is less distinct, and compartments are harder to separate. Around 1% of frames with a single predominant localization are highlighted with larger markers to intuitively annotate clusters.

You can see that, once the embeddings are labeled with ground-truth localization categories, the model has clearly learned to distinguish different subcellular compartments based on raw pixel patterns. In the UMAP plot on the left (with the `ClassificationHead`), compartments like chromatin, mitochondria, and nucleolus form fairly tight, distinct clusters—indicating that the model has learned consistent visual features for these structures. Vesicles sometimes form a separate group but often show overlap with ER and cytoplasmic frames, likely reflecting their more diffuse and variable appearance in the raw images.

Removing the `ClassificationHead` has a clear visual impact. In the righthand UMAP, the model still produces structured embeddings, but the clusters are less distinct. Localizations such as cytoplasm, ER, and vesicles are more intermixed, suggesting that without the auxiliary classification task, the model learns a weaker or more entangled representation of subcellular identity. This comparison highlights how architectural choices—even auxiliary objectives—can meaningfully shape the structure of learned representations.

> Since we want to compare how the presence or absence of the `ClassificationHead` affects the learned embeddings, it's important that both UMAP visualizations are directly comparable. By default, techniques like UMAP involve some randomness in initialization and optimization, which can lead to inconsistent visual layouts across runs. To address this, we use an *aligned UMAP*, a variant of UMAP that tracks shared data points across embeddings to ensure consistent alignment and make visual comparisons meaningful.

While these UMAP plots are useful for qualitative interpretation, it's also possible to quantify clustering quality more formally; for example, by comparing the average distances between embeddings from the *same* localization versus those from *different* localizations. If the model has learned a meaningful embedding space, we'd expect intra-class distances (within a localization category) to be small, and inter-class distances to be larger. This basic intuition underpins several clustering quality metrics and was used by Kobayashi et al. to evaluate their model quantitatively. In this chapter, we'll focus on visual exploration, but you can always extend the code to include such metrics if you would like to.

Inspecting Feature Spectrums

To better understand what our model has learned, we can analyze how it uses its codebook across different proteins. Remember, each image is encoded using entries from a learned set of code vectors (the codebook). Over time, some entries may become specialized, for instance, consistently firing for mitochondrial proteins, or for nuclear-localized ones.

A *feature spectrum* is a simple but powerful way to summarize this behavior. For each protein, we:

- Collect all its image frames.
- Count how often each codebook entry is used across those frames.
- Aggregate these counts into a histogram. This is the protein's feature spectrum.

Each feature spectrum essentially acts like a "fingerprint" of how the model encodes that protein's spatial patterns.

To explore how specialized the codebook has become, we compare these spectra across proteins. If certain codebook entries are consistently used for similar proteins (e.g., those localized to the ER), we expect those spectra to be correlated.

The following function computes a Pearson correlation matrix across all protein spectra and then applies hierarchical clustering to group similar patterns. This reveals which sets of codebook entries (features) tend to be co-used and may correspond to broader localization categories:

```python
def cluster_feature_spectrums(
    protein_histograms: np.ndarray, n_clusters: int
) -> tuple[np.ndarray, np.ndarray, np.ndarray]:
    """Cluster proteins based on similarity in codebook usage patterns."""
    corr_idx_idx = np_pearson_cor(protein_histograms, protein_histograms)
    tree = linkage(
        corr_idx_idx,
        method="average",
        metric="euclidean",
        optimal_ordering=True,
    )
    encoding_clusters = fcluster(tree, n_clusters, criterion="maxclust")
    return corr_idx_idx, tree, encoding_clusters
```

We first compute how correlated the spectra are across all proteins using Pearson correlation. Then, we build a tree (dendrogram) from that correlation matrix and cluster it to find groups of related proteins based on how similarly they use the codebook. These clusters often reflect shared localization patterns or functional relationships.

The resulting heatmap in Figure 6-15 gives us a bird's-eye view of which parts of the codebook are used together and by which proteins:

```python
from dlfb.localization.inspect.embeddings.feature_spectrum import (
    plot_encoding_corr_heatmap,
)
from dlfb.localization.inspect.embeddings.utils import aggregate_proteins

protein_ids, protein_histograms = aggregate_proteins(
    dataset_splits["valid"], **frame_embeddings["with_head"]
)
```

```
corr_idx_idx, tree, encoding_clusters = cluster_feature_spectrums(
    protein_histograms, n_clusters=8
)
plot_encoding_corr_heatmap(corr_idx_idx, tree, encoding_clusters);
```

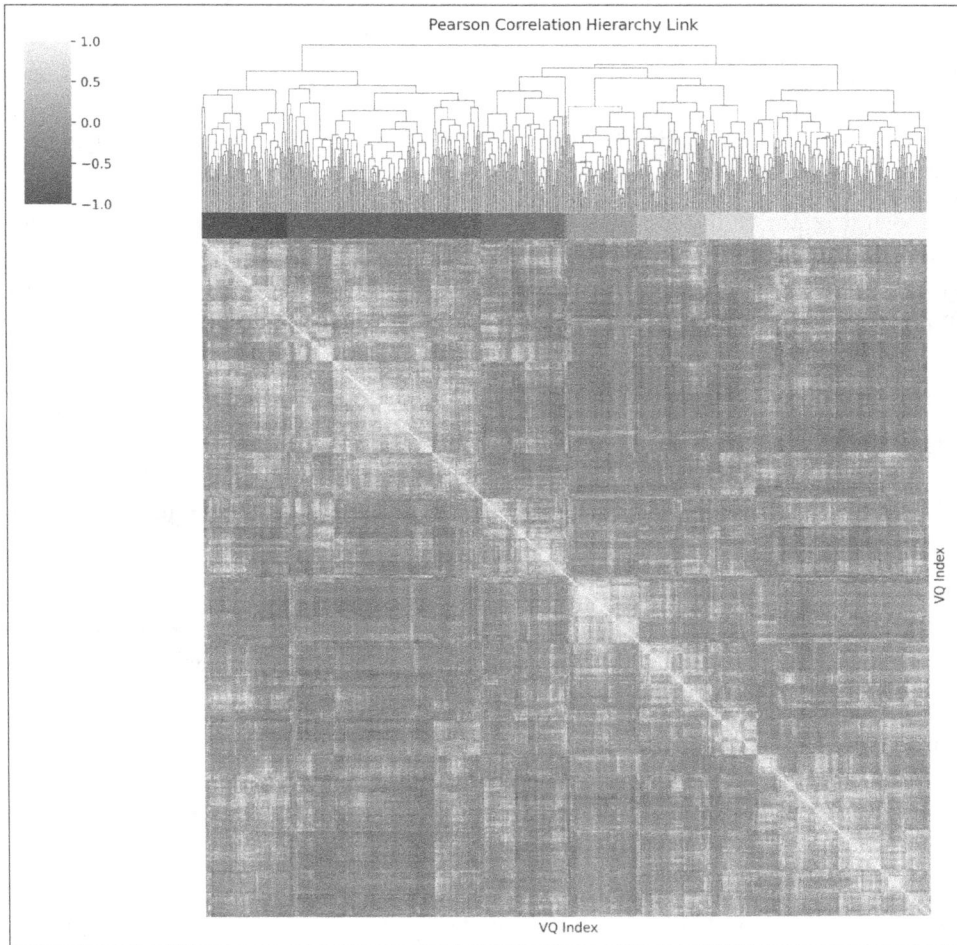

Figure 6-15. Correlation heatmap between codebook entries, based on how often they co-occur across proteins. Each axis shows the index of a vector quantization (vq) codebook entry (not all 512 indices are written out due to space constraints). The shading indicates the Pearson correlation between code usage patterns across proteins: lighter denotes strong positive correlation, and darker strong negative correlation. The dendrogram here clusters codebook entries into groups that represent shared spatial features learned by the model.

We can now take a deep dive into what the model's feature spectra actually represent. Recall that for each protein, we computed a histogram of how often each codebook entry is used—essentially, a fingerprint of localization patterns learned by the model. Then, we computed the Pearson correlation between these fingerprints across proteins to see which codebook entries tend to be used together.

The resulting plot in Figure 6-15 shows a correlation heatmap between all 512 codebook entries, with the color scale ranging from –1 (strong anticorrelation, darker) to +1 (strong correlation, lighter). Codebook entries that are frequently used together—for example, to describe a specific organelle—appear as bright blocks.

The dendrogram at the top reflects a hierarchical clustering of these correlations. Based on this structure, the clustering procedure has grouped the codebook entries into eight broader encoding clusters, each of which represents a recurring spatial motif captured by the model. The color bar just below the dendrogram indicates these groupings.

Here's what this heatmap tells us:

Diagonal blocks
    The visible lighter blocks along the diagonal show groups of codebook entries that are highly correlated, suggesting that they work together to represent similar spatial features across proteins.

Off-diagonal near-zero correlations
    The lack of strong off-diagonal structure implies that these clusters are relatively distinct. The model has learned specialized regions of the codebook for different spatial contexts.

Unsupervised insight
    Crucially, this entire structure emerged without any localization labels. It reflects how the model has self-organized its representation space purely from image similarity.

This is a powerful window into the internal structure of the learned representations, revealing that even individual quantized codes fall into larger functional groups that track meaningful biological variation.

To take this analysis a step further, we can examine how these codebook entries are used across known subcellular compartments. If the model has learned meaningful localization features, we should see that certain codebook vectors are enriched for specific compartments or structures like chromatin, vesicles, or ER.

This is shown in Figure 6-16, where we average the feature spectra across all proteins within each localization class. Even though the model was never told anything about localization labels during training, we can now see clear signatures that align with known biology:

```
from dlfb.localization.inspect.embeddings.feature_spectrum import (
  plot_stacked_histrograms,
)
from dlfb.localization.inspect.embeddings.utils import aggregate_localizations

localizations, localization_histograms = aggregate_localizations(
  dataset_splits["valid"], protein_ids, protein_histograms
)
plot_stacked_histrograms(
  localizations, localization_histograms, tree, encoding_clusters
);
```

Although this is our initial, relatively simple model trained on a small subset of proteins, we already see some evidence that it is learning meaningful biological structure. Each row in the figure represents the average codebook usage (feature spectrum) for a given localization class. For instance, nucleolus (first row) shows peaks mostly within group VIII, suggesting that only a few specific codebook entries are consistently used to represent that compartment. This kind of narrow, high-signal spectrum is typical for highly structured and easily distinguishable compartments.

In contrast, mitochondria (third row), er (fourth row), and especially cytoplasm (fifth row) display broader, more distributed activation across various groups—perhaps reflecting their more heterogeneous or variable visual features, or simply lacking in learned separation due to our limited dataset. Still, each compartment has its own distinctive fingerprint: for instance, mitochondria shows much higher peaks in group VI. Interestingly, while nucleoplasm is spatially adjacent to nucleolus in the cell, it exhibits a markedly different activation pattern—whereas nucleoplasm appears much more similar to chromatin, possibly due to less structural separation or shared staining characteristics.

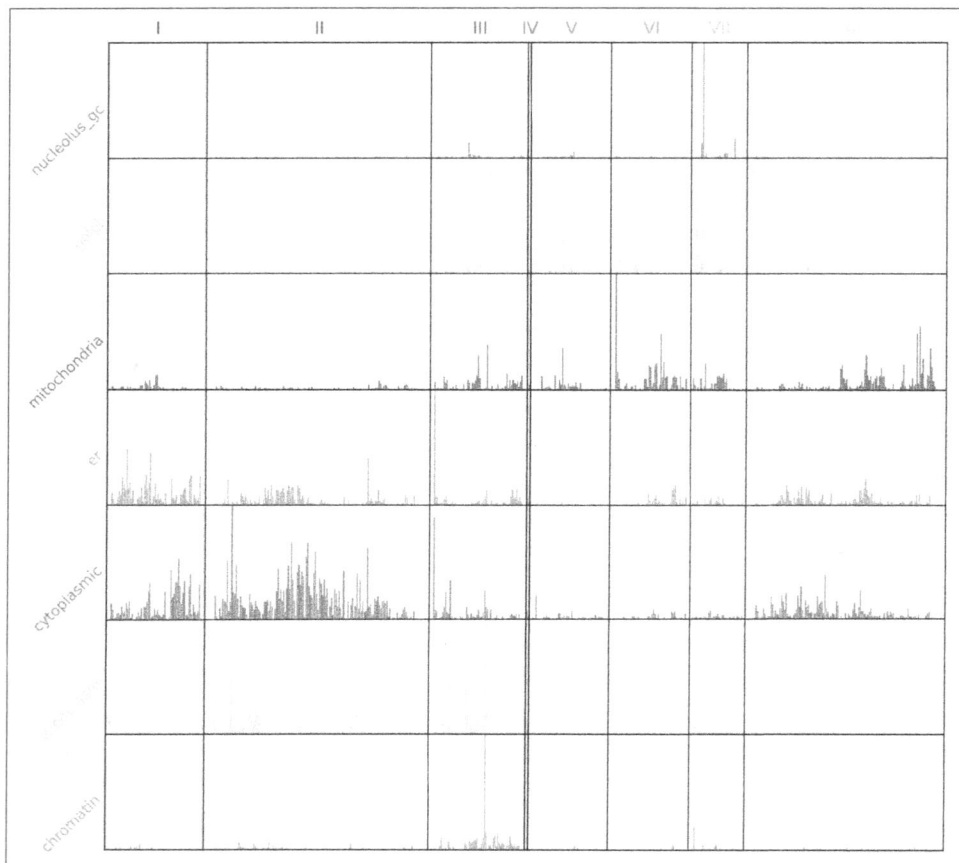

Figure 6-16. Stacked histograms of codebook usage for different protein localization classes. Each row shows the average feature spectrum for a specific subcellular compartment (e.g., ER, nucleoplasm). Vertical groupings (I–VIII) reflect clusters of related codebook entries, as defined in the correlation heatmap shown previously. Despite never being trained with localization labels, the model has learned to associate certain codebook vectors with specific biological structures, revealing meaningful and interpretable signatures.

Looking at Individual Proteins

Much of our analysis so far has focused on aggregate behavior—examining feature spectra, UMAP projections, and clustering across many proteins. While this provides a useful global picture, it can obscure how the model treats individual examples.

A valuable complementary approach is to pick a *single protein* and investigate it more closely:

- Review its raw input images across training frames.
- Visualize which codebook vectors are most active (its feature spectrum).
- Find its most similar neighbors using cosine similarity between embeddings.
- Inspect the classification output. If it's misclassified, are the top predicted alternatives biologically reasonable?

This kind of deep-dive analysis grounds abstract metrics in biological intuition and often reveals surprising model behavior. As an exercise, try applying this to proteins with contrasting properties—disease-associated versus not, structured versus disordered, or well-annotated versus unknown.

We've only scratched the surface using a small dataset and a simple architecture. In the next section, we'll scale up both the model and the data to see how far this approach can go.

Improving the Model

We've seen that even a relatively simple model with a small dataset of 50 proteins can learn meaningful representations of protein localization. The next step is to scale up the dataset to push performance further.

Scaling Up the Data

We now increase the number of proteins and their corresponding imaging frames significantly (from 50 to 500). This presents new challenges, especially in terms of training time and resource management. To make this process more manageable, we'll save the final training state so that it can be reloaded for further inspection and evaluation:

```
dataset_splits = DatasetBuilder(
  data_path=assets("localization/datasets")
).build(
  rng=rng_dataset,
  splits={"train": 0.80, "valid": 0.10, "test": 0.10},
  n_proteins=500,  # a larger number of proteins
```

```
)
model = LocalizationModel(
  num_classes=count_unique_proteins(dataset_splits),
  embedding_dim=64,
  num_embeddings=512,
  commitment_cost=0.25,
  dropout_rate=0.45,
  classification_head_layers=2,
)

state, metrics = train(
  state=model.create_train_state(
    rng=rng_init,
    dummy_input=dataset_splits["train"].get_dummy_input(),
    tx=optax.adam(0.001),
  ),
  rng=rng_train,
  dataset_splits=dataset_splits,
  num_epochs=10,
  batch_size=256,
  classification_weight=1,
  eval_every=1,
  store_path=assets("localization/models/large"),
)
```

This model takes considerably longer to train, but the payoff is a more expressive and well-structured representation space. Figure 6-17 shows the aggregated codebook usage (feature spectra) across different localization classes, showing clearer localization signatures across a broader range of proteins (this plot also includes an expanded number of location classes as rows):

```
from dlfb.localization.inspect.embeddings.feature_spectrum import (
  plot_stacked_histrograms,
)
from dlfb.localization.inspect.embeddings.utils import (
  aggregate_localizations,
  aggregate_proteins,
  cluster_feature_spectrums,
  get_frame_embeddings,
)

frame_embeddings = get_frame_embeddings(state, dataset_splits["valid"])
protein_ids, protein_histograms = aggregate_proteins(
  dataset_splits["valid"], **frame_embeddings
)
_, tree, encoding_clusters = cluster_feature_spectrums(
  protein_histograms, n_clusters=12
)
localizations, localization_histograms = aggregate_localizations(
  dataset_splits["valid"], protein_ids, protein_histograms
)
```

```
plot_stacked_histrograms(
  localizations, localization_histograms, tree, encoding_clusters
);
```

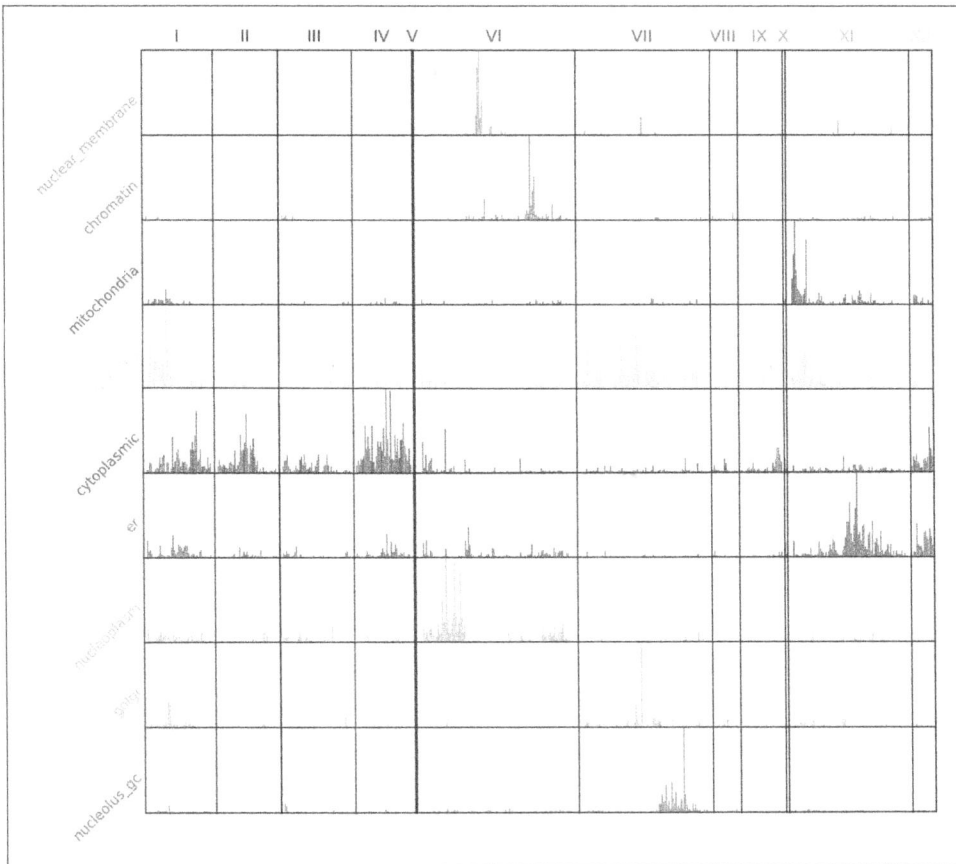

Figure 6-17. Histograms of embedding codebook usage (feature spectra) for different subcellular localizations. Each row shows the average distribution of codebook features across all image frames annotated with a given localization (e.g., chromatin, vesicles, nuclear membrane). Distinct localization categories exhibit unique spectral "signatures"—for instance, nuclear membrane and chromatin show sharp, narrow peaks. This representation provides a quantitative "signature" of protein localization and allows predictions for unannotated proteins based on their similarity to known spectra.

Now we're cooking; we see more clearly differentiated spectral signatures compared to the earlier, smaller model. Many compartments display sharp, concentrated peaks, indicating that the model consistently relies on a small set of highly specific codebook vectors for these localizations. The rows—each representing a different subcellular compartment—now show more distinct and characteristic patterns. For example, comparing the "before" (Figure 6-16) and "after" (Figure 6-17) spectra for `mitochon dria` and `er`, we see that in the "after" state, the signatures are less diffuse and more sharply defined, suggesting that the model has learned to focus more precisely on key features for these compartments.

We could speculate that compartments showing broader activation across several codebook regions, such as the `er` and `cytoplasm`, do so because of their inherently more variable or sprawling visual structure. These compartments tend to span large regions of the cell and appear with more morphological diversity. For instance, the ER forms a network that spans the cell, while cytoplasmic proteins can exhibit diffuse or context-dependent patterns. These differences likely lead the model to spread their representations across a wider set of features.

> These improvements to the spectra came purely from scaling up the dataset; no changes were made to the model architecture. Try exploring the model side of things next: increase capacity by adjusting `num_embeddings`, `embedding_dim`, or `classification_head_layers`, or experiment with your own architecture changes. There's a lot of room to get creative here.

Next, we can replot a UMAP projection to visualize the model's learned representation space (see Figure 6-18):

```
from dlfb.localization.inspect.embeddings.clustering import (
    calculate_projection,
    plot_projection,
)

projection = calculate_projection(frame_embeddings)
plot_projection(
    projection,
    dataset_splits["valid"],
    subset_mode="single",  # Only show frames with single localization
    titles=["Localization UMAP Projection"],
);
```

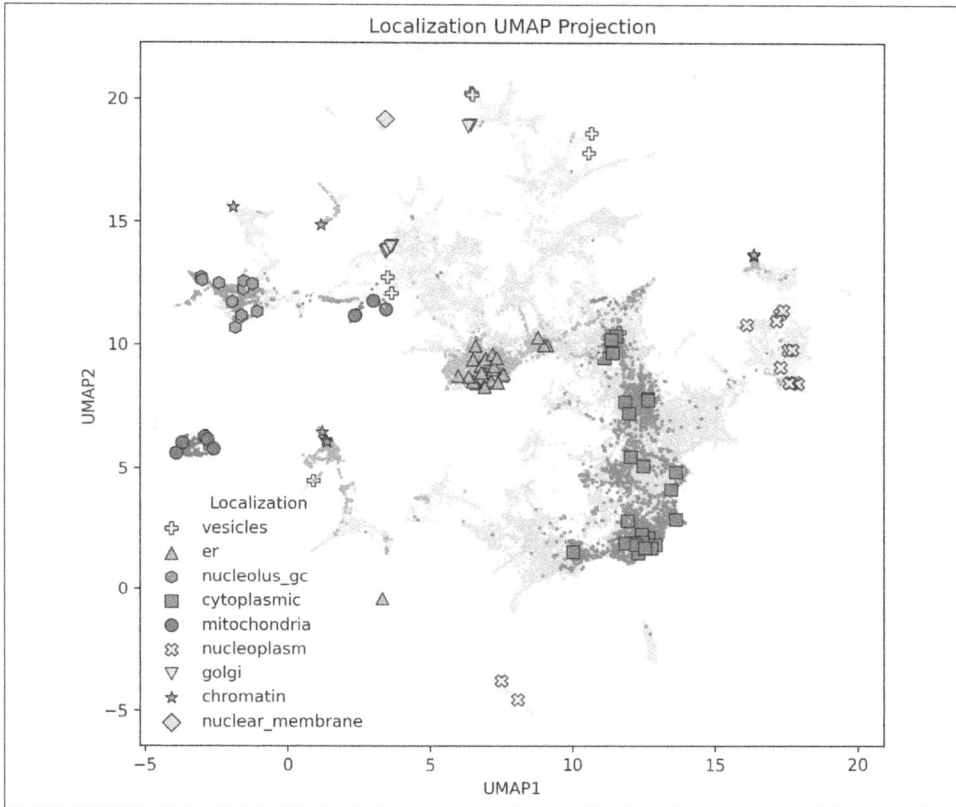

Figure 6-18. UMAP projection of embeddings produced by the larger model and dataset. The tight clustering of similar labels (e.g., nucleolus, ER, chromatin) shows that the model has learned to distinguish complex spatial patterns—without using labels during training.

This projection offers a compelling view into how well the larger model has learned to distinguish subcellular compartments. Each point represents an individual image frame, colored by its known localization. But remember, the model was trained in a self-supervised manner and never had access to these labels.

First, we see more points in the plot because the dataset is larger. We also observe that the separation between groups is now more distinct compared to earlier runs. Compartments such as the cytoplasm, vesicles, and nucleoplasm now form well-defined groups and are less intermixed, suggesting that the model has learned highly consistent visual signatures for these localizations. In some cases—such as vesicles—the model even appears to discover distinct subcategories, which are cleanly separated in the embedding space. Previously diffuse or heterogeneous categories like cytoplasm now exhibit more internal structure, with clearer spatial separation.

Explore Rich Embeddings

This kind of embedding space is rich with both biological and technical signals. Here are a few ideas for exploring it further:

- If certain localization classes (like `vesicles`, `chromatin`, `nucleoplasm`) form multiple separate clusters, inspect representative examples from each. Do the clusters correspond to specific cell cycle stages, subtypes, or imaging artifacts? Is there any sort of visible pattern that distinguishes them?

- Examine *outliers*, which are individual frames that sit far from any cluster. These might reflect mislabeled data, rare biological phenomena, or technical noise.

- Leverage known protein complexes. Do proteins that physically interact also lie close together in embedding space? Could you rank candidate complexes using dot products or distances between embeddings, potentially discovering novel interactions? (This is broadly similar to what was done in the cytoself paper.)

It's also worth checking how well the test set embeddings separate, to further confirm that the model has learned a generalizable representation.

Embedding spaces like these are not just for visualization. They offer a powerful tool for downstream discovery.

Going Further

This is as far as we'll go in this chapter, but there's much more you can explore from here. The original cytoself paper introduces several architectural innovations that improve both performance and biological insight:

Split quantization
> Instead of quantizing full feature vectors, cytoself divides them into smaller chunks and quantizes each independently. This improves codebook usage and the richness of the learned representations.

Global and local representations
> The model processes each image at two spatial scales:

- A *coarse* ($4 \times 4 \times 576$) representation that captures high-level localization patterns.
- A *fine* ($25 \times 25 \times 64$) representation that preserves detailed spatial features. Each is quantized using a separate codebook.

These additions enabled cytoself to achieve several impressive results, all in an unsupervised setting:

Predicting localization of previously uncharacterized proteins

For instance, it correctly inferred that FAM241A localized to the endoplasmic reticulum—a prediction that was later confirmed through colocalization experiments and mass spectrometry.

Distinguishing subtle subcellular differences

The model learned to separate visually similar compartments like lysosomes and endosomes. Though hard to distinguish by eye, these compartments differ in function: lysosomes degrade cellular material, while endosomes serve as transport vesicles en route to lysosomes.

Discovering protein complexes

Perhaps most impressively, cytoself grouped proteins belonging to the same complexes, such as ribosomes or the proteasome, purely from embedding similarity. By comparing protein-level embeddings, the authors showed that known complex members cluster tightly, even without any labels or prior knowledge. In some cases, this outperformed previous supervised methods, and it even hinted at previously uncharacterized complex members.

If you're interested in learning more, we encourage you to read the original paper and explore the OpenCell resource (*https://oreil.ly/eX8XH*), which hosts the data and interactive tools used in the study. The paper also outlines exciting future directions like 3D imaging, label-free microscopy, and cross-species generalization.

Summary

In this chapter, we built a self-supervised deep learning model to learn spatial organization within human cells, all without relying on any manual annotations. This kind of approach is especially powerful for large-scale biological imaging, where manual curation not only is expensive and time-consuming but also can introduce human bias. You tackled a real-world challenge involving complex data, biological nuance, and custom neural architectures, all while drawing on key ideas from across the book.

This was also the final end-to-end project chapter of the book. If you made it this far, congratulations. We hope this book helped you build intuition, fluency, and confidence in applying deep learning to biology. Whether you're exploring new datasets, designing new models, or testing the boundaries of what today's models can (and can't) do, we're excited to see where you go next.

Happy modeling—and keep exploring!

## CHAPTER 7

Tips and Tricks for Deep Learning in Biology

This final chapter brings together common themes from earlier chapters and distills practical strategies for applying deep learning techniques to biological problems. In machine learning, it's rare for things to work perfectly on the first try—or even the tenth. Debugging is an expected part of the process, not a sign of failure. Don't get discouraged.

Here, we share a collection of tips that have helped us (and others) navigate the challenges of deep learning in biology. Some were learned the hard way, and others emerged from writing this book. This list isn't exhaustive, but we hope it shortens your path to developing working models—and sharpens your instincts for when things go wrong.

> Don't expect steady, incremental improvements in your project. Progress in deep learning—especially with biological data—is often highly nonlinear. You might spend weeks debugging with no clear gains, only to make one small change that suddenly unlocks everything. This is normal—and not a cause for concern.

Simplify

When things stop making sense, simplify. Strip your problem back to the bare minimum—a smaller dataset, a shallower model, or a simpler loss function. It's easy to get lost in complex pipelines, but debugging is much easier when you can isolate one thing at a time.

Once you've got something working again, you can reintroduce complexity gradually. Think of this as turning the knobs one by one instead of all at once. It's not glamorous, but it's one of the most reliable strategies for making progress.

Simplify Your Model

When your model has too many bells and whistles, it becomes difficult to pinpoint where things are going wrong. Often, the most effective debugging strategy is to strip it down to the bare essentials.

Simplify your architecture
Complex architectures can make it difficult to reason about what's going wrong. To help with this:

Eliminate unnecessary layers
If a layer doesn't contribute directly to the input–output mapping—such as layers that simply add model capacity without altering dimensions—it's best to remove it during debugging.

Call basic layers directly
Use layers like nn.Conv and nn.Dense instead of custom blocks, which can obscure bugs and internal behavior.

Reduce depth and width
If your model has many layers or units per layer, consider reducing both. A shallower model is easier to debug and understand, especially in the early stages of development.

Remove residual connections
These can complicate debugging by introducing dependencies between layers and by masking issues in the layers they connect (like poor initialization or gradient problems).

The ultimate simplification is to reduce your model to a direct mapping from inputs to outputs. For example, pass your input through a single nn.Dense layer and see if it can overfit a tiny batch of data. If that fails, the issue is likely not with the architecture, but with your data pipeline, loss function, or optimizer.

Turn off extras, like normalization and dropout
These add complexity that's often unnecessary during early debugging. In Flax, you must explicitly manage both model state (e.g., batch norm statistics) and random number generators (RNGs), which can easily lead to subtle bugs.

Don't use batch norm
The complexity of batch norm is threefold. First, it behaves differently during training and inference. Second, it introduces additional state (running mean and variance) that must be updated outside standard gradient updates.

Third, it breaks a key assumption: most layers operate independently on each batch element, but batch norm computes statistics across the batch. This makes it incompatible with tools like vmap or sharded training (pmap, pjit) unless you take special care to synchronize statistics across devices.

Skip dropout

Dropout introduces stochasticity, making it harder to determine whether poor performance is due to randomness or a deeper issue.

Simplify your optimizer

Don't worry about experimenting with different optimizers or learning rate schedules until the basics are working. Pick a sensible default—like good old Adam with a learning rate of 1e-3, and focus on solving more fundamental issues first.

Avoid mixed precision

While lower-precision data types like bfloat16, float16, or TensorFloat32 can improve performance and memory usage (out of scope for this book), they can lead to subtle numerical instability that's extremely hard to debug. Note that even if you pass in float32 inputs, JAX may default to lower-precision matmuls. To force full float32 precision globally (especially during debugging), add jax.config.update("jax_default_matmul_precision", "float32").

Although turning off these more advanced features may lead to worse performance (due to underfitting or overfitting) and feel like you're going backward, once the basic setup is functioning correctly, you can systematically reenable features to assess their impact on model performance.

Simplify and Control Your Environment

Your model might be fine, but your technical environment could be introducing unexpected issues. Many bugs that look like deep learning failures are really just environmental gremlins—this means that cleaning up your setup is often the fastest way forward. Here are some tips to keep in mind:

Sort out determinism and reproducibility

It's easier to isolate issues if your experiments are reproducible. In addition to turning off stochastic components like dropout, consider:

Setting explicit random seeds

For reproducibility in JAX, you need to set the seed for jax.random.PRNGKey(...)—this controls all randomness in model initialization, dropout, and other JAX-based operations. Note that you don't need to set

Python's `random.seed(...)` or NumPy's `np.random.seed(...)` unless you're also using them separately (e.g., in your data preprocessing or non-JAX components).

Turning off dataset shuffling
Don't shuffle your training data, or shuffle with a fixed random seed, to maintain a consistent order of examples across runs.

Keeping the environment constant
Avoid inconsistencies caused by external factors (e.g., use the same hardware, library versions, and configurations).

Strip down your training loop
Especially in JAX and Flax, training loops require manual control over RNGs, state, and updates—making them powerful but easy to get wrong.

Train on a single batch for just a few steps
This is often enough to catch major issues.

Disable extras like logging, metrics, or learning rate schedules
These can obscure what's actually happening during training.

Use fixed inputs and random seeds
Run your training step on the same batch every time. This eliminates variability due to changing data and makes bugs easier to isolate.

Avoid high-level abstractions temporarily
If you're using tools like `TrainState`, try replacing them with raw variable updates until the core logic works.

Make your code self-contained
Especially when working in interactive environments like Colab or Jupyter notebooks, it's easy for your environment to get cluttered with old variables or states without you realizing it. Restarting the kernel and organizing your code into self-contained functions can really help with debugging.

Turn off JIT compilation (with caution)
Disabling `@jax.jit` can help identify issues in your training step more easily by making stack traces clearer and behavior more explicit. However, be aware: turning off JIT can cause huge memory usage spikes and drastically slower execution, making this approach impractical for larger runs.

Train on a single GPU instead of multiple GPUs
If you are using GPUs, start with just one. Multi-GPU training via sharding (e.g., with `jax.pjit`) introduces a lot of additional complexity. (Note: `pmap` is being deprecated in favor of `pjit`, so it's best not to rely on it.)

Use CPU for simpler debugging setups
    If your model is small and you're trying to isolate a bug, using CPU can remove the complexity of device placement and driver issues. It's slower, but often easier to reason about. Just be aware that some bugs may appear only on accelerators.

Go back to NumPy
    While `jax.numpy` mimics NumPy closely, the original NumPy library has simpler internals and often better error messages. You can't train models or use autodiff, but for testing data transformations or verifying calculations, it can be useful to isolate and debug numpy-only code outside of JAX.

These steps may feel tedious, but a clean and controlled environment is often the difference between spinning your wheels for days and finding a bug in 10 minutes.

> A quick note: different hardware backends behave slightly differently when running JAX code, especially when it comes to reproducibility.
>
> - *TPUs* are fully deterministic in JAX. If you use the same random seed, you'll get the same results every time.
> - *GPUs* are mostly deterministic, but some operations (like convolutions or matrix multiplications) may vary slightly across runs unless you disable certain performance optimizations.
> - *CPUs* are generally deterministic, but subtle sources of nondeterminism (like thread scheduling) can still appear.
>
> For early debugging in JAX, running on CPU can simplify things—but just be aware that bugs might show up differently when you move to GPU or TPU.

Simplify the Data and Problem

Make your dataset and prediction task easier to simplify debugging. Here are some ways to make the data more manageable during early experimentation:

Visualize individual examples
    Actually plot or print raw inputs and labels—not just summaries. You'll often catch issues like incorrect encodings, off-by-one errors, or mismatched image-label pairs this way. It's surprisingly tempting to skip this step—don't. Simply looking at the raw data can reveal issues that save you hours later.

Check class balance
    An imbalanced dataset can make the model appear broken when it's just doing the naive thing (predicting the majority class). Consider subsampling or rebalancing during debugging.

Remove data augmentation

Augmentations like cropping, flipping, or adding noise can hide underlying issues or make the task unnecessarily hard. Turn them off until you're confident the core pipeline works.

Reduce the number of classes

Instead of predicting many categories, reframe your task as binary classification to focus on the clearest signal first.

Simplify the output space

On a similar note, if your target is complex (e.g., a regression label or structured output), try reducing it to something simpler. For instance, predict a binary class or thresholded version of the label instead. This can help verify the pipeline before tackling the full problem.

Make your dataset smaller

Large datasets slow everything down. Use a small, representative subset that captures the key structure.

Limit the scope of your data

Use a natural slice of your dataset. For example, restrict to a single species, tissue type, year, or patient group. This cuts variability and helps isolate bugs.

Check for label leakage

Especially in biological datasets, label leakage can creep in through metadata like patient ID, batch number, or experiment date. This can cause your model to perform suspiciously well by learning shortcuts. Double-check that no features or splits accidentally leak target-related information.

Work with synthetic or simulated data

If feasible, start with synthetic data that mimics key characteristics of your real dataset but is easier to understand and trace. You can also add controlled noise (e.g., `np.random.normal(0, 1)`) to test the model's robustness.

Simplifying the data isn't about giving up—it's about creating a controlled testbed where you can debug quickly, eliminate uncertainty, and build confidence before scaling back up.

Overfit to a Single Batch of Data

We've mentioned this briefly before, but it's worth stating explicitly: if your model doesn't seem to be learning much at all, a classic debugging strategy is to try to overfit to a single batch—meaning, rerun the training loop on the same batch repeatedly and see if the model can memorize it. This test helps confirm that the training loop, loss function, and optimizer are wired up correctly.

Instead of the usual training loop:

```
for step in range(num_steps):
    batch = next(training_data)
    state, loss = update_step(state, batch)
```

you can try this:

```
num_debugging_steps = 10
batch = next(training_data)

for step in range(num_debugging_steps):
    state, loss = update_step(state, batch)
    print(f"Loss at step {step}: {loss}")
```

If your setup is working correctly, the model should rapidly memorize the batch, and the loss should decrease significantly after a few steps. If not, consider checking for:

Learning rate issues:
> The learning rate might be too high (causing divergence) or too low (causing no learning).

Frozen parameters or bad gradients
> Sometimes parameters are not being updated at all—for example, if they were accidentally excluded from the params dict due to naming mismatches or scoping issues. Also inspect gradients—if they're all zero or NaN, that's a clue.

Loss function bugs
> Make sure you're using the correct loss function for your task (e.g., cross-entropy for classification) and that it's behaving numerically as expected. Prefer standard implementations over custom home-made ones, at least during debugging.

Model initialization problems
> Poor or inconsistent weight initialization can prevent learning, especially in deeper networks. If you're using custom modules, double-check their initializations.

Batch size too small
> Very small batches (e.g., 1–2 examples) can lead to noisy gradients and unstable updates. For debugging, use a small but reasonable batch size like 8–32.

Silent shape mismatches or broadcasting errors
> These won't always crash your code, but they can silently mess up your loss or gradients. Print tensor shapes and inspect intermediate outputs to confirm everything lines up as expected.

This is one of the fastest and most informative debugging steps—if your model can't learn a tiny batch, don't bother scaling up yet.

Go Back to Basics

If none of these debugging tips works and you are rapidly losing your mind, one of the most effective strategies is to revert to a simple, well-understood example that you know works:

Start with an established example
Train a simple model on a well-known dataset—like a basic CNN on MNIST for image problems. These examples are widely used and well-documented, making them a reliable way to get something working end-to-end.

Reproduce known results
Make sure your setup can successfully train the model and reach the expected performance (e.g., ~99% accuracy on MNIST). This confirms your training loop, model, and loss function are functioning correctly.

Swap in your dataset
Once the baseline works, begin replacing the dataset with your own. Proceed gradually and check that everything still functions as expected.

Iteratively add complexity
With your data integrated, incrementally introduce more complex components—like deeper architectures or new training strategies. Watch for breakage after each change.

There's absolutely nothing wrong with going back to a tutorial and training a simple linear model—sometimes that's the fastest way to confirm your setup and get your bearings.

Log Everything

Good logging is the difference between productive debugging and blindly guessing. When something goes wrong, clear logs can help you trace back exactly what happened—and when things go right, they help you understand why.

Log training loss and key metrics over time
At a minimum, track loss, accuracy, and any relevant task-specific metrics (like auROC or auPRC). This makes it easier to spot overfitting, instability, or underperformance.

Log validation performance at regular intervals
Seeing how your model generalizes during training helps detect overfitting early and can catch bugs where validation performance diverges for no obvious reason.

Log inputs, predictions, and errors
> Save a few input samples, predicted outputs, and errors at each step (or epoch). This is especially useful for spotting systematic failures (e.g., always misclassifying a certain class).

Record configuration and hyperparameters
> Save the learning rate, batch size, optimizer type, and model architecture along with each run. You will forget. Everyone forgets.

Use a structured logger or tracking tool
> Tools like TensorBoard, Weights & Biases, or even just structured JSON logs can make it easier to compare runs and understand what changed.

> Logging might feel like overhead in the moment, but it's one of the best time investments you can make. Even minimal logs can help you debug faster and avoid retraining unnecessarily.
>
> There's a rule of thumb: every new log reveals a bug you didn't know you had.

Ask for Help

If you're still stuck, don't be afraid to reach out. Community forums like Stack Overflow or GitHub Discussions are valuable resources. Talking to a colleague or friend can also help—sometimes just explaining the problem out loud (rubber ducking) leads to breakthroughs.

You can also use LLMs like ChatGPT, Gemini, or Claude to help troubleshoot or explore ideas. Just remember that while these models can be very helpful, their suggestions aren't always correct and can introduce new bugs—so double-check any code they generate.

Common Data Issues

Often, it's not your model that has a bug—it's the data. Subtle issues in your dataset can quietly break your learning pipeline, and you may find yourself spending hours debugging the model setup when the problem is actually upstream. This section covers common data pitfalls that are worth checking before tearing your architecture apart.

Data Leakage

As physicist Richard Feynman famously said, "The first principle is that you must not fool yourself—and you are the easiest person to fool." It's all too easy to believe your model is doing well when it's actually cheating. Data leakage happens when information that should be hidden during training is inadvertently accessible to the model, leading to overly optimistic performance metrics.

- Obvious cases
  - Evaluating the model on the training set itself. This sounds slightly silly, but it's surprisingly common—especially in informal settings like Kaggle notebooks.
  - Evaluating on a validation or test set where some examples overlap with the training set.
- Subtle cases
  - Leakage through preprocessing, for example, normalizing the full dataset before splitting into train/valid/test sets.
  - Features that leak future information—correlated with the target only because they wouldn't be available at prediction time.

Here are some real-world examples of subtle leakage:

- You want to classify skin lesions as malignant or benign. Patients with cancer are photographed in one clinic and healthy patients in a different clinic. If one clinic uses brighter lighting, your model may learn to use brightness as a proxy for cancer.
- You're predicting protein binding to a gene, and you include gene expression level as an input to the model. Since a gene's expression may be affected by binding, your model might learn to rely on this proxy instead of the DNA sequence features you intended it to learn.

> If your model performs well on a test set but fails to generalize to new datasets or real-world settings, data leakage is one of the first things to investigate.

To avoid data leakage:

- Always ensure that test data is fully isolated and untouched by the training pipeline.
- When adding new training data partway through a project, check whether it already appears in your validation or test sets.

- Ask yourself: *Would this feature be available at inference time?* If not, don't use it.
- Use model interpretation tools to see what aspects of your data your model is relying on when making its predictions. Does what you see match your expectations, or is the model picking up on artifacts?

Incorrect Data Labels

It seems obvious, but incorrectly labeled data is one of the most common and frustrating causes of model underperformance. Some high-risk scenarios include:

Labels stored separately from inputs
If labels are in a separate file (e.g., a CSV with filenames and classes), they can easily be mismatched or misjoined during preprocessing. We certainly have several shameful anecdotes along these lines.

Shuffling inputs and labels independently
If you shuffle data and labels separately, they'll fall out of sync—silently.

Shape mismatches in TensorFlow datasets:
`tf.data.Dataset` won't necessarily complain if your labels and inputs have mismatched shapes like data of shape `(100, 10)` but labels of shape `(43,)`. This can result in silent failures that only manifest much later.

Merging tabular datasets incorrectly
Joining datasets without verifying alignment (e.g., via `merge` in pandas) can mislabel data rows without throwing errors.

Data augmentation pipelines modifying labels incorrectly
Augmentation effectively increases your dataset—so it's also a high-risk area for introducing label corruption.

> A common warning sign of label issues: the training loss does goes down slightly during training but plateaus early at a high value, and accuracy (or other metrics) stays near random chance level.
>
> For example, if the labels are scrambled, your accuracy would hover around random baseline values like 50% in balanced binary classification.

The best antidote to label issues is simply spending time inspecting your input-label pairs, both at the beginning as raw data, and at different points in your data pipeline. Check a few batches by hand. Plot examples and verify the label. It may feel tedious—but it's one of the fastest ways to catch silent bugs.

Imbalanced Classes

In biology, it's common for one class to vastly outnumber others—like detecting rare mutations or identifying diseased cells. This class imbalance is not an issue in itself, but a model trained on imbalanced data might just learn to always predict the majority class, achieving deceptively high accuracy.

Warning signs:

- Accuracy is high, but precision or recall on the minority class is poor.
- Confusion matrix shows the model rarely predicts the minority class.

To address this:

- Use class weighting or focal loss to penalize the model more for mistakes on the minority class. Focal loss down-weights easy examples and focuses learning on hard, misclassified ones—especially useful when the rare class is easily overwhelmed.
- Resample the data—either oversample the minority class or undersample the majority class—to reduce imbalance. Oversampling is often safer when data is scarce but can lead to overfitting if not done carefully.
- Use stratified sampling to ensure class balance is preserved across your train, validation, and test splits. This means splitting the data so each subset maintains the original class proportions, avoiding skewed performance estimates.

Distribution Shifts

Training and test data can often come from different sources—different labs, species, sequencing protocols, or imaging setups. These shifts can cause models to learn dataset-specific artifacts instead of generalizable biology.

Warning signs:

- Strong validation performance doesn't transfer to real-world or external datasets.
- A model trained to predict dataset labels (e.g., lab or batch ID) performs surprisingly well.

To catch and correct for this:

- Visualize embeddings (e.g., via PCA or UMAP) colored by data source to spot clustering.
- Use batch effect correction or domain adaptation methods if needed.

- Be cautious when mixing data from different sources—explicitly test generalization to new settings.

Biology-Specific Gotchas

Biology is a vast field, filled with complex systems and ever-evolving datasets. We can't cover every pitfall here, but the following are some common sources of bugs and errors that we've encountered repeatedly—and that are worth keeping in mind when building models on biological data:

Versioning issues

Many biological datasets are tied to reference versions (e.g., genome builds, gene IDs, transcript annotations). It's dangerously easy to mix up genome versions (e.g., GRCh37 vs GRCh38), gene ID versions, or even organism accessions. Make sure all parts of your pipeline are using consistent versions—or explicitly map between them.

Data integration challenges

Combining datasets from different sources is common in biology but can lead to subtle inconsistencies: mismatched identifiers, differing file formats, or incompatible measurement units (e.g., read counts versus TPM versus RPKM). Carefully check alignment before merging datasets.

Biological heterogeneity

Biological systems vary widely—across individuals, cell types, populations, and species. A model trained on European ancestry samples may not generalize to other ancestries. Likewise, models trained on data from immortalized cancer cell lines can fail when applied to normal primary cells. Always consider the scope and limitations of your training data.

Ambiguous or soft labels

Biological categories are often not cleanly defined: cell types can be graded or transitional, and protein binding is often a continuous score, not a binary yes/no. Hard labels, where present, may oversimplify what is actually a spectrum. In these cases, performance ceilings may reflect label ambiguity, not model failure.

Experimental noise

Just adding more data isn't always better—low-quality experimental data can introduce noise that overwhelms signal. Look for ways to filter or denoise.

Use quality metrics

Many experiments include built-in quality scores. Filtering on these can help.

Leverage replicates
> Use experimental replicates (same exact setup, multiple times) or biological replicates (same protocol, but different samples) to reduce variance. You can average replicate signals or use them to quantify uncertainty.

Batch effects
> Differences in lab conditions, reagent lots, sequencing machines, or protocols can introduce strong confounding signals. These technical artifacts often dominate the true biological signal if not accounted for. Visualize your data (e.g., with PCA or UMAP colored by batch) to assess how much batches cluster apart. You can also train a model to predict batch labels—if it performs better on this than your actual task, your model is probably learning batch-specific noise. If needed, apply normalization techniques like quantile normalization to mitigate these effects.

Common Model Issues

Not all training failures come from bad data—sometimes, the model itself is the problem. In this section, we highlight common issues that arise during model training, from overfitting to gradient instability.

There are more detailed deep learning debugging guides out there—for example, the Deep Learning Tuning Playbook (*https://oreil.ly/-t96r*) by Google. But here we'll recap some of the most common and practical failure modes, with a focus on how to identify and fix them efficiently.

Overfitting and Poor Generalization

Overfitting is one of the most common issues in deep learning. Deep neural networks typically have high capacity and are only told to minimize training loss—without any built-in notion of generalization. As a result, they often perform well on training data but poorly on unseen data.

Fortunately, there's a well-established set of regularization techniques to help reduce overfitting—many of which we've already discussed throughout this book:

Dropout
> Randomly disables units in the network during training to prevent over-reliance on any one path.

Weight decay
> Penalizes large weights (L1 or L2 regularization) to encourage simpler models that generalize better.

Early stopping

Monitors validation performance and stop training when performance starts to degrade, even if training loss continues to drop.

Data augmentation

Expands your dataset by applying small, meaningful transformations (e.g., rotations, flips in images; jittering or cropping in sequences).

Ensembling

Combines predictions from multiple models trained with different seeds or splits as a form of error correction. Even ensembling the same architecture can significantly improve robustness.

Also, check whether your validation or test data differs fundamentally from your training data (see the previous section). If the distribution has genuinely shifted, then the model might not be overfitting so much as encountering data it was never trained to handle.

> A model that performs well on training data but poorly on validation is likely overfitting. A model that performs poorly on both might be underfitting or struggling with a broken setup.

Vanishing or Exploding Gradients

Vanishing gradients, where gradient values approach zero, and exploding gradients, where they become excessively large, can severely disrupt training. Fortunately, these issues are relatively easy to detect.

A simple way to monitor gradients is to compute their L2 norm (also called the Euclidean norm), which summarizes the overall magnitude of the gradient as a single scalar. You can log this value alongside the loss during training.

Here's how to compute the L2 norm of gradients in Flax:

```
@jax.jit
def compute_gradients_l2_norm(grads):
    """Compute L2 norm of gradients."""
    grads_flat = jax.tree_util.tree_leaves(grads)  # Flatten.
    return jnp.sqrt(sum([jnp.sum(jnp.square(g)) for g in grads_flat]))

# Example usage inside a training step:
loss, grads = jax.value_and_grad(loss_fn)(state.params)
grad_norm = compute_gradients_l2_norm(grads)
```

You can log this `grad_norm` over time and visualize it alongside the loss to examine gradient behavior:

- If `grad_norm` is close to 0, gradients are likely vanishing.
- If it grows rapidly or spikes erratically, you may be seeing exploding gradients.

Common fixes to try out include:

- Lower the learning rate or use a learning rate schedule.
- Use better weight initializers: try Xavier (Glorot) or He initialization, depending on your activation function.
- Normalize activations: Batch normalization or layer normalization helps stabilize the flow of gradients.
- Add residual connections: These help gradients propagate through deep networks without degradation.
- Clip gradients: This is a blunt but effective tool to cap extreme values and prevent instability.

The fixes to this issue tend to be ones we've already mentioned, like reducing the learning rate or using a learning schedule, using different weight initializers, and adding either batch normalization or layer normalization. Adding residual connections between blocks can also be helpful. Finally, explicitly clipping gradients to a fixed threshold to avoid excessive values might sound like a really crude approach but is common and very effective. Here is an example implementation:

```
def clip_gradients(grads, threshold):
    """Clip gradients."""
    return jax.tree_map(lambda g: jnp.clip(g, -threshold, threshold), grads)
```

Or, if you are using `optax`, you can also clip gradients with:

```
tx = optax.chain(optax.clip(threshold), optax.adam(learning_rate))
```

Training Instability

A related issue to gradient issues is *training instability*, which can manifest in several ways, including erratic training losses, sudden spikes in validation loss, or even full-blown divergence. By *divergence*, we mean that the model fails to *converge* toward a stable solution; instead, the loss may oscillate wildly or become `NaN`.

Training instability typically arises from a few common causes:

Learning rate is too high
A high learning rate can cause the optimizer to overshoot minima, leading to instability. Try lowering the learning rate or using a warmup schedule that starts small and ramps up gradually.

Using a nonadaptive optimizer
Adaptive optimizers like Adam, RMSProp, or Adagrad adjust learning rates per parameter and tend to be more robust out-of-the-box. While vanilla stochastic gradient descent (SGD) can be effective, it typically requires more careful tuning, especially with larger models or noisy data.

Exploding gradients
In deep networks, gradients can grow too large and destabilize updates. As discussed earlier, apply *gradient clipping* or use normalization layers (like batch norm or layer norm) to control this.

Inappropriate batch size
Very small batches can lead to noisy gradient estimates that make training unstable. Larger batches offer more stable gradients—generally, try using the largest size your hardware allows, especially during early debugging.

Poor weight initialization
Improper initialization can cause gradients to vanish or explode. Flax uses LeCun normal as the default initializer for `nn.Dense` and `nn.Conv`, which works well with ReLU activations. But for very deep networks or specific architectures, Xavier or He initialization may perform better.

Activation blowup
As networks deepen, intermediate activations can grow excessively large, especially with ReLUs or unnormalized inputs. To prevent this, keep activations centered and bounded, most commonly by applying batch normalization.

Poor Model Performance

The model trains and the dataset looks good. You've squashed overfitting. Everything generally looks sane. The only problem is that the model is just not that good.

How Well Should You Do?

We touched on this point in the introduction, but it's worth restating: to judge performance, you need context. Here are some ways to anchor your expectations:

Random chance performance
 What would random guessing achieve? For regression, how well would you do by always predicting the mean or median of the training set?

Baseline models
 Try a simple linear regression or logistic regression. Sometimes these models perform surprisingly well—and if your deep model doesn't beat them, something's wrong.

Other published models
 If others have worked on this task, check what performance they report. You can often get architectural or preprocessing ideas from their work. But beware—published metrics aren't always trustworthy, and they may not be directly comparable to your setup.

Human performance
 Can a human do this task? How well would an expert do it? This can help you calibrate expectations.

Experimental replicates
 Biological measurements often contain noise due to sampling variability, measurement error, or biological variability itself. One way to estimate the ceiling for your model's performance is to check how consistent the raw signal is across replicate experiments. If two replicates have a correlation of 0.85, your model is unlikely to exceed that. Don't expect your model to be more consistent than biology itself.

Addressing Poor Model Performance

While there's no universal fix for a model that just isn't performing well, the following strategies can help identify what's wrong and suggest paths forward:

Check data quality
 Many model issues are actually data issues. Dive into specific examples—especially ones the model gets wrong—and look for inconsistencies, noise, or labeling errors. If the data is highly domain specific and you're not an expert, ask someone who is.

Run error analysis

Where is the model doing well? Where does it consistently fail? Are there patterns to its mistakes—specific classes, edge cases, or confounding conditions? Systematic errors often point to missing features or broken assumptions.

Add more data

More data can help if the model is underfitting or struggling with rare cases. You can also try synthetic augmentation, bootstrapping from known examples, or generating simulations. Watch how performance scales with dataset size—plateaus may indicate other bottlenecks.

Tune hyperparameters

Some hyperparameters matter more than others—start with learning rate, batch size, model depth, and regularization strength. Use grid or random search over a small range to find better-performing settings.

Try transfer learning

If similar datasets or tasks exist, use pretrained models as a starting point. You can either fine-tune the whole model or freeze its feature extractor and train a smaller model on top. Alternatively, use learned embeddings from a related model as input features.

As always, if you're stuck, revisit the basics: simplify the model, overfit a single batch, sanity-check your labels, and compare against baseline performance. Many of the strategies that help fix broken models can also clarify why a working model isn't yet a good one.

Final Thoughts

Deep learning in biology is hard. The data is messy, the goals are often open-ended, and training useful models can be finicky. But that's exactly what makes it exciting. With every experiment, you're not just solving a technical challenge—you're helping push the boundaries of how we understand life itself.

The journey won't always be smooth—models will fail in surprising ways, you'll write catastrophic bugs, and the data will contain monumental errors. But if you stay curious, keep things modular, simplify when in doubt, and stay patient, you'll find your way through.

Whether you're building models to decode genomes, predict protein structures, or interpret microscopy images, we hope this book has helped you approach the work more confidently—and maybe even enjoy the process a little more.

Good luck, and keep going!

Index

About the Authors

Charles Ravarani is chief technology officer at biotx.ai and holds a PhD in computational biology. He brings together deep academic expertise and software engineering to drive innovation in drug discovery and machine learning education.

Natasha Latysheva is a research engineer at Google DeepMind and has a PhD in computational biology. She specializes in deep learning for genomics and is passionate about making machine learning education accessible and engaging.

Colophon

The animal on the cover of *Deep Learning for Biology* is the paper nautilus (*Argonauta argo*). Even though it's in the name, the paper nautilus isn't a nautilus (a cephalopod with a hard, coiled external shell), but is a pelagic octopus. The name "paper nautilus" comes from the delicate, paper-thin shell females create to protect their eggs. They can be found all over the globe in tropical and subtropical waters, preferring to stay close to the surface, rather than on the seabed like other octopuses.

Male and female paper nautili vary significantly in size and appearance. Females are much larger, growing to about 11 inches long (including their shell), while males barely grow to 1 inch and do not have a shell. Not only does the female's shell provide protection for her eggs, but it's used to help her move through the water. The shell captures air bubbles, allowing her to float through the water while conserving her energy.

Paper nautili have an interesting way of reproducing. The males have a specialized "arm" (the hectocotylus) containing sperm that they transfer to the female during reproduction, which then detaches from the male. The female then carries and protects the eggs in her shell until they hatch. Females can reproduce multiple times during their lifespan, whereas males will only do so once.

The cover illustration is by José Marzan Jr., based on an antique line engraving from *Meyers Kleines Lexikon*. The series design is by Edie Freedman, Ellie Volckhausen, and Karen Montgomery. The cover fonts are Gilroy Semibold and Guardian Sans. The text font is Adobe Minion Pro; the heading font is Adobe Myriad Condensed; and the code font is Dalton Maag's Ubuntu Mono.

O'REILLY®

Learn from experts.
Become one yourself.

60,000+ titles | Live events with experts | Role-based courses
Interactive learning | Certification preparation

**Try the O'Reilly learning platform
free for 10 days.**

www.ingramcontent.com/pod-product-compliance
Lightning Source LLC
Chambersburg PA
CBHW080138220326
41598CB00032B/5099